METHODS IN MOLECULAR BIOLOGY™

Series Editor
John M. Walker
School of Life Sciences
University of Hertfordshire
Hatfield, Hertfordshire, AL10 9AB, UK

For other titles published in this series, go to
www.springer.com/series/7651

METHODS IN MOLECULAR BIOLOGY™

DNA Microarrays for Biomedical Research

Methods and Protocols

Edited by

Martin Dufva

Department of Micro and Nanotechnology
Technical University of Denmark, 2800 Kgs. Lyngby, Denmark

Humana Press

Editor
Martin Dufva
Department of Micro and Nanotechnology
Technical University of Denmark
2800 Kgs. Lyngby, Denmark
Martin.Dufva@nanotech.dtu.dk

Series Editor
John M. Walker
University of Hertfordshire
Hatfield, Herts.
UK

ISSN 1064-3745 e-ISSN 1940-6029
ISBN 978-1-934115-69-5 e-ISBN 978-1-59745-538-1
DOI 10.1007/978-1-59745-538-1

Library of Congress Control Number: 2008938537

Printed on acid-free paper

springer.com

Preface

DNA microarray technology has revolutionized research in the past decade. In the beginning microarray technology was mostly used for mRNA expression studies, but soon spread to other applications such as comparative genomic hybridization, SNP and mutation analysis. These applications are now in everyday use in many laboratories and therefore the focus of this volume. It is clear from the protocols in this volume that DNA microarray assays are very complicated to perform even if fabrication of microarray is not considered. It is also clear that there are many different ways to perform microarray assays even if the basic concept is the same, i.e. hybridization of sample DNA (or RNA) to immobilized single stranded capture DNA. Minute changes to a protocol can be pivotal between success and failure in a microarray assays.

DNA microarrays fabrication can be divided into two broad categories: on chip synthesis and spotting off chip synthesized DNA. The latter is by far the most common method in house for fabrication of DNA arrays in house. In house fabrication of microarray is necessary when microarrays are not commercially available or is not an economical possibility. The largest providers of microarray are Affymetrix, Illumina and Agilent and all are exemplified in this volume on different kinds of applications. Commercial arrays are typically targeted towards popular organisms and application such as SNP, gene expression analysis, and microarray user that have other requirements are left to fabricate arrays themselves. This volume therefore addresses fabrication issues theoretically as well as giving examples of practical detailed methods.

The main advantage of DNA microarray is that many reactions are taking place in parallel on the surface of microarrays. This advantage is also microarray technology's greatest weakness because all these hybridization reactions need to operate at one single condition applied to the array which put large demands on probe's choice. Furthermore, we have little knowledge about what is taking place on the surface of microarrays that complicates array development. This volume provides robust protocols for performing microarray assays reproducibly. However, reproducible does not necessarily mean that data obtained correctly reflects what is going on in a cell or an organism.

DNA microarray technology is slowly filtering into diagnostic applications that presumably will benefit from miniaturization and highly multiplex assays just like the research community has been doing and will be doing for a considerable time yet. Before microarray comes into clinical use though, we need to find new short and efficient protocols based on the current state-of-the-art protocols provided here.

Contents

Contributors

JUSTIN BISHOP • *University of Utah, Salt Lake City, Utah, USA*
STEVE BLAIR • *University of Utah, Salt Lake City, Utah, USA*
DIETMAR H. BLOHM • *Department of Biotechnology and Molecular Genetics, Center for Applied Genesensor-Technology (CAG), University of Bremen, Bremen, Germany*
NIGEL P. CHARTER • *Wellcome Trust, Sanger Institute, Cambridge, UK*
ALEXANDER CHAGOVETZ • *University of Utah, Salt Lake City, Utah, USA*
MARTIN DUFVA • *Technical University of Denmark, Kgs. Lyngby, Denmark*
MICKAËL DURAND-DUBIEF • *Karolinska Institute /NOVUM, Huddinge, Sweden*
KARL EKWALL • *University College Södertörn, Huddinge, Sweden.*
TOMAS FITZGERALD • *Wellcome Trust, Sanger Institute, Cambridge, UK*
KEVIN L. GUNDERSON • *Illumina, Inc., San Diego, CA, USA*
BAOCHUAN LIN • *The Center for Bio/Molecular Science and Engineering, Naval Research Laboratory, Washington, DC, USA*
ANTHONY P. MALANOSKI • *The Center for Bio/Molecular Science and Engineering, Naval Research Laboratory, Washington, DC, USA*
LILI MILANI • *Uppsala University, Uppsala, Sweden.*
PETER NILSSON • *School of Biotechnology, UTH-Royal Institute of Technology, Stockholm, Sweden*
DANIEL A. PEIFFER • *Illumina, Inc., San Diego, CA, USA*
JESPER PETERSEN • *Technical University of Denmark, Kgs. Lyngby, Denmark*
LENA POULSEN • *Technical University of Denmark, Kgs. Lyngby, Denmark*
RICHARD REDON • *Wellcome Trust, Sanger Institute, Cambridge, UK*
DIANE RIGLER • *Wellcome Trust, Sanger Institute, Cambridge, UK*
MICHAEL STANGEGAARD • *University of Copenhagen, Copenhagen, Denmark*
ANN-CHRISTINE SYVÄNEN • *University Hospital, Uppsala, Sweden*
SASHA TODT • *Center for Applied Genesensor-Technology (CAG), University of Bremen, Bremen, Germany*
ANNELIE WALDÉN • *School of Biotechnology, UTH-Royal Institute of Technology, Stockholm, Sweden*
RASMUS WERNERSSON • *Center for Biological Sequence Analysis, Technical University of Denmark, Kgs. Lyngby, Denmark*
LAYNE WILLIAMS • *University of Utah, Salt Lake City, Utah, USA*

Chapter 1

Introduction to Microarray Technology

Martin Dufva

Abstract

DNA microarrays can be used for large number of application where high-throughput is needed. The ability to probe a sample for hundred to million different molecules at once has made DNA microarray one of the fastest growing techniques since its introduction about 15 years ago. Microarray technology can be used for large scale genotyping, gene expression profiling, comparative genomic hybridization and resequencing among other applications. Microarray technology is a complex mixture of numerous technology and research fields such as mechanics, microfabrication, chemistry, DNA behaviour, microfluidics, enzymology, optics and bioinformatics. This chapter will give an introduction to each five basic steps in microarray technology that includes fabrication, target preparation, hybridization, detection and data analysis. Basic concepts and nomenclature used in the field of microarray technology and their relationships will also be explained.

Key words: Microarray, application, method.

1. Introduction to Microarray Assays

Microarray assays originate from traditional solid phase assays—DNA/RNA dot blot assays and Enzyme Linked Immuno Sorbent Assays (ELISA)—that have been used for decades in laboratories. A solid phase assay has molecules attached to a solid support and these molecules are designated 'capture molecules' or 'probes'. The capture molecules are *probing* the sample for the presence of *target* molecules. The probe should demonstrate as high specificity and affinity for the target molecule as possible. The probes can be PCR products *(1)*, oligonucleotides *(2)*, or plasmids or bacterial artificial chromosomes *(3)* for analysis of genomes and transcriptomes. Though not discussed in this volume, probes can be made of many other types

Martin Dufva (ed.), *DNA Microarrays for Biomedical Research: Methods and Protocols, vol. 529*

DOI 10.1007/978-1-59745-538-1_1 Springerprotocols.com

of molecules such as proteins *(4)*, antibodies *(5, 6)*, DNA/RNA aptamers *(7, 8)*, small molecules *(9)* and carbohydrates *(10, 11)* for analysis of proteomes.

The key advantage of microarray technology is that minute amounts of many different probes are immobilized onto a solid support yielding tremendous parallel analysis capacity needed to analyse whole 'oms', such as the transcriptome, in one single batch process. Numerous different probes are typically immobilized in arrays of spots on a solid support where each spot contains multiple copies of a particular capture molecule/probe. For example, when fabricating an array it is known that the spot at coordinates (x1, y1) contains copies of the probe for gene 'G', the target in this case. The identity of probes is therefore encoded by a position in a 2D array (**Fig. 1.1**). If each spot contains a different probe, a single sample can be probed for the presence of many different target molecules. Typical microarrays contain thousands to million probes on a single 'chip' (substrates usually with other dimensions than microscope slide) or microscope slide used for

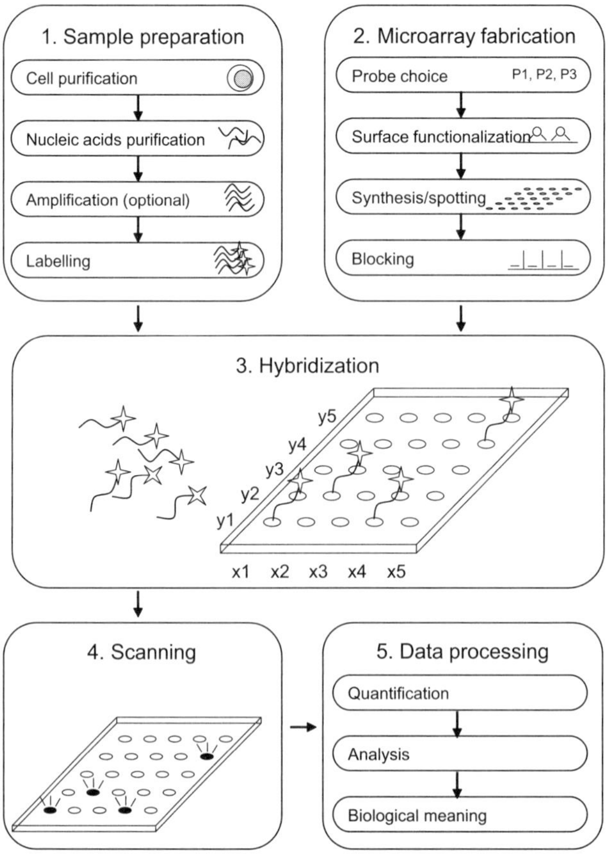

Fig. 1.1. Layout of the process step for making and using DNA microarray.

analysis of the genomes and transcriptomes. Significantly smaller microarrays encompassing 10–100 spots exist as well for sensitive diagnostics of viral and bacterial infections and cost efficient genotyping.

The operation whereby sample/target is allowed to react with the probes to generate probe–target interactions is referred to as hybridization. To maximize the sensitivity of the assay, the target should be highly concentrated. Compared to other methods this is not a disadvantage of microarrays because the very small size of microarrays means fairly small quantities of sample/target are required. The target and the array or probes are then left to react under conditions that facilitate hybridization; typically long hybridization times in high ionic strength buffers at relatively high temperatures. Mixing can be used to increase hybridization kinetics decrease the background binding and obtain homogeneous hybridization over the array. After hybridization, arrays are usually subjected to stringent washing procedures to remove cross-hybridizations, i.e. target molecules that have bound to the wrong spot (probes) during the hybridization reaction.

Target molecules are labelled using fluorescent molecules or other dyes either pre- or post-hybridization so that probe–target hybridizations can be detected via the generation of a signal. For example, a signal obtained at spot (x1, y1) indicates that gene G (target) is present in the sample. The signal does not provide any other information other than the presence of the target. The size or length of the captured target molecules or the complete sequence/composition of the captured target is not known. This is one weakness of microarray technology as compared to Serial analysis of gene expression *(12)* and Northern blot analysis. These methods yield target sequence frequencies and target size information, respectively.

2. Early Development and Origin of Arrays

Microarrays offer high-throughput and miniaturized versions of the assay formats they were based on: microtitre plates and dot blot assays. Microtitre plate solid phase assays are based on immobilizing specific capture molecules in the wells of a microtitre plate. Each well contains only one type of probe. Thus a well in a microtitre plate assay can be viewed as equivalent to a spot in a microarray. Typically, the probes are immobilized by adding 100–200 μL of probe solution to a well. In comparison, a microarray spot is typically produced by spotting 0.1–1 nL probe solution. After removing the excess probe solution, 100–200 μL sample is added to the well. Although microtitre plate assays are highly sensitive, the reaction conditions are not optimal during the assay to obtain maximum sensitivity according to Ekins et al. *(13, 14)*. The reason is the large amount of immobilized probes used in microtitre plate assays capture so many target molecules from the sample that the concentration of the target molecules is decreased. The result is that the density of immobilized target molecules is decreased leading to a lower signal to noise ratio. In microarray assays however, the spots are tiny and contain very small amounts of probe molecules. These small amounts of probe do not affect the concentration of target molecules in the sample. Therefore, microspot assays yield higher density of immobilized target molecules and results in higher signal to noise ratios compared to microtitre plate assays. Typically, microarray assays use less than one percent of the target molecules present in the sample *(15)* .

Even though immunoassays were the first to be explored for increased sensitivity and decreased sample requirement *(13, 14)*, microarray-based protein assays did not show promising results until late 1990s *(4, 16)*. DNA arrays started the revolution in the early 1990s where biology transitioned from mainly hypothesis driven research to also include discovery driven research. Microarrays are powerful tools for answering questions such as 'which genes are up-regulated by drug X...'. In contrast, classical hypothesis driven research attempted to answer questions such as 'is drug X regulating Gene Y'.

Discovery grade array requires large number of spots per surface area in order to be useful. Arraying DNA onto membranes was introduced in 1979 and was referred to as 'dot blot' *(17)*. At the time, because of the porosity of membranes and the large spotting volumes, a limitation of the spotting equipment, the density of spots in the arrays was quite low. In other words, only a few different probes were immobilized within a given area of the membrane. Since the spots were at least 1 mm large and the distance between each spot was 1 mm, only 25 different spots could be fit in each cm^2. It was clear that the throughput of dot blot was incapable of extracting the huge amount of information contained within cells. For example, a dot blot assay to determine gene expression of the 40,000 different mRNA in the cell would require a membrane of the size 1600 cm^2 corresponding to 160 microscope slides. It is clear that using 160 microscope slides per experiment is cumbersome, expensive and would require 160-fold more samples. Moving from a porous solid support to rigid solid support allowed for the emergence of high density arrays that could be used for discovery driven research *(18, 19)*.

3. Applications

3.1. Gene Expression Arrays

Gene expression profiling using DNA microarrays gives information about the relative differences in gene expression between two different cell populations, e.g. 'treated' cells compared to 'untreated' cells or cancer cells compared to normal cells. The degree of up- and down-regulation can be estimated for each gene but not the amount (absolute number of molecules) of mRNA expressed in treated cells vs. untreated cells. The first pan-transcriptome arrays contained probes towards all the genes in yeast (slightly more than 6,000 transcripts) and were used in several ground breaking publications that demonstrated the usability of DNA microarray technology for genome wide gene expression analysis. Yeast pan-transcriptome arrays were used to determine which genes are involved in the metabolic shift from fermentation to respiration *(18)*, cell cycle *(20, 21)*, sporulation *(22)* and ploidy *(23)*. After the human genome was sequenced, it was possible to design probes towards known as well as predicted transcripts/genes in order to get as complete a transcriptome analysis as possible. With arrays capable of analyzing the whole human transcriptome, gene expression analysis has been widely used for research on cell physiology and to find diagnostic markers/mechanisms for diseases such as cancer *(24)*. However, gene expression analysis also has other applications including the determination of water pollution by examining expression profiles in mussels *(25)*, of biocompatibility of surfaces and microchips for cell culture *(26, 27)* and responses to irradiation *(28)*. Examples of other types of gene expression arrays are exon arrays *(29)* and siRNA arrays *(30, 31)*

It is not by chance that gene expression analysis was and still is the most used application of microarray technology. Biologically, mRNA levels usually reflect gene function though function of a gene is ultimately determined at the protein activity level. Technically, the transcriptome is easily accessible. Probes made of DNA are easy to obtain and manipulate during fabrication. The transcriptome is relatively well described and limited in size as compared to the corresponding genome. Furthermore, probes can be designed based on expressed sequence tags (EST). The ESTs are generated by sequencing clones of poly A+ molecules in the cell. They can be viewed as a collection of sequences representing the mRNAs that are expressed in a cell at a given time. Conveniently, the whole genome does not need to be sequenced to collect EST data and the first large microarrays were based on ESTs *(32, 33)*. Therefore, transcriptional maps can be generated even for organisms for which there is limited genome sequence data.

3.2. Genome Wide SNP Analysis and Mutation Analysis of Genes

It is estimated that there are about 10 million single nucleotide polymorphisms (SNPs) in the human population. The SNPs are spread throughout the whole genome and can be used as genomic markers for finding links between genes and diseases. A SNP is typically a substitution of one base for another specific base. For example, a G is substituted with a C while all other bases in close proximity of the SNP are unchanged. As SNPs can be located inside as well as outside of genes, target needs to be prepared in such a way that the whole genome is represented. This can be major obstacle for such genome wide analysis. Affymetrix and Illumina provide arrays for genotyping 2.8 million and 1 million SNPs, respectively. The technologies of these companies are based on allele-specific hybridization *(34)* and allele-specific primer extension, respectively *(35)*. Allele-specific hybridization is based on probes that are centred over the mutation site so that the variant base is approximately in the middle of probe. Centering the variant base to the middle of the probe destabilizes mismatch hybrids maximally and the probe will therefore be highly sensitive to mutations in the target. At least two probes are used for detecting a particular SNP: one probe is perfectly matched with one allelic variant and another probe is specific for the other allelic variant. The relative signal strength between the two different probes after stringent hybridization and/or washing is used to assign genotype. Though two probes suffice in principle, Affymetrix uses about 20 probes for each SNP analyzed to obtain enough specificity in the assay *(34)*. Allele-specific primer extension is based on placing the probes so that last nucleotide of the probe is placed over the site of the SNP. A polymerase reaction can be initiated if the probe ends with a perfect match while mismatch hybridization will give a 'flapping' 3' end that cannot serve as an initiation structure for polymerization.

The same technologies used for SNP genotyping can be used to genotype mutations that cause monogenetic diseases *(36, 37)*. The drive of array-based assay is to replace automatic sequencing for diagnostics. Depending on the fabrication and detection methods used microarrays can be cheaper, faster and less laborious than automated sequencing. In monogenetic diseases, the gene is mutated in ways that modulate the activity of the corresponding protein. Protein activity can be modulated by mutations in the promoter, exon and introns. Typically many mutations can be found near sequences that encode regulatory motifs or catalytic sites of the protein product. The consequence is an increase in the number and complexity of the probes required to genotype a single mutation within such regions as the probe usually overlaps many mutations simultaneously.

3.3. Comparative Genomic Hybridization

Comparative genomic hybridizations (CGH) are used to find large deletions and amplifications within genomes *(38)*. CGH was originally based on immobilizing whole chromosomes on glass slides and co-hybridizing different fluorescent labelled controls and sample DNAs to the chromosomal preparation. The control DNA originates from cells with normal karyotype while the sample can derive for example from a tumour. The different genetic content in the sample and the control DNA is then resolved using the immobilized condensed chromosomes. The resolution of the original approach is quite low, about 20 Mb in range, due to the use of condensed chromosomes as probes. Array CGH is currently a very popular technique and is based on an immobilized array of probes, much like gene expression arrays. As in the original assay, differently labelled DNAs from sample and control is hybridized to the arrays. In array CGH, the resolution is mainly determined by the number of probes on the array. For example, 32,000 different probes evenly distributed throughout the human genome gives array CGH a resolution of about 0.1 M base *(3)*. CGH arrays can consist of rather large probes produced using BAC clones *(3, 39)* or smaller cDNA clones *(40)*. Alternatively, SNP arrays can be utilized where loss of heterozygocity is taken as proof of a deletion *(3, 41)*. CGH can be used for finding insertion and deletions of chromosomal material *(42)* or copy number variation analysis *(43)*.

3.4. Array Based Chromatin Immunoprecipitation Assays (ChIP or Chip Assays)

Chromatin immunoprecipitation or ChIP Assays are used to find the promoters that bind a specific transcription factor *(44)* . The principle of ChIP is to crosslink the DNA and proteins together and subsequently isolate DNA fragments that have bound a particular transcription factor using immunoprecipitation with antibodies specific to the transcription factor of interest. After amplification, different fragments can be identified on DNA microarrays consisting of probes towards the promoter regions. The ChIP assay is not limited to transcription factors and can also be used for other DNA binding proteins such as histones.

3.5. Other Assays

3.5.1. Sequencing

Microarrays can be successfully used for re-sequencing purposes *(45, 46)*. Re-sequencing arrays are in principle the same as SNP array with the exception that four probes are used to determine the bases in a particular site. The variant base is centred in the middle of the probes as described above. There are therefore four probes for each base investigated in a sequence and for example re-sequencing 10 bases in a row requires 40 probes (ten times four). Therefore, sequencing of one million bases requires four million probes. Such re-sequencing can be used for identification of pathogens *(45–47)* and mutational analysis of mitochondria *(48)* and genetic variability of genome segments *(49)*.

3.5.2. Transfection Arrays

DNA microarrays can also be used for experiments that are more complex than hybridization reactions. Plasmids contained in gelatin or other similar matrices can be arrayed by spotting these onto glass microscope slides. Cells are subsequently plated and grown over the surface of the array and a transfection of the plasmids within the array is mediated simultaneously by liposomes *(50)*. Genes involved in apoptosis were efficiently mapped by transfection arrays using 1959 different plasmids spotted on a microscope slide. Subsequent utilization of gene expression arrays on transfectants gave information about which proteins were regulators and which were effectors of apoptosis *(51)*.

3.5.3. Template for Protein Array Synthesis

Plasmids can also be used as templates for the 'just in time' in situ creation of protein microarrays. In such arrays, plasmids carrying different genes, cloned in-frame with the GST gene under the control of a T7 promoter, are spotted together with an antibody towards GST. The arrays of plasmids are transcribed and translated simultaneously using a cell-free lysate such a reticulocyte lysate. The fusion protein is retained on the respective spots by the co-spotted antibody *(52)*. Such arrays can subsequently be used for protein–protein interaction studies.

4. Detailed Description of Microarray Technology

Though the principals behind microarray technology seem simple, it is far from easy to perform a 'complete' microarray experiment from start to finish. The outline of what is required for a microarray experiment is shown in **Fig. 1.1** and discussed in detail below.

4.1. Fabrication

Most users are likely not concerned with details concerning the fabrication of DNA microarrays such as probe choice and chemistry, as in most cases a complete hybridization-ready microarray can be purchased. Manufacturers offering pre-made arrays include Affymetrix, Illumina and Agilent. Microarrays for gene expression, comparative genomic hybridization and detailed SNP analysis are commercially available for a number of popular organisms. However, microarrays for gene expression investigation for the majority of organisms are not commercially available and in these cases, choices regarding probe design, chemistry and fabrication are required.

4.1.1. Probe Choice

The production of a DNA microarray that is ready for hybridization is a complex process. First, probes for microarrays are selected from nucleotide sequence databases such as Genebank. Probe choice strategy is highly dependent on the application and microarray fabrication platform used but a probe should be specific and be able to efficiently capture the target. For gene expression arrays based on 25–60 nucleotide long probes, probe sequences are often chosen from nucleotide sequences found in the 3'end of the transcript. The reason is that cDNA synthesis is often initiated from the 3' end of the transcript using polyT primers. The results are cDNA fragments that mostly represent the 3'end of the transcripts. In order to maximize signal, the probes are then placed towards the 3'end of the transcript. This results in a bias towards the presence of 3'end sequences. Alternatively, polyT sequences attached to a solid support can be used to select Poly A transcripts before random primer extension is initiated. If possible gene expression array probes should have the same theoretical melting temperature (Tm) in order to function at the same stringency and similar Gibbs free energy to yield similar hybridization signals between spots *(53)*. cDNA arrays for expression analysis utilize probes that are >1000 nt and the requirement for Tm matching is less of a problem since there is little variation in base composition between >1000 nt sequences.

In contrast to probes used for gene expression studies that can be placed 'somewhere' in the 3' end, probes for analysis of mutation need to be placed at the site of mutation whether or not it is in a GC rich or a AT rich region. This can put severe restraints on probe selection because it can be difficult to Tm match probes. Most sensitive to this is allele-specific hybridization that requires precise Tm match of probes to discriminate between single base changes. Mutation analysis using allele-specific primer extension or mini-sequencing is not dependent on precise Tm matching but only requires that the probes end either at the nucleotide being investigated or at the nucleotide just before the nucleotide being investigated. Tm matching of probes is easier for SNP analysis than mutation analysis of specific genes. The reason is that SNPs can be chosen with the only criterion that it is a marker for a specific locus. Thus SNP in GC rich and AT rich regions can be avoided. This is not possible to do for mutation analysis of genes since each mutation has been described to have a phenotype and must thus be genotyped.

4.1.2. Immobilization of DNA

The solid support and the chemistry used to immobilize probes is very important and may influence the background signal, stability of the bond between the probes and the solid support, probe density, hybridization efficiency, DNA hybrid characteristics, spot morphology, spot density and spot reproducibility (**Table 1.1**). Typically DNA microarrays are fabricated on a solid glass support because glass is rigid, allows for fluorescent detection as it is transparent, and can easily be chemically modified. Polymeric materials have also been considered as solid supports because of the

Table 1.1
Factors influencing different parameters of a microarray assay

		Specificity	Sensitivity	Probe density	Morphology	Spot density	Geometry
Microarray fabrication	Robotics (X-Y precision)					+++	+++
	Probe sequence	+++	+++				
	Spotter type (inkjet etc)		+	+	+++	+++	+
	Spotting conditions		++	++	+++	+++	+
	Immobilization chemistry	++	+++	+++	+++	++	++
	Probe conc	+	++	+++	+		+
	Spotting buffer	+	+	++	++	++	+
Target prep	Cell purity	+++	+++				
	Nucl. Acids purification	++	+++				
	Amplification	+	+++		+		
	Labelling	+	+++				
Hyb cond.	Mixing	+	+++		+		
	Stringency	+	++				
S/N	Background	+	+++				
	Detection methods	++	+++				

possibilities to incorporate other approaches, including the use of microfluidic structures within the solid support itself. The flexibility of polymeric materials is a drawback for detection purposes (see below) but the advantage of these is that they are not brittle like glass is. Silicon solid supports can also be utilized for microarray fabrication. There are protocols for making arrays of pre-synthesized oligonucleotides (reviewed in *(54)* and in Chapter 6) as well as for on-chip synthesis of oligonucleotides on polymeric, glass and silicon solid supports *(55–57)*.

4.2. Target Preparation

Target preparation is a complex procedure that is always the responsibility of the end user of the microarray. Inappropriate target preparation limits the potential of the microarray experiment even if high quality microarrays and advanced bioinformatic systems are used. Target preparation is usually a multi-step process that can be divided into cell and nucleic acids purification, amplification and labelling (**Fig. 1.1**). The number of steps required is dependent on the application and the biological material at hand.

4.2.1. Cell Preparation

Cells can be selected from complex matrixes such as tissue or blood using laser microdissection *(58)* or antibody affinity purification, respectively. For gene expression applications, the selection of target cells is very important because analyzing a mixture of different cells will result in an average gene expression profile of the mixture(normal and treated cells) and important gene regulations can be missed. CGH analysis may also require selection of cells. Analysis of complex tissues such as tumours requires purification of cancer cells from the surrounding healthy tissue prior to analysis. Normal diploid cells will reduce the amplitude of the signal coming from amplifications and/or deletions of the tumour cells. Analysis of inherited chromosomal changes is by contrast not dependent at all on the type of diploid cell analyzed and requires no purification.

4.2.2. Nucleic Acid Purification

The aim of nucleic acid purification is to prepare sufficient amount of either DNA or RNA to levels of such purity that it can be used in enzymatic reactions. DNA is the least sensitive nucleic acid and can readily be prepared from fresh or frozen tissue materials in sufficient amounts using a number of different methods. Archived material such as paraffin embedded tissue slices can also be used but yield DNA of lesser quality *(59)*. By contrast, RNA is more sensitive as it is easily degraded by endogenous nucleases (RNases). RNA preparation methods must inhibit RNase activity. Often, guanidinium isothiocyanate (GITC)-based methods are used in the RNA preparation protocols to avoid RNase activity *(60)*. Though RNase activity is inhibited during the preparation protocol, RNA may be degraded after the sample is taken if it is not snap frozen or directly lysed in GITC containing lysis buffer. In this buffer, RNA is stable for long-term storage at −20 °C *(61)*.

4.2.3. Amplification

Often there is a need to assay from limited amounts of sample material. Large arrays for the investigation of genome wide SNP analysis, CGH or gene expression analysis typically require large amounts of target in order to give sufficient hybridization signal. As such, there is a need to amplify the DNA or RNA prior to labelling of the target.

The genome can first be cut into fragments with restriction enzymes and primer sequences subsequently ligated to the fragments. The complex DNA can then be amplified using PCR. This amplification method requires little starting material, 250 ng, but reduces the complexity of the resulting target *(34)*. Alternatively, Phi29- based random primed isothermal wide genome amplification is a non-PCR-based method that gives better coverage of complex genome and allows genome wide genetic analysis from 100 ng DNA samples *(35)*.

mRNA can also be amplified but only indirectly. The most popular method is to reverse transcribe mRNA into cDNA using polyT primers modified with T7 promoter sequences in the 5'end. Double stranded DNA is then generated where each fragment carries a T7 viral promoter that can be utilized for T7 in vitro transcription reaction (IVT). An 80-fold amplification of the mRNA can be achieved using this method and the amplified product is designated 'aRNA' *(62)*. The method has since been modified to use two rounds of amplification resulting in an approximately millionfold amplification of the target *(63)*. This is sufficient for generating gene expression profiles from single cells *(64)*. Besides the large amplification achieved by this method, another attractive feature is that it generates single stranded targets. This in part explains the efficiency of this amplification method for microarray analysis. Even though the target is generated using random priming and is by nature a linear amplification form, it usually represents the original mRNA population qualitatively and quantitatively less well than cDNA directly reverse transcribed *(65)*. The reason is that there is always a selection during enzymatic reactions and the selection is enhanced by the amplification process. It is therefore not unexpected that the correlation between the gene expression profile obtained using different target preparation methods is gradually decreased with increasing amplification of the target *(65)*. Though reverse transcription is the gold standard for target preparation for microarray experiments, it has been shown that reverse transcription can also introduce systematic biases in array experiment as compared to hybridization of labelled mRNA *(66)*.

4.2.4. Labelling

Typically, target hybridized to an array is labelled with fluorescent molecules or biotins for post-hybridization staining. For gene expression analysis, the target is usually labelled during cDNA synthesis from RNA by spiking labelled nucleotides into the reverse transcription reaction. Similarly, aRNA ready for hybridization contains labelled ribonucleotides that are incorporated during the IVT reaction. Alternative approaches have been used to avoid enzymatic treatment of mRNA prior to hybridization. These involve direct labelling of RNA prior to hybridization *(66)* or to stain the RNA post-hybridization with gold nanoparticles covered with poly-T oligonucleotides *(68)*.

There are similar ways to label DNA prior to hybridization for genetic tests. Direct labelling of PCR primers is a rapid and convenient method to introduce a label during the amplification process. Alternatively, labelled nucleotides can be spiked into a labelling reaction to give random labelling of the fragments. The latter method typically introduces several labels per strands compared to end labelled primers.

4.3. Hybridization Reactions

Microarrays were first hybridized under cover slips using 2 μL of highly concentrated target solution per cm^2 *(1)*. This hybridization method relies solely on diffusion of target molecules to the corresponding spot. Since target molecules are fairly large, the diffusion time from one side of the array to the other takes many years, it can be expected that the spot only reacts with targets that are present in a fraction of the total hybridization volume. Hybridization without mixing often results in heterogeneous reaction conditions over the array surface. It would therefore be advantageous to perform mixing on arrays. A literary survey indicates several solutions including; cavitation micro streaming *(69)*, magnetic bar stirring *(70)*, air driven bladders *(15)*, centrifugal mixing *(71)* and shear driven mixing *(72)*. A drawback of most mixing strategies is that the sample is significantly diluted in order to be mixed using the above described methods compared to static hybridization using very small volumes. Despite dilution, mixing often gives a 2–10-fold increase in signal, provides homogeneous hybridization conditions over the entire array and lowers background signals.

4.4. Detection

By far the most popular method for detection of hybridized array is fluorescence. Fluorochromes offer high sensitivity, large dynamic range, are easy to work with and a single array can be stained with up to four different fluorochromes, each with distinct spectral properties. The drawbacks of fluorescence are photo bleaching during exposure and decomposition of fluorochromes over time. These make fluorescent stains less suitable for long-term archiving of hybridized slides. Since a microarray assay is generally not quantitative, each sample must be compared with a control. Gene expression arrays can be used with either one or two fluorescent dyes. The use of a single fluorescent dye for detection requires that the sample and the control are hybridized on two different slides, whereas the use of two different fluorescent dyes allows the use of a single slide to probe both the sample and control targets.

Utilization of three fluorochromes gives opportunities for quality control. One of the dyes is only used as an indicator of the presence and relative quantity of immobilized probes while the other two fluorochromes can be used for sample and control target labelling and detection *(73, 74)*. Missing spot or poor quality spots can then be easily filtered out prior to analysis.

Four different fluorochromes might be used to detect the four different nucleotides that can be incorporated in DNA as in minisequencing reactions *(75)*.

Although fluorescence is popular, the method requires fairly expensive scanners and many applications such as gene expression profiling and CGH also need better assay detection limits. Therefore many other approaches for microarray detection have been proposed. Light scattering of silver enhanced gold nanoparticles has several orders of magnitude better detection limits compared to traditional fluorescent detection and allows for SNP analysis from unamplified genomic DNA *(76)* and gene expression analysis from as little as 0.5 μg of unamplified total RNA *(68)*. Hesse et al. have recently demonstrated a novel but not yet commercially available fluorescent-based scanner system that gives similar or better sensitivity to scattering of light by gold/silver particles for the detection of single molecules hybridized to microarray spots *(77)*.

Measuring absorbance provides inexpensive solution for detection of analyte binding to DNA microarray. Suitable staining methods include gold nano particles *(78)*, gold/silver particles *(79)* and enzyme based stains such as alkaline phosphatase BCIP/NBT reactions *(80, 81)*. Such stains are visible provided that the spot is sufficiently large but can conveniently be digitized with an inexpensive 1200 dpi or better flatbed scanner. The drawbacks of these staining methods are significantly higher detection limit and limited dynamic range of the assays.

4.5. Data Analysis

4.5.1. Quantification

After digitalization using a scanner, the images (usually TIF files) are analyzed in specialized software. The relative fluorescence of a spot is quantified by calculating the 'whiteness' of the spot. This is simply done by calculating the pixel values within a spot where black is equal to no signal, white is maximum signal and the different gray scales correspond to everything in between minimum and maximum signal. Defining a spot and the pixels that should be counted is not easy. Some software requires that each spot is defined by the user by encircling the spot (the freeware Scanalyze *(82)* is an example). Once the spot is defined, the pixel value of every pixel within the spot is summed to give a total spot signal or summed and divided by the number of pixels in order to give the density of the fluorescence. Spots of different sizes constitute a problem and usually require user intervention, a cumbersome and time intensive process. Spotfinder, another freeware, only allows pixels above the background signal level to be included in the spot. In Spotfinder, spots with severely malformed morphologies such as 'coffespots' and 'halfmoon' shapes can be still be quantified.

The background signal originating from dark currents of the instrument, substrate chemistries as well as unspecific binding of target to the substrate surface can be calculated in different ways.

Local background signal is calculated by examining pixels surrounding each spot. The fluorescent value of the spot is then calculated to be (signal from spot – signal from the background). This is a very good method to use if the microarray slide has uneven background. A simpler approach is to subtract the average background value of the entire microarray slide from each of the spots on the array.

4.5.2. Normalization

A drawback of co-hybridization in microarray-based CGH and gene expression assays is that different fluorescent dyes are not perfectly matched in terms of quantum yields and sensitivity to light and ozone. For instance, there is a non-linear relationship between the signals from Cy3 and Cy5, the most popular dyes for detection on arrays. These differences must be compensated for by normalization software such as QSPLINE *(83)*. Using the same dye for both the sample and control does not have the above problem. In this case, differences between hybridization reactions must be compensated for in silico. After normalization, the data set can be analyzed by various statistical methods.

5. Parameters Used to Describe Microarray Assays

5.1. Geometry

Geometry of the array refers to how well the spots are ordered into an array; i.e. how even/equal are the distances between each spot. Even spacing between the spots is important because misaligned spots are difficult to quantify automatically. Geometry of the array is mainly determined by the precision of the machinery used to fabricate the arrays. For spotted arrays, geometry can be affected when spotting on hydrophobic surfaces because the droplet can 'move' from the point where it was deposited.

5.2. Spot Density

Spot density is defined as the number of spots per unit area. Spot density is determined by the precision of the machinery to localize a spot in (x,y) co-ordinate system, probe solution ejection system, immobilization chemistry and the composition of spotting solution (probe concentration and the spotting buffer). In most cases it is the size of the spots that determines spot density. Spot size can be altered by changing the spotting volume. Spot volumes are determined by the deposition technique used to deliver the droplets to the surface, the hydrophobicity of the surface (determined by the chemistry) and the composition of the spotting buffer. For instance, spots on hydrophobic surfaces can be made larger by adding appropriate amount of detergents in the spotting buffer.

Increased spot density is required in recent years to meet the demands of the ever increasing complexity of microarray experiments. The present drivers for increasing spot density are genome and proteome analysis and not gene expression arrays. Chapter 5 discusses methods to fabricate arrays with higher spot density/ more compact arrays.

5.3. Probe Density

Probe density is defined as the number of probe molecules per unit area. It is a measurement used to characterize immobilization chemistries that link DNA to microarray surfaces. The probe density is determined by the chemistry of the surface, modifications made to the DNA to increase immobilization, the size of the molecules to be immobilized, the spotting buffer and the probe concentration (**Table 1.1**).

The chemistry of the surface and the probes determines the efficiency of immobilization of the probes. Spotting buffers also need to be chosen correctly so that the spotting buffer does not interfere with the chemistry. For instance, TRIS buffers need to be avoided if the surface contains epoxy or aldehyde funtionalization because TRIS contains amines that can react with the surface before the DNA has a chance to bind. The spotted probe concentration is very important and too low a concentration will give spots with few capture molecules and thus low maximum signal is usually translated to low sensitivity. Optimal spotted probe concentration often needs to be titrated for each surface chemistry. Critical to high probe density is the deposition of the correct drop volume on the surface (*see* also **Section 5.2**) so that the spots, when dried, contain at least a monolayer of molecules that can be attached to the surface. Probe density is usually optimized to maximize hybridization signal but not hybridization efficiency. This is to produce spots with as large a dynamic range and as high a sensitivity as possible. Probe density also determines the upper limit of the possible hybridizations within a spot. In many cases, only a fraction of the probes immobilized to the surface can undergo hybridization even under saturated conditions. The density of hybridized targets to a surface is referred to as 'hybridized density'.

5.4. Sensitivity

The sensitivity of a microarray assay is defined as the lowest concentration of target molecules that can be detected on a spot. Sensitivity is affected by all factors of a microarray experiment and is the parameter that most users have problems with. In particular, users that set up their own microarray assays have to consider all factors that could influence sensitivity (**Table 1.1**). Users who buy ready made arrays are limited to optimize and/or select appropriate target preparation methods, detection systems and/or hybridization conditions to increase the sensitivity.

Probe density and organization on the surface as well as the affinity of the probes is determined during the fabrication phase. Probe affinity should be high. This is predicted using the calculated ΔG values of the probes *(84)*. The probes must also be immobilized in the correct density to obtain maximum signal. Factors affecting probe density are discussed above (**Section 5.3**). Low signal on arrays can result from suboptimal probe function, which can be caused by the molecular organization of the probes on the surface. It is well known that the use of molecular spacers, to move short probes away from the substrate surface gives better signal than short probes directly linked to surfaces.

Target preparation is very critical to produce highly sensitive assays. As previously discussed, the target preparation method used determines the concentration of target molecules to be hybridized to the array which in turn determines the sensitivity of the assay. Mixing benefits sensitivity because it moves target molecules from one end of the microarray to the other; something not possible by diffusion alone.

Finally, the detection system and method has a large impact on sensitivity (*see* **Section 4.4**). Usually, instrumentation is fixed because of the high cost associated with acquiring new equipment. However, the sensitivity of the detector in the instruments can, in most cases, be adjusted to appropriate levels to obtain the highest assay sensitivity. For example, for weakly fluorescent arrays the sensitivity of instruments can easily be adjusted so that the array source is exposed to more excitation light resulting in more light emission from the arrays. However, it should be noted that the background signal usually increases as well when the sensitivity of the instrument is increased.

5.5. Specificity

Specificity (or selectivity) is defined as how selective the probes are to capture the intended target in a complex background of other target molecules. Targets with similar but not identical sequences may bind to the probe intended for the intended target. The binding of an 'unintended' target to a probe is called cross-hybridization. Cross-hybridization must be minimized as it decreases the diagnostic power of genotyping arrays and the resolution of up-and down-regulation of genes in gene expression microarray experiments. In traditional Northern blot assays specificity is obtained by the sequence and length of the target. Thus unspecific hybridization to other fragments than the intended one is easily observed because these fragments typically are of different lengths. The specificity of microarrays is based on the sequence of the probes, therefore probe selection is critical. Closely related to probe selection is the stringency level of the assay. Probes should be chosen so that all the probes on the array function optimally at a single stringency condition, i.e. on determined buffer concentration and temperature. Optimal condition for probe function can be predicted by calculating the melting temperature of the probes. Choosing appropriate stringency is usually a balance between having high signals on the arrays and sufficient specificity. Approaches other than the probe sequence can be used to maximize the specificity of the assay. One is to remove unwanted nucleic acids from a complex target. This can be achieved for some applications by selecting only cells of interest using immuno capture and thereby removing the 'contaminating' cells from the assay. For monogenetic diseases or viral/bacterial diagnostics, it is very convenient to 'select' nucleic acids to be analyzed using PCR.

Acknowledgements

Thanks are due to David Sabourin, Lena Poulsen and Jesper Petersen for reading the manuscript and helpful discussions.

References

1. Schena, M., Shalon, D., Davis, R.W. and Brown, P.O. (1995) Quantitative monitoring of gene expression patterns with a complementary DNA microarray. *Science*, **270**, 467–470.
2. Zammatteo, N., Jeanmart, L., Hamels, S., Courtois, S., Louette, P., Hevesi, L. and Remacle, J. (2000) Comparison between different strategies of covalent attachment of DNA to glass surfaces to build DNA microarrays. *Anal Biochem*, **280**, 143–150.
3. Ishkanian, A.S., Malloff, C.A., Watson, S.K., DeLeeuw, R.J., Chi, B., Coe, B.P., Snijders, A., Albertson, D.G., Pinkel, D., Marra, M.A. et al. (2004) A tiling resolution DNA microarray with complete coverage of the human genome. *Nat Genet*, **36**, 299–303.
4. Lueking, A., Horn, M., Eickhoff, H., Bussow, K., Lehrach, H. and Walter, G. (1999) Protein microarrays for gene expression and antibody screening. *Anal Biochem*, **270**, 103–111.
5. Huang, R.P. (2001) Detection of multiple proteins in an antibody-based protein microarray system. *J Immunol Methods*, **255**, 1–13.
6. Huang, R.P., Huang, R., Fan, Y. and Lin, Y. (2001) Simultaneous detection of multiple cytokines from conditioned media and patient's sera by an antibody-based protein array system. *Anal Biochem*, **294**, 55–62.
7. Li, Y., Lee, H.J. and Corn, R.M. (2006) Fabrication and characterization of RNA aptamer microarrays for the study of protein-aptamer interactions with SPR imaging. *Nucleic Acids Res*, **34**, 6416–6424.
8. Lee, M. and Walt, D.R. (2000) A fiber-optic microarray biosensor using aptamers as receptors. *Anal Biochem*, **282**, 142–146.
9. Bradner, J.E., McPherson, O.M., Mazitschek, R., Barnes-Seeman, D., Shen, J.P., Dhaliwal, J., Stevenson, K.E., Duffner, J.L., Park, S.B., Neuberg, D.S. et al. (2006) A robust small-molecule microarray platform for screening cell lysates. *Chem Biol*, **13**, 493–504.
10. Wang, D., Liu, S., Trummer, B.J., Deng, C. and Wang, A. (2002) Carbohydrate microarrays for the recognition of cross-reactive molecular markers of microbes and host cells. *Nat Biotechnol*, **20**, 275–281.
11. Horlacher, T. and Seeberger, P.H. (2006) The utility of carbohydrate microarrays in glycomics. *Omics*, **10**, 490–498.
12. Velculescu, V.E., Zhang, L., Vogelstein, B. and Kinzler, K.W. (1995) Serial analysis of gene expression. *Science*, **270**, 484–487.
13. Ekins, R., Chu, F. and Biggart, E. (1990) Fluorescence spectroscopy and its application to a new generation of high sensitivity, multi-microspot, multianalyte, immunoassay. *Clin Chim Acta*, **194**, 91–114.

14. Ekins, R., Chu, F. and Biggart, E. (1990) Multispot, multianalyte, immunoassay. *Ann Biol Clin (Paris)*, **48**, 655–666.
15. Adey, N.B., Lei, M., Howard, M.T., Jensen, J.D., Mayo, D.A., Butel, D.L., Coffin, S.C., Moyer, T.C., Slade, D.E., Spute, M.K. et al. (2002) Gains in sensitivity with a device that mixes microarray hybridization solution in a 25-microm-thick chamber. *Anal Chem*, **74**, 6413–6417.
16. MacBeath, G. and Schreiber, S.L. (2000) Printing proteins as microarrays for high-throughput function determination. *Science*, **289**, 1760–1763.
17. Kafatos, F.C., Jones, C.W. and Efstratiadis, A. (1979) Determination of nucleic acid sequence homologies and relative concentrations by a dot hybridization procedure. *Nucleic Acids Res*, **7**, 1541–1552.
18. DeRisi, J.L., Iyer, V.R. and Brown, P.O. (1997) Exploring the metabolic and genetic control of gene expression on a genomic scale. *Science*, **278**, 680–686.
19. Iyer, V.R., Eisen, M.B., Ross, D.T., Schuler, G., Moore, T., Lee, J.C., Trent, J.M., Staudt, L.M., Hudson, J., Jr., Boguski, M.S. et al. (1999) The transcriptional program in the response of human fibroblasts to serum. *Science*, **283**, 83–87.
20. Cho, R.J., Campbell, M.J., Winzeler, E.A., Steinmetz, L., Conway, A., Wodicka, L., Wolfsberg, T.G., Gabrielian, A.E., Landsman, D., Lockhart, D.J. et al. (1998) A genome-wide transcriptional analysis of the mitotic cell cycle. *Mol Cell*, **2**, 65–73.
21. Spellman, P.T., Sherlock, G., Zhang, M.Q., Iyer, V.R., Anders, K., Eisen, M.B., Brown, P.O., Botstein, D. and Futcher, B. (1998) Comprehensive identification of cell cycle-regulated genes of the yeast Saccharomyces cerevisiae by microarray hybridization. *Mol Biol Cell*, **9**, 3273–3297.
22. Chu, S., DeRisi, J., Eisen, M., Mulholland, J., Botstein, D., Brown, P.O. and Herskowitz, I. (1998) The transcriptional program of sporulation in budding yeast. *Science*, **282**, 699–705.
23. Galitski, T., Saldanha, A.J., Styles, C.A., Lander, E.S. and Fink, G.R. (1999) Ploidy regulation of gene expression. *Science*, **285**, 251–254.
24. Perez-Diez, A., Morgun, A. and Shulzhenko, N. (2007) Microarrays for cancer diagnosis and classification. *Advances in experimental medicine and biology*, **593**, 74–85.
25. Venier, P., De Pitta, C., Pallavicini, A., Marsano, F., Varotto, L., Romualdi, C., Dondero, F., Viarengo, A. and Lanfranchi, G. (2006) Development of mussel mRNA profiling: Can gene expression trends reveal coastal water pollution? *Mutat Res.*
26. Stangegaard, M., Wang, Z., Kutter, J.P., Dufva, M. and Wolff, A. (2006) Whole genome expression profiling using DNA microarray for determining biocompatibility of polymeric surfaces. *Molecular Biosystems*, **2**, 421–428.
27. Stangegaard, M., Petronis, S., Jorgensen, A.M., Christensen, C.B. and Dufva, M. (2006) A biocompatible micro cell culture chamber (microCCC) for the culturing and on-line monitoring of eukaryote cells. *Lab Chip*, **6**, 1045–1051.
28. Tusher, V.G., Tibshirani, R. and Chu, G. (2001) Significance analysis of microarrays applied to the ionizing radiation response. *Proc Natl Acad Sci U S A*, **98**, 5116–5121.
29. Frey, B.J., Mohammad, N., Morris, Q.D., Zhang, W., Robinson, M.D., Mnaimneh, S., Chang, R., Pan, Q., Sat, E., Rossant, J. et al. (2005) Genome-wide analysis of mouse transcripts using exon microarrays and factor graphs. *Nat Genet*, **37**, 991–996.
30. Castoldi, M., Schmidt, S., Benes, V., Noerholm, M., Kulozik, A.E., Hentze, M.W. and Muckenthaler, M.U. (2006) A sensitive array for microRNA expression profiling (miChip) based on locked nucleic acids (LNA). *Rna*, **12**, 913–920.
31. Beuvink, I., Kolb, F.A., Budach, W., Garnier, A., Lange, J., Natt, F., Dengler, U., Hall, J., Filipowicz, W. and Weiler, J. (2007) A novel microarray approach reveals new tissue-specific signatures of known and predicted mammalian microRNAs. *Nucleic Acids Res.*
32. Schena, M., Shalon, D., Heller, R., Chai, A., Brown, P.O. and Davis, R.W. (1996) Parallel human genome analysis: microarray-based expression monitoring of 1000 genes. *Proc Natl Acad Sci U S A*, **93**, 10614–10619.
33. Loftus, S.K., Chen, Y., Gooden, G., Ryan, J.F., Birznieks, G., Hilliard, M., Baxevanis, A.D., Bittner, M., Meltzer, P., Trent, J. et al. (1999) Informatic selection of a neural crest-melanocyte cDNA set for microarray analysis. *Proc Natl Acad Sci U S A*, **96**, 9277–9280.

34. Matsuzaki, H., Dong, S., Loi, H., Di, X., Liu, G., Hubbell, E., Law, J., Berntsen, T., Chadha, M., Hui, H. et al. (2004) Genotyping over 100,000 SNPs on a pair of oligonucleotide arrays. *Nat Methods*, **1**, 109–111.
35. Gunderson, K.L., Steemers, F.J., Lee, G., Mendoza, L.G. and Chee, M.S. (2005) A genome-wide scalable SNP genotyping assay using microarray technology. *Nat Genet*, **37**, 549–554.
36. Chou, W.H., Yan, F.X., Robbins-Weilert, D.K., Ryder, T.B., Liu, W.W., Perbost, C., Fairchild, M., de Leon, J., Koch, W.H. and Wedlund, P.J. (2003) Comparison of two CYP2D6 genotyping methods and assessment of genotype-phenotype relationships. *Clin Chem*, **49**, 542–551.
37. Gemignani, F., Perra, C., Landi, S., Canzian, F., Kurg, A., Tonisson, N., Galanello, R., Cao, A., Metspalu, A. and Romeo, G. (2002) Reliable detection of beta-thalassemia and G6PD mutations by a DNA microarray. *Clin Chem*, **48**, 2051–2054.
38. Kallioniemi, A., Kallioniemi, O.P., Sudar, D., Rutovitz, D., Gray, J.W., Waldman, F. and Pinkel, D. (1992) Comparative genomic hybridization for molecular cytogenetic analysis of solid tumors. *Science*, **258**, 818–821.
39. Fiegler, H., Redon, R., Andrews, D., Scott, C., Andrews, R., Carder, C., Clark, R., Dovey, O., Ellis, P., Feuk, L. et al. (2006) Accurate and reliable high-throughput detection of copy number variation in the human genome. *Genome Res*, **16**, 1566–1574.
40. Pollack, J.R., Perou, C.M., Alizadeh, A.A., Eisen, M.B., Pergamenschikov, A., Williams, C.F., Jeffrey, S.S., Botstein, D. and Brown, P.O. (1999) Genome-wide analysis of DNA copy-number changes using cDNA microarrays. *Nat Genet*, **23**, 41–46.
41. Barrett, M.T., Scheffer, A., Ben-Dor, A., Sampas, N., Lipson, D., Kincaid, R., Tsang, P., Curry, B., Baird, K., Meltzer, P.S. et al. (2004) Comparative genomic hybridization using oligonucleotide microarrays and total genomic DNA. *Proc Natl Acad Sci U S A*, **101**, 17765–17770.
42. Pinkel, D. and Albertson, D.G. (2005) Array comparative genomic hybridization and its applications in cancer. *Nat Genet*, **37 Suppl**, S11–17.
43. Redon, R., Ishikawa, S., Fitch, K.R., Feuk, L., Perry, G.H., Andrews, T.D., Fiegler, H., Shapero, M.H., Carson, A.R., Chen, W. et al. (2006) Global variation in copy number in the human genome. *Nature*, **444**, 444–454.
44. Ren, B., Robert, F., Wyrick, J.J., Aparicio, O., Jennings, E.G., Simon, I., Zeitlinger, J., Schreiber, J., Hannett, N., Kanin, E. et al. (2000) Genome-wide location and function of DNA binding proteins. *Science*, **290**, 2306–2309.
45. Malanoski, A.P., Lin, B., Wang, Z., Schnur, J.M. and Stenger, D.A. (2006) Automated identification of multiple micro-organisms from resequencing DNA microarrays. *Nucleic Acids Res*, **34**, 5300–5311.
46. Lin, B., Wang, Z., Vora, G.J., Thornton, J.A., Schnur, J.M., Thach, D.C., Blaney, K.M., Ligler, A.G., Malanoski, A.P., Santiago, J. et al. (2006) Broad-spectrum respiratory tract pathogen identification using resequencing DNA microarrays. *Genome Res*, **16**, 527–535.
47. Lin, B., Blaney, K.M., Malanoski, A.P., Ligler, A.G., Schnur, J.M., Metzgar, D., Russell, K.L. and Stenger, D.A. (2007) Using a resequencing microarray as a multiple respiratory pathogen detection assay. *J Clin Microbiol*, **45**, 443–452.
48. Chee, M., Yang, R., Hubbell, E., Berno, A., Huang, X.C., Stern, D., Winkler, J., Lockhart, D.J., Morris, M.S. and Fodor, S.P. (1996) Accessing genetic information with high-density DNA arrays. *Science*, **274**, 610–614.
49. Wang, D.G., Fan, J.B., Siao, C.J., Berno, A., Young, P., Sapolsky, R., Ghandour, G., Perkins, N., Winchester, E., Spencer, J. et al. (1998) Large-scale identification, mapping, and genotyping of single-nucleotide polymorphisms in the human genome. *Science*, **280**, 1077–1082.
50. Ziauddin, J. and Sabatini, D.M. (2001) Microarrays of cells expressing defined cDNAs. *Nature*, **411**, 107–110.
51. Palmer, E.L., Miller, A.D. and Freeman, T.C. (2006) Identification and characterisation of human apoptosis inducing proteins using cell-based transfection microarrays and expression analysis. *BMC Genomics*, **7**, 145.
52. Ramachandran, N., Hainsworth, E., Bhullar, B., Eisenstein, S., Rosen, B., Lau, A.Y., Walter, J.C. and LaBaer, J. (2004) Self-assembling protein microarrays. *Science*, **305**, 86–90.

53. Bruun, G.M., Wernersson, R., Juncker, A.S., Willenbrock, H. and Nielsen, H.B. (2007) Improving comparability between microarray probe signals by thermodynamic intensity correction. *Nucleic Acids Res.*
54. Dufva, M. (2005) Fabrication of high quality microarrays. *Biomol Eng*, **22**, 173–184.
55. Moorcroft, M.J., Meuleman, W.R., Latham, S.G., Nicholls, T.J., Egeland, R.D. and Southern, E.M. (2005) In situ oligonucleotide synthesis on poly(dimethylsiloxane): a flexible substrate for microarray fabrication. *Nucleic Acids Res*, **33**, e75.
56. Southern, E.M., Case-Green, S.C., Elder, J.K., Johnson, M., Mir, K.U., Wang, L. and Williams, J.C. (1994) Arrays of complementary oligonucleotides for analysing the hybridisation behaviour of nucleic acids. *Nucleic Acids Res*, **22**, 1368–1373.
57. Fodor, S.P., Read, J.L., Pirrung, M.C., Stryer, L., Lu, A.T. and Solas, D. (1991) Light-directed, spatially addressable parallel chemical synthesis. *Science*, **251**, 767–773.
58. Ohyama, H., Zhang, X., Kohno, Y., Alevizos, I., Posner, M., Wong, D.T. and Todd, R. (2000) Laser capture microdissection-generated target sample for high-density oligonucleotide array hybridization. *Biotechniques*, **29**, 530–536.
59. Gilbert, M.T., Haselkorn, T., Bunce, M., Sanchez, J.J., Lucas, S.B., Jewell, L.D., Van Marck, E. and Worobey, M. (2007) The isolation of nucleic acids from fixed, paraffin-embedded tissues-which methods are useful when? *PLoS ONE*, **2**, e537.
60. Chomczynski, P. and Sacchi, N. (1987) Single-step method of RNA isolation by acid guanidinium thiocyanate-phenol-chloroform extraction. *Anal Biochem*, **162**, 156–159.
61. Gilleland, R.C. and Hockett, R.D., Jr. (1998) Stability of RNA molecules stored in GITC. *Biotechniques*, **25**, 944–946, 948.
62. Van Gelder, R.N., von Zastrow, M.E., Yool, A., Dement, W.C., Barchas, J.D. and Eberwine, J.H. (1990) Amplified RNA synthesized from limited quantities of heterogeneous cDNA. *Proc Natl Acad Sci U S A*, **87**, 1663–1667.
63. Eberwine, J., Yeh, H., Miyashiro, K., Cao, Y., Nair, S., Finnell, R., Zettel, M. and Coleman, P. (1992) Analysis of gene expression in single live neurons. *Proc Natl Acad Sci U S A*, **89**, 3010–3014.
64. Kamme, F., Zhu, J., Luo, L., Yu, J., Tran, D.T., Meurers, B., Bittner, A., Westlund, K., Carlton, S. and Wan, J. (2004) Single-cell laser-capture microdissection and RNA amplification. *Methods Mol Med*, **99**, 215–223.
65. Hughes, T.R., Mao, M., Jones, A.R., Burchard, J., Marton, M.J., Shannon, K.W., Lefkowitz, S.M., Ziman, M., Schelter, J.M., Meyer, M.R. et al. (2001) Expression profiling using microarrays fabricated by an ink-jet oligonucleotide synthesizer. *Nat Biotechnol*, **19**, 342–347.
66. Gupta, V., Cherkassky, A., Chatis, P., Joseph, R., Johnson, A.L., Broadbent, J., Erickson, T. and DiMeo, J. (2003) Directly labeled mRNA produces highly precise and unbiased differential gene expression data. *Nucleic Acids Res*, **31**, e13.
67. Stangegaard, M., Dufva, I.H. and Dufva, M. (2006) Reverse transcription using random pentadecamer primers increases yield and quality of resulting cDNA. *Biotechniques*, **40**, 649–657.
68. Huber, M., Wei, T.F., Muller, U.R., Lefebvre, P.A., Marla, S.S. and Bao, Y.P. (2004) Gold nanoparticle probe-based gene expression analysis with unamplified total human RNA. *Nucleic Acids Res*, **32**, e137.
69. Liu, R., Lenigk, R., Druyor-Sanchez, R., Yang, J. and Grodzinski, P. (2003) Hybridization enhancement using cavitation microstreaming. *Anal Chem*, **75**, 1911–1917.
70. Yuen, P., Li, G., Bao, Y. and Muller, U. (2003) Microfluidic devices for fluidic circulation and mixing improve hybridization signal intensity on DNA arrays. *Lab on a Chip*, **3**, 46–50.
71. Bynum, M.A. and Gordon, G.B. (2004) Hybridization enhancement using microfluidic planetary centrifugal mixing. *Anal Chem*, **76**, 7039–7044.
72. Pappaert, K., Vanderhoeven, J., Van Hummelen, P., Dutta, B., Clicq, D., Baron, G.V. and Desmet, G. (2003) Enhancement of DNA micro-array analysis using a shear-driven micro-channel flow system. *J Chromatogr A*, **1014**, 1–9.
73. Hessner, M.J., Wang, X., Khan, S., Meyer, L., Schlicht, M., Tackes, J., Datta, M.W., Jacob, H.J. and Ghosh, S. (2003) Use of a three-color cDNA microarray platform to measure and control support-bound probe for improved data quality and reproducibility. *Nucleic Acids Res*, **31**, e60.

74. Hessner, M.J., Wang, X., Hulse, K., Meyer, L., Wu, Y., Nye, S., Guo, S.W. and Ghosh, S. (2003) Three color cDNA microarrays: quantitative assessment through the use of fluorescein-labeled probes. *Nucleic Acids Res*, **31**, e14.
75. Lindroos, K., Sigurdsson, S., Johansson, K., Ronnblom, L. and Syvanen, A.C. (2002) Multiplex SNP genotyping in pooled DNA samples by a four-colour microarray system. *Nucleic Acids Res*, **30**, e70.
76. Bao, Y.P., Huber, M., Wei, T.F., Marla, S.S., Storhoff, J.J. and Muller, U.R. (2005) SNP identification in unamplified human genomic DNA with gold nanoparticle probes. *Nucleic Acids Res*, **33**, e15.
77. Hesse, J., Jacak, J., Kasper, M., Regl, G., Eichberger, T., Winklmayr, M., Aberger, F., Sonnleitner, M., Schlapak, R., Howorka, S. et al. (2006) RNA expression profiling at the single molecule level. *Genome Res*, **16**, 1041–1045.
78. Han, A., Dufva, M., Belleville, E. and Christensen, C. (2003) Detection of analyte binding to microarrays using gold nano particles labels and a desktop scanner. *Lab Chip*, **3**, 336–339.
79. Alexandre, I., Hamels, S., Dufour, S., Collet, J., Zammatteo, N., De Longueville, F., Gala, J.L. and Remacle, J. (2001) Colorimetric silver detection of DNA microarrays. *Anal Biochem*, **295**, 1–8.
80. Petersen, J., Stangegaard, M., Birgens, H. and Dufva, M. (2007) Detection of mutations in the beta-globin gene by colorimetric staining of DNA microarrays visualized by a flatbed scanner. *Anal Biochem*, **360**, 169–171.
81. Chen, J.J., Wu, R., Yang, P.C., Huang, J.Y., Sher, Y.P., Han, M.H., Kao, W.C., Lee, P.J., Chiu, T.F., Chang, F. et al. (1998) Profiling expression patterns and isolating differentially expressed genes by cDNA microarray system with colorimetry detection. *Genomics*, **51**, 313–324.
82. Eisen, M.B. and Brown, P.O. (1999) DNA arrays for analysis of gene expression. *Methods Enzymol*, **303**, 179–205.
83. Workman, C., Jensen, L.J., Jarmer, H., Berka, R., Gautier, L., Nielser, H.B., Saxild, H.H., Nielsen, C., Brunak, S. and Knudsen, S. (2002) A new non-linear normalization method for reducing variability in DNA microarray experiments. *Genome Biol*, **3**, research0048.
84. Luebke, K.J., Balog, R.P. and Garner, H.R. (2003) Prioritized selection of oligodeoxyribonucleotide probes for efficient hybridization to RNA transcripts. *Nucleic Acids Res*, **31**, 750–758.

Chapter 2

Probe Design for Expression Arrays Using OligoWiz

Rasmus Wernersson

Abstract

Since all measurements from a DNA microarray is dependant on the probes used, a good choice of probes is of vital importance when designing custom microarrays. This chapter describes how to design expression arrays using the *OligoWiz* software suite. The desired general features of good probes and the issues which probe design must address are introduced and a *conceptual* (rather than mathematical) description of how OligoWiz scores the quality of the potential probes is presented. This is followed by a detailed step-by-step guide to designing expression arrays with OligoWiz.

The scope of this chapter is exclusively on expression arrays. For an in-depth review of the entire field of probe design (including a comparison of different probe design packages) as well as instructions on how to produce special purpose arrays (e.g., splice detection arrays), please refer to *(1)*.

Key words: Probe design, probe selection, expression array, oligonucleotide array, DNA microarray, software, bioinformatics, transcripts.

1. Introduction

A good choice of probes is vital to the usefulness of a microarray since the probes determine what signal will be detected (from both intended and non-intended targets). In summary a good probe must fulfill the following criteria:

- An ideal probe must discriminate well between its intended target and all other potential targets in the target pool.
- The probe must be able to detect concentration differences under the applied hybridization conditions.

OligoWiz website: http://www.cbs.dtu.dk/services/OligoWiz/

Martin Dufva (ed.), *DNA Microarrays for Biomedical Research: Methods and Protocols, vol. 529*

DOI 10.1007/978-1-59745-538-1_2 Springerprotocols.com

These two points are the ultimate goal to achieve for all probe design software packages, even if the actual algorithms used can be quite different *(1)*. The following sections describe how this is handled in the OligoWiz software package.

1.1. Introducing OligoWiz

Since the computational burden of performing the scoring of all possible probe positions is substantial, OligoWiz has been implemented as a *client–server* solution.

The workflow is as follows: The user interfaces with the Graphical User Interface (the "client" – written in Java for platform-independent use,*see* **Fig. 2.1**), and selects a dataset and a set of parameters for the probe design project. Next the

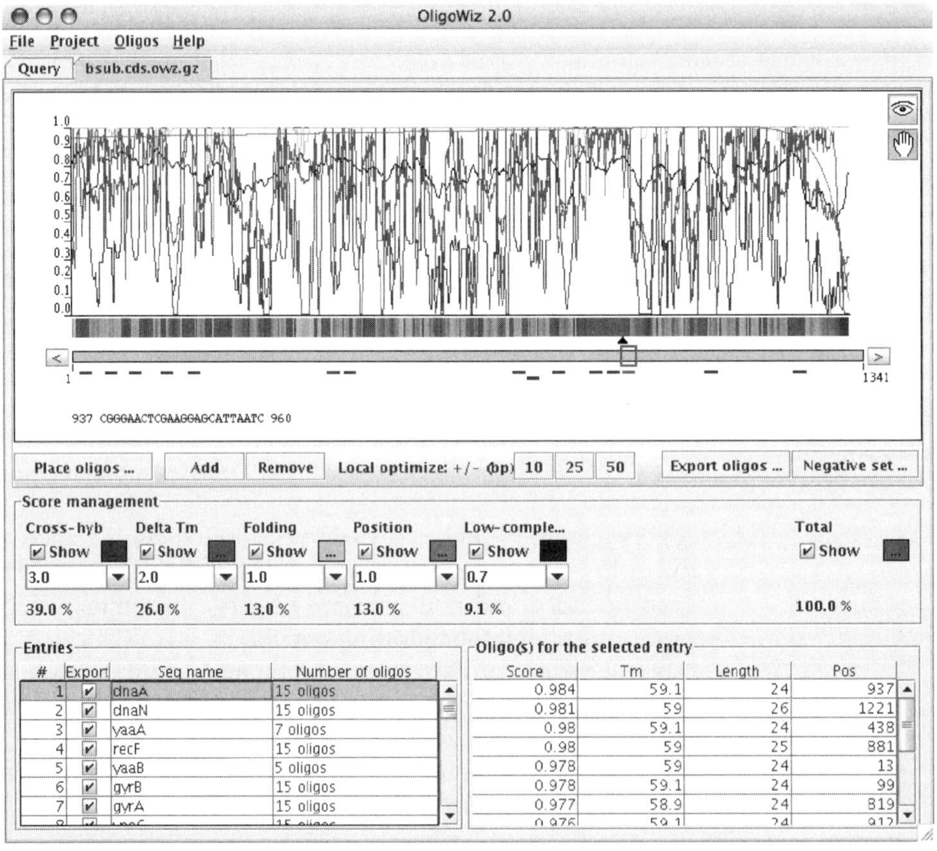

Fig. 2.1. **OligoWiz 2.0 screenshot.** This screenshot shows the main functionality of the software – including the graphical representation of the probe-"goodness" scores and the placement of probes along the selected transcript. The orange bar below the curves represents the currently selected transcript (dnaA). In this example a short-mer (24–26 bp) probe design for *Bacillus subtilis* is in progress, and up to 15 probes per transcript have been placed. The probes are visualized as lines below the transcript, and details are provided in the lower right-hand corner. Please note that the five scores are color-coded (cannot be seen here) – examples of the color coding is found at the OligoWiz website and in OligoWiz publications *(1–3)*.

data is uploaded to the server (hosted at a multi-CPU supercomputer located at the Center for Biological Sequence Analysis at the Technical University of Denmark) where all the computationally heavy algorithmic processing takes place. Once calculation of a particular dataset is completed a datafile with scoring information about *each potential* probe along all transcripts in the dataset is returned to the user, and all further work on the actual probe selection happens in a completely off-line fashion using the GUI.

1.2. Probe Suitability Scores in OligoWiz

OligoWiz uses a scoring-scheme that works as follows: For each position along all transcripts in the input dataset the suitability of placing a probe here is evaluated according to five criteria: Cross-hybridization, ΔT_m, Folding (self-annealing), Position (within the transcript) and "Low-complexity." Each individual score has a value between **0.0** (not suited – a bad position for placing a probe) to **1.0** (well suited – no problems detected). The individual scores are then combined with different weights (e.g., Cross-Hybridization is more important than Low-Complexity, *see* **Fig. 2.1** for the default values) to form a **Total score** which is also normalized to be between **0.0** and **1.0**. The actual selection of the best position for probe placement is based on the Total score.

In the following sections the *conceptual* workings of the individual scores will be described. The actual formulas for the calculations are found in the two main OligoWiz publications *(2, 3)*.

1.2.1. Cross-Hybridization

As mentioned previously a vital property for a probe is to pick up only the intended signal. A way to ensure this is to avoid probes that may hybridize (partially) to other transcripts. It has been shown *(4)* that a 50-mer will detect a significantly false signal from an unintended target that has more than 75–80% identity at the sequence level. Also, short stretches (> 15 bp) of complete complementarity will give rise to a signal from cross-hybridization. Similar result for short oligos (23–27 bp) has recently been shown by *(1)*.

The perfect way to get around this problem is to calculate the actual hybridization energy between all probes and all targets at the correct individual concentrations. However, since the concentrations of the targets are not known, and since such calculations are very time-consuming we have opted for an approximate solution: screen the entire genome (for prokaryotes and small eukaryotes) or transcriptome (Unigene collection for large eukaryotes, like mammals) using BLAST *(5, 6)* for regions with substantial similarity to the transcripts in question. By default regions with more that 75% similarity over at least 15 bp is considered to be problematic.

1.2.2. ΔT_m

Another important aspect of probe design is to ensure uniform hybridization conditions throughout the array. Traditionally this has been done by controlling the GC ratio within the probe. OligoWiz addresses this issue by forcing the distribution of T_m (melting temperature) to be as narrow as possible.

This is done in two ways:

- A "ΔT_m" score[1] that evaluates how far the T_m of a potential probe is from the mean T_m of all potential probes.
- Allowing the length of the probes to vary. **Fig. 2.2** shows how the T_m distribution of a set of oligonucleotides becomes increasingly narrow, if the most optimal length (within an interval) can be chosen. Working with short probes it is the experience of the OligoWiz authors that even allowing the length to vary just between 24–26 bp will improve the T_m profile. *Finding the optimal length is the very first step performed by OligoWiz*: For each position the most optimal length within the user-specified interval is determined, and this length is used for the calculation of all other scores.

1.2.3. Folding

To ensure uniform hybridization conditions for all probes on the microarray, the probes should avoid self-annealing (folding). The classical way of investigating this issue is to calculate the free

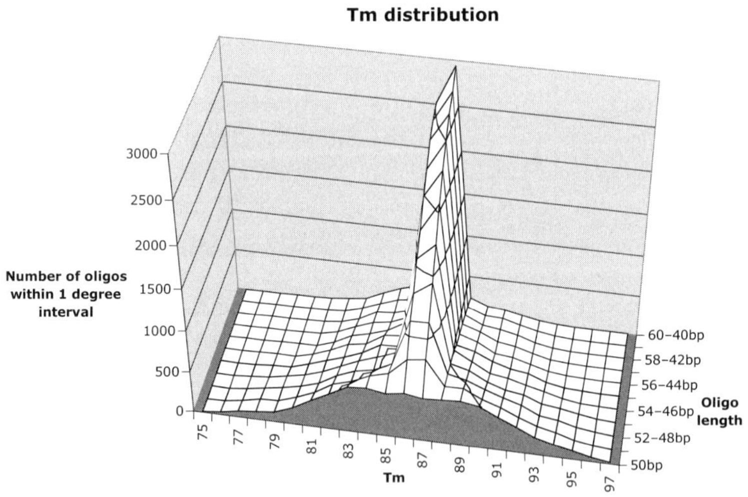

Fig. 2.2. **T_m distribution in optimized length intervals of oligonucleotides.** This figure shows how the T_m-distribution of a large set of oligonucleotides (based on all 50 mers within the Yeast genome) can be made increasingly narrow by allowing the length to vary and selecting the most optimal length within each interval. (Based on data from *(2)*).

[1] Listed as "Delta-Tm" in the interface.

energy of potential secondary structures using programs such as MFOLD *(7)*. However, using MFOLD is very time consuming[2] and for this reason approximate methods that are two orders of magnitude faster was developed for OligoWiz *(3)*. Briefly, this method is based on the idea of aligning the oligo to itself using a dinucleotide alphabet using dynamic programming *(8)* and a subsitution matrix based on the dinucleotide binding energies. The resulting alignment will represent the lowest folding energy state given the input sequence. This approximate method is in good agreement with MFOLD (*see (3)*, **Fig. 2.2**) – especially for the sequences with strong secondary structure, which are the most important to avoid when designing probes for DNA microarray. Since all possible probe positions along all target transcripts must be scored, the calculations can be done in a sliding window fashion, where most of the dynamic programming matrix from the previous position can be reused, this contributes significantly to the speed-up. Please *see (3)* for further details on the implementation.

1.2.4. Low Complexity

In order to avoid picking up background signal, probes that contain a lot of "sub-words" that are common in the genome/transcriptome should be avoided. This can be illustrated with the following example (human DNA):

Oligo with low-complexity:

AAAAAAAGGAGTTTTTTTCAAAAAACTTTTAAAAAAGCTTTAGGTTTTTA

Oligo without low-complexity:

CGTGACTGACAGCTGACTGCTAGCCATGCAACGTCATAGTACGATGACT

In OligoWiz, this problem is addressed by counting the occurrence of all 8 bp "words" in the genome/transcriptome and scoring the degree to which a probe consist of frequent sub-words.

1.2.5. Position

The optimal position within the transcript for placing a probe depends on the labeling and/or amplification method used. When using standard poly-T priming (targeting the poly-A tail of eukaryotic transcripts) the labeling starts from the 3′ end of the transcript. Since there is a certain probability that the reverse transcriptase will not complete the synthesis of cDNA in full length, most signals are detected using probes targeting the 3′ end of the transcript. In OligoWiz the following position preference models are built in.

[2] 2 seconds for a 30-mer and 16 minutes for all 30-mers in a 500 bp transcript at OligoWiz reference platform at the time.

- **Poly-T priming:** Push probes towards the 3′ prime end (Probabilistic model of the labeling from the 3′ prime end).
- **Random priming:** Avoid probes at the extreme 3′ prime end (Probabilistic model of the labeling using random hexamers).
- **Linear 5′ preference:** 1.0 at the 5′ end and decreases linearly to 0.0 over 2000 bp.
- **Linear 3′ preference:** As the 5′ preference, but counting from the 3′ end instead.
- **Linear mid preference:** 1.0 at the midpoint decreasing to 0.0 over 1000 bp to each side.

Observe that it is possible to completely ignore the position score, by setting its weight to **0.0**. This is especially useful in situations like placing splice-junction probes, where the position is constrained by the gene structure.

1.3. Rule Based Placement of Probes

As mentioned previously, the OligoWiz server returns a datafile to the client (the graphical interface) which contains *scoring of all possible probes.* At this point no decisions about the actual placement (how many per transcript, spacing etc.) of the probes have been made. All the computations on the placement of the probes is performed solely on the user's own computer in a completely off-line manner. This means that once the data file has been created it contains everything needed for further work, and can be stored on the user's own computer/network or be shared with collaborators using email, for instance.

The actual placement of the probes is done using a rule based method (*see* **Fig. 2.3** for an overview of the options). The placement algorithm is as follows (repeated for each transcript):

1. **Apply filters:** If any filters have been defined (e.g., requiring the total-score to be above a certain value), start by masking out the regions disallowed by the filters. For the advanced optional use of filter please *see (1)*.
2. **Place probe:** Select the currently available position with the highest Total score for probe placement.
3. **Mask out surrounding positions:** Positions within the desired minimum distance are masked out.
4. **Repeat/terminate:** Terminate if the maximum total number of probes has been reached or if no more positions are available. Otherwise, go to step 2.

Since the computationally heavy calculations (scoring of all probe position) have already been performed on the server, the placement algorithm is fast. This makes it possible to

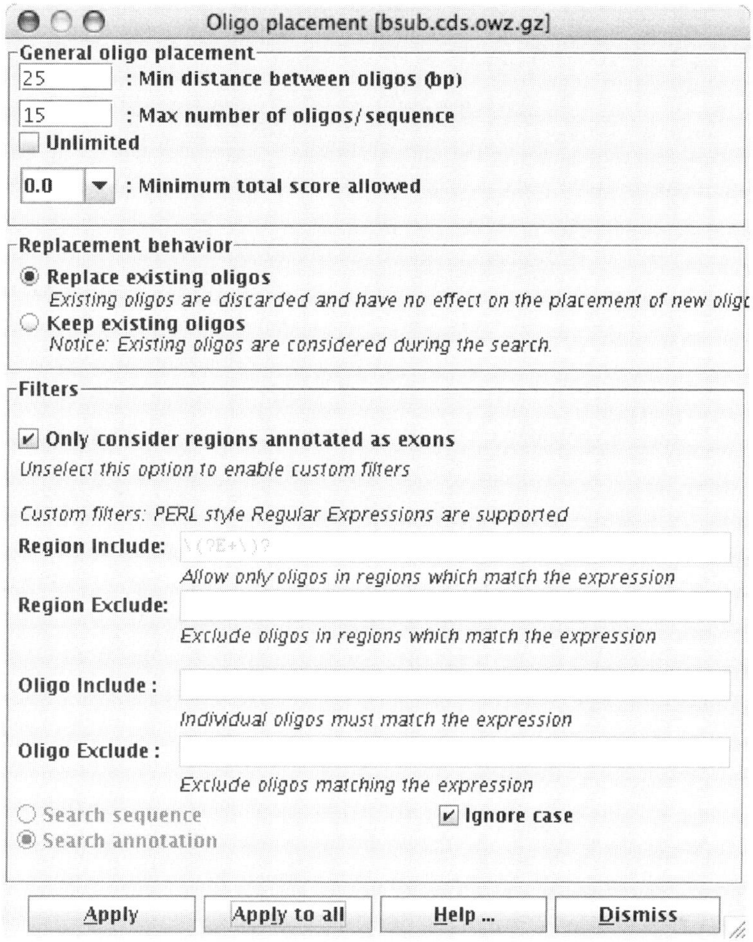

Fig. 2.3. **Probe selection dialog.** The spacing criteria are specified in the topmost box. The use of filters and sequence feature annotation (e.g., intron/exon structure) are not described here. For further details please refer to the OligoWiz website and *(1)*.

experiment with the probe placement parameters, evaluate the result, and refine the parameter in a real-time iterative fashion.

1.4. Exporting the Probe Sequences

The final step in the probe design process will be to actually order the array (e.g., NimbleExpress) or the oligonucelotides to be spotted. In order to make this step easy, OligoWiz support exporting the probe sequence to both FASTA and TAB format, and has the option of reverse-complimenting the probes (if needed) and automatically creating PM/MM probe pairs, if that is desired. Furthermore, it should be noted that a Material and Methods section describing the parameters used in the probe design is auto-generated and added to the file,

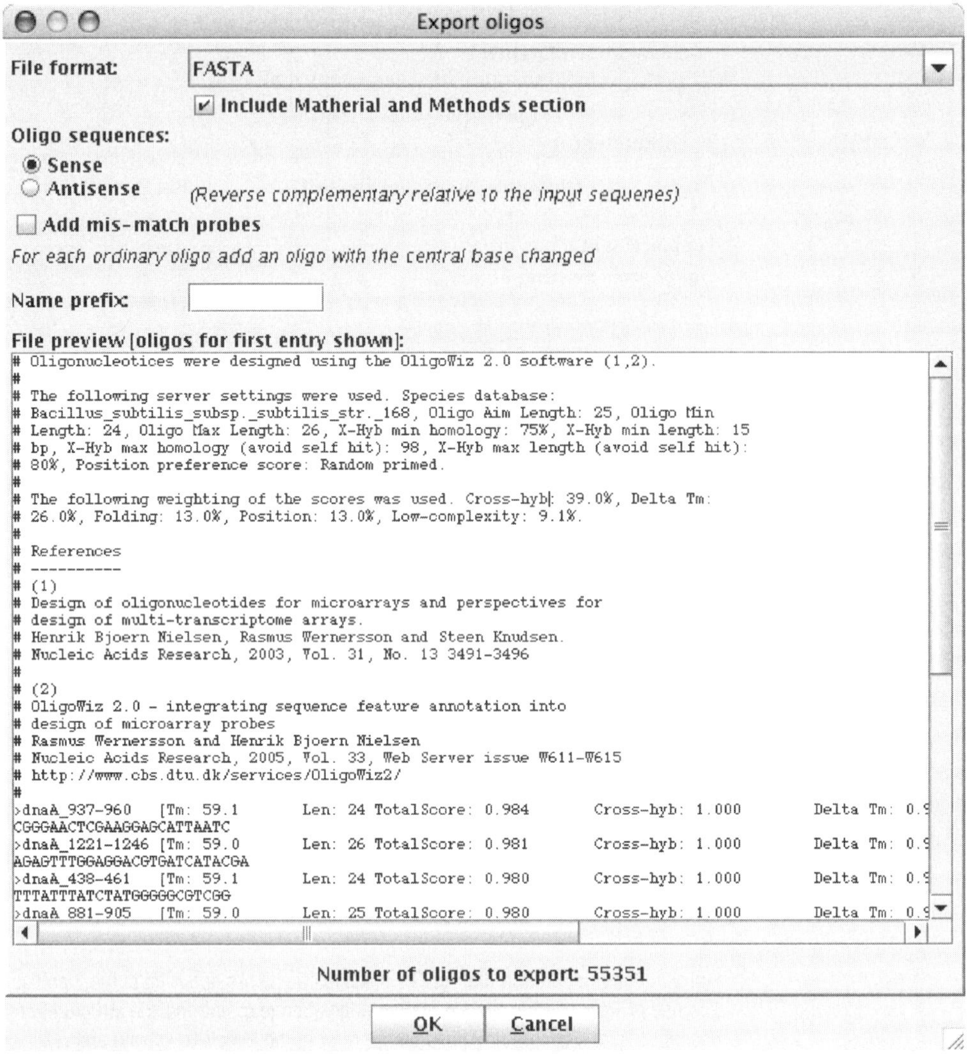

Fig. 2.4. **OligoWiz probe sequence export options.**

documenting the probe design process (*see* **Fig. 2.4** and step 10 in the step-by-step guide).

2. Materials

An internet connected computer with Java 1.4 (or newer) installed. The OligoWiz client is tested on Windows, Mac OS X, Irix, Solaris and Linux – it is written with cross-platform use in mind and should work on virtually any operating system for which a Java Runtime Environment (JRE) exists. The optional use of a local installation of the OligoWiz server software is not covered here, please *see (1)* and the OligoWiz website for further details.

3. Methods

This section summarizes the steps the user has to go through to select probes for an expression array.

1. **Prepare target sequences in FASTA format.** (For instructions on how to use TAB files please *see (1, 9)* – or the descriptions on the OligoWiz website). The very first step is to identify the sequences that the array should detect. This could for example be an entire prokaryotic genome or a set of transcripts from the human genome/transcriptome. For prokaryotic sequences a file prepared from the CDS (protein coding genes) regions of the full genomic sequence is recommended. In many cases a FASTA file with only the transcripts/CDSs can be downloaded from the same datasource as the full genomic builds. For higher eukaryotes (e.g., Human or Mouse), sequences from the UNIGENE collections are recommended. Observe that it is important to also include control targets/genes – since most normalization algorithms used in the downstream processing assumes that only a minor (10%) proportion of the transcript vary from array to array *(10)*. *See* **Note 1** for further details about the input data.
2. **Launch the OligoWiz client.**
   - 2.1. Download the most recent version of the OligoWiz client from the OligoWiz website: www.cbs.dtu.dk/services/OligoWiz/.
   - 2.2. Download Java version 1.4 (or newer) if it is not already installed on the local computer. Instruction on how to do this on various platforms (Windows/Linux/Mac) is detailed on the webpage.
   - 2.3. Launch the OligoWiz client by double-clicking on the JAR file (Windows and Mac) or from the command-line (Linux and UNIX). *See* **Note 2** for issues relating to the memory usage of the program.
3. **Select input file.** Click the "..." button next to the *Input FASTA or TAB file* field (*see* **Fig. 2.5**), and select the FASTA file prepared in Step 1. The OligoWiz client will suggest a unique filename for the result file (not generated yet) – accept this, or customize the filename/placement if desired.

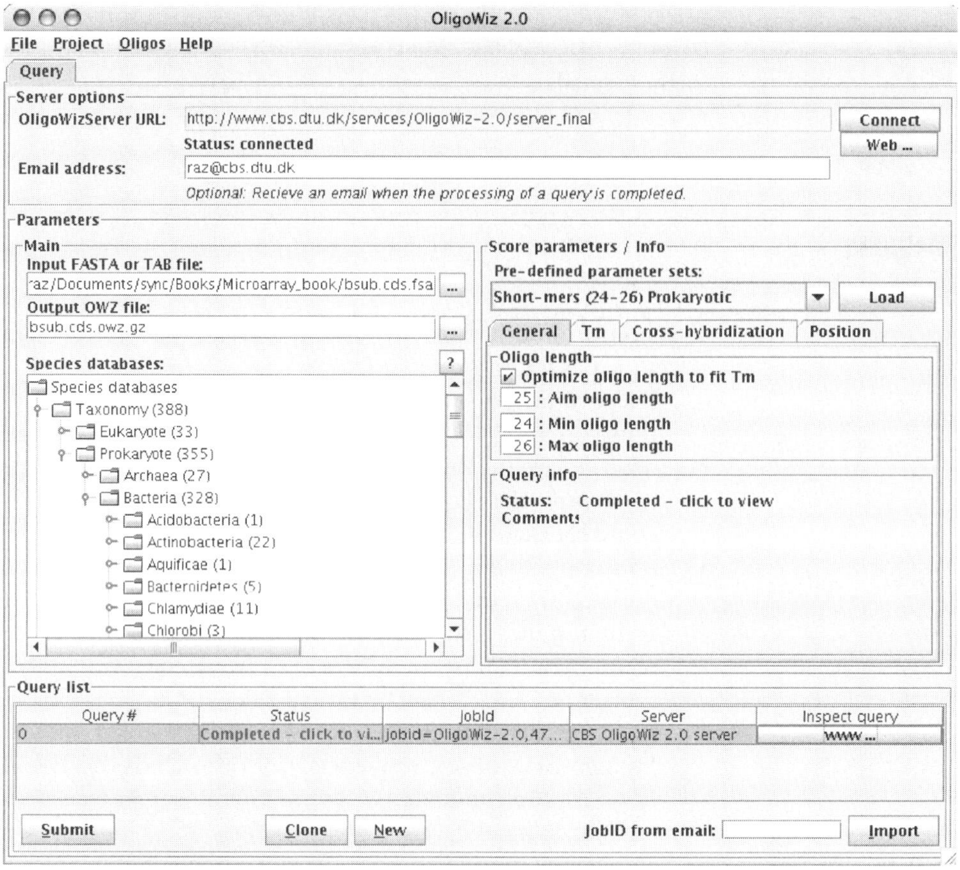

Fig. 2.5. **OligoWiz query launch page.**

4. **Select species database.** Select the species database that will be used for calculating the Cross-hybridization and Low-Complexity scores. A full description of all the databases[3] is available on the OligoWiz website. (If the species-tree is empty, please refer to **Note 3** describing how to trouble-shoot network issues).

5. **Customize score parameters.** Select the best fitting predefined parameter set in the *Score parameters/info* box and press *Load* (*see* **Fig. 2.5**). The predefined parameter sets can be customized further, as described below:

   5.1. **Oligo Length:** Determines if OligoWiz should aim at a fixed length or allow the length to vary within an interval in order to optimize T_m (recommended).

   5.2. **Tm**

      5.2.1. Select if OligoWiz should determine the optimal T_m (recommended) – alternatively a specific T_m to aim for can be specified.

5.2.2. Select if OligoWiz should use a DNA:DNA or RNA:DNA model for calculating the T_m. Select DNA:DNA if DNA is to be hybridized to the array and RNA:DNA if RNA is used (this is typically the situation).

5.3. **Cross-Hybridization**

5.3.1. Set the cut-off values of when a BLAST hit is to be considered: % minimum similarity and minimum length. Hits below this threshold will be completely ignored. It is recommended to use the default values.

5.3.2. Set the cut-off when a BLAST hit is considered a "self-hit" (the target sequence it self). For prokaryotic arrays the default values are recommended – if the input data is transcripts for a complex eukaryotic organism with a large degree of alternative splicing, the issue of detecting self-hits is more complicated. In this case it is recommended to lower the self-hit criteria. A pragmatic solution is to lower the self-hit length criteria to ~40% (0.4) – *see (1)* for a detailed discussion.

5.4. **Select position model.** For labeling protocols using poly-T (usually the case for running eukaryotic arrays) select the *Poly-T* option. For labeling protocols using random hexamers (usually the case for prokaryotic arrays) select the *Random priming* option.

6. **Submit the query**

6.1. **Optional step:** Enter your email address in the *Email address* field – this will make the server send you an email once the processing is completed with a link to direct download of the result data file. This is especially useful for long running queries.

6.2. **Press the "Submit" button**

7. **Wait for the server to finish processing the query**. The status of the processing can be seen in the *Query List* table. Once the processing has completed, the data file (file type: .owz.gz) will automatically be downloaded and stored on the local computer.

8. **Load the data file.** Double-click on the downloaded query in the "Query List" table to load in the data. This will load in the data and launch the main interface for placing probes (*see* **Fig. 2.1**).

Notice: If the data file has been downloaded manually by following the link in the server-generated email, the data can be loaded by using the File -> Open menu option.

9. **Place probes**
   9.1. **Adjust score weights (if needed).** It is recommended to keep the default settings. However, notice that it's possible to disable a score by setting its weight to **0.0**.
   9.2. **Bring up the Oligo Placement window.** Press the "*Place Oligos...*" button to launch the probe selection dialog (*see* **Fig. 2.3**).
   9.3. **Select probe placement criteria.** For short probes (~25 bp) 8 probes or more per target sequence is recommended, for long probes (50–70 bp) 2–4 (or more) is recommended *(1)*.
   9.4. **Apply selection criteria.** Press the *Apply to all* button to search for probes fulfilling the criteria in the entire data set. (The *Apply* button can be used to test the criteria on a single sequence).
   9.5. **Inspect the placement of the probes.** Keep the probe placement window open, and inspect the placement of the probes in the main window. Notice that both the *Entries* and *Oligos* lists can be sorted by clicking on the header elements. This makes it easy to identify target sequences for which no or few probes have been selected.
   9.6. **Repeat step b-e if needed.**
10. **Export probe sequences.** Press the "*Export oligos...*" button to bring up the Probe Export window (*see* **Fig. 2.4**). The sequences can be exported in FASTA and TAB format. Optionally the probe sequences can be exported as anti-sense probes and/or pairs or PM/MM (perfect match/Mis-match) probes can be generated. In most cases the probes should be saved as "sense" probes in FASTA format – however, it is important to make sure that the strandness is correct for the protocol to be used in the lab.
11. **Optional: Export negative set.** If a sub-set of the target sequence proves to be difficult to design probes for, this sub-set can be extracted from the full set of target sequences, by pressing the *Export negative set* button. This makes it possible to isolate the troublesome cases, and re-run the entire probe-design process for these sequences only with more relaxed settings (or alternatively deciding NOT to target these sequences in the array design).

4. Notes

1. **Problems related to input data**: The most common source of problems with running OligoWiz is problems with the input data:

1.1. Please make sure that the data is in a supported file format (TAB or FASTA). Notice that the file must be a text-only file (an otherwise correctly formatted FASTA file within a MS-Word document will NOT work).

1.2. Please make sure that the file contains the sequences of the transcripts/genes which should be targeted. Submitting a file with a single large DNA sequence representing an entire prokaryotic genome will not work. OligoWiz is designed to work in a gene/transcript oriented way (for comments on how to design a chromosomal tiling array please *see (1)*).

1.3. Please make sure that the input sequences are of a sufficient length. Entries that are shorter than the minimum probe length will be discarded.

2. **Memory problems:** For very large datasets, the default amount of memory available to Java may become a problem. As a rule of thumb more memory may be needed if a FASTA file with more than 10,000 sequences (average prokaryotic CDS length) is submitted. The OligoWiz webpage contains detailed instruction of how to start the OligoWiz client with more memory on various platforms.

3. **Network problems:** If the OligoWiz client fails to connect to the OligoWiz server (the species database list remains empty, and the connection status remains "not connected") it is most likely to be due to problems with the network setup. The OligoWiz client communicates with the server using HTTP (like a web browser), and it needs a direct connection rather than going through a HTTP proxy. If the local network setup uses a HTTP proxy (inspect the browser proxy settings – or ask the local system administrator), this is likely to be the cause of the problem. The OligoWiz website contains a description of a work-around of this issue.

References

1. Wernersson, R., Juncker, A.S. and Nielsen, H.B. (2007) Probe Selection for DNA Microarrays using OligoWiz. *Nature Protocols*, 2, 2677–2691.
2. Nielsen, H.B., Wernersson, R. and Knudsen, S. (2003) Design of oligonucleotides for microarrays and perspectives for design of multi-transcriptome arrays. *Nucleic Acids Res*, 31, 3491–3496.
3. Wernersson, R. and Nielsen, H.B. (2005) OligoWiz 2.0-integrating sequence feature annotation into the design of microarray probes. *Nucleic Acids Res*, 33, W611–W615.
4. Kane, M.D., Jatkoe, T.A., Stumpf, C.R., Lu, J., Thomas, J.D. and Madore, S.J. (2000) Assessment of the sensitivity and specificity of oligonucleotide (50 mer) microarrays. *Nucleic Acids Res*, 28, 4552–4557.
5. Altschul, S.F., Gish, W., Miller, W., Myers, E.W. and Lipman, D.J. (1990) Basic local alignment search tool. *J Mol Biol*, 215, 403–410.

6. Altschul, S.F., Madden, T.L., Schäffer, A.A., Zhang, J., Zhang, Z., Miller, W. and Lipman, D.J. (1997) Gapped BLAST and PSI-BLAST: a new generation of protein database search programs. *Nucleic Acids Res*, 25, 3389–3402.
7. Zuker, M. (1994) Prediction of RNA secondary structure by energy minimization. *Methods Mol Biol*, 25, 267–294.
8. Needleman, S.B. and Wunsch, C.D. (1970) A general method applicable to the search for similarities in the amino acid sequence of two proteins. *J Mol Biol*, 48, 443–453.
9. Wernersson, R. (2005) FeatureExtract-extraction of sequence annotation made easy. *Nucleic Acids Res*, 33, W567–W569.
10. Workman, C., Jensen, L.J., Jarmer, H., Berka, R., Gautier, L., Nielsen, H.B., Saxild, H.-H., Nielsen, C., Brunak, S. and Knudsen, S. (2002) A new non-linear normalization method for reducing variability in DNA microarray experiments. *Genome Biol*, 3, research0048.

Chapter 3

Comparative Genomic Hybridization: Microarray Design and Data Interpretation

Richard Redon and Nigel P. Carter

Abstract

Microarray-based Comparative Genomic Hybridization (array-CGH) has been applied for a decade to screen for submicroscopic DNA gains and losses in tumor and constitutional DNA samples. This method has become increasingly flexible with the integration of new biological resources generated by genome sequencing projects. In this chapter, we describe alternative strategies for whole genome screening and high resolution breakpoint mapping of copy number changes by array-CGH, as well as tools available for accurate analysis of array-CGH experiments. Although most methods listed here have been designed for microarrays comprising large-insert clones, they can be adapted easily to other types of microarray platforms, such as those constructed from printed or synthesized oligonucleotides.

Key words: Probe design, clone selection, normalization, outlier detection, CNV calling, Comparative Genomic Hybridization, array-CGH.

1. Introduction

Comparative Genomic Hybridization (CGH) was developed in the early 1990s to screen for chromosomal deletions and duplications along whole genomes *(1, 2)*. Originally, CGH consisted of co-hybridizing one test and one reference labeled probe DNA onto metaphase chromosomes spread on glass slides in the presence of Cot-1 DNA to suppress high repeat sequences (*see* **Chapter 17**). During the 1990s, CGH on chromosomes was widely used by research laboratories, in particular to screen for chromosome numerical aberrations associated with the progression of solid tumors *(3)*: chromosome analysis by G-banding was

Martin Dufva (ed.), *DNA Microarrays for Biomedical Research: Methods and Protocols, vol. 529*

DOI 10.1007/978-1-59745-538-1_3 Springerprotocols.com

technically challenging with tumor cells, due to the frequency of highly rearranged karyotypes and difficulties in culturing cells in vitro to obtain good quality metaphase chromosomes.

However, although CGH became widely used in cancer research, it did not prove to be particularly valuable as a standard method in diagnostic laboratories for the analysis of genomic imbalance in patients with developmental disorders. This was firstly due to the poor spatial resolution of metaphase CGH, which is limited to 5–15 Mb by the image acquisition of probe signals on metaphase spreads using fluorescence microscopy. Secondly, metaphase CGH is technically challenging, requiring expertise for preparation of suitable metaphase chromosomes as well as image acquisition and analysis.

From the mid 1990s, the International Human Genome Sequencing Project released new information on the human genome sequence, which was derived from the construction and characterization of libraries comprising large-insert clones such as bacterial artificial chromosomes (BACs) *(4)*. These resources allowed the CGH method to be modified such that metaphase chromosomes could be replaced by arrayed DNA fragments representing precise chromosome coordinates. This strategy was initially called matrix-CGH *(5)* and then array-CGH *(6)*, and it is this name that is now in common usage. The development of array-CGH improved significantly the potential of CGH for the analysis of small chromosomal imbalances. Initial arrays provided a more than tenfold increase in resolution such that micro rearrangements that were invisible previously on chromosome preparations became detectable. Also, for the first time, deletion and duplication breakpoints could be localized directly on the human genome sequence assembly.

The large insert clones used for the first array-CGH applications – in particular BACs and fosmids – have since become widely available. This has facilitated the construction of microarrays covering the whole genome at increasingly higher resolution. However, the relatively large size of these clones (~170 kb for BACs, ~ 40 kb for fosmids) limits the ultimate resolution of these types of arrays. In the past couple of years, small-insert clones, PCR products, and oligonucleotides have been developed for use in array-CGH *(7, 8)* allowing a greater degree of flexibility and higher resolution (down to just a few base pairs) in the design of microarray experiments, which can be tailored to the specific biological question. This chapter describes many critical factors that should be considered when designing new array-CGH experiments and discusses different possible strategies for data analysis. It focuses on microarrays comprising cloned DNA printed on slides, though some strategies and tools described here can also apply for the design of microarrays composed of printed or synthesized oligonucleotides.

2. Array-CGH Design

2.1. Clone Selection

The first step in array-CGH is the design or choice of the microarray to be used for interrogating test genomes. There are two common strategies: (i) the design or the selection of one microarray covering the whole genome in order to screen for every deletion or duplication in a given test genome compared to a reference DNA; (ii) the construction and use of one microarray targeted to one part of the genome only, such as one chromosome or one region.

The design of a whole genome microarray is dependent on the resources available to construct the array. Construction of arrays from large insert clones requires physical spotting of the clone DNA onto microscope slides, which typically limits the number of elements on the array to less than 50,000. For this reason, many laboratories used BAC clones for whole genome coverage, because with an average length of 170 kb coverage of the whole genome with overlapping clones requires approximately 30,000 BACs while it would require more than 120,000 fosmids (40 kb in length). Covering the whole genome at tiling path resolution is an important investment in time and resources, which may not be suitable for many laboratories. For this reason, most BAC microarrays used for whole genome screening comprise only approximately 3,000 clones. They cover the whole genome with clones regularly interspaced, each single clone positioned at an interval of approximately 1 Mb apart. Although this strategy is not efficient for the detection of copy number changes below 1–2 Mb in size, it has proved to be valuable for the screening of most large-scale deletions or duplications, such as those responsible for severe congenital anomalies.

Several sets of clones designed specifically for the construction of CGH microarrays are publicly available. The Wellcome Trust Sanger Institute has developed two sets of large-insert clones for the construction of microarrays covering the whole genome at 1-Mb and tiling path resolutions (1Mb and 30k TPA sets, respectively). The coverage of the human genome by these two sets of clones can be visualized on the Ensembl browser (www.ensembl.org, *see* **Fig. 3.1A**) and clones are available through GeneService (www. geneservice.co.uk). Another selection of 32,000 overlapping BAC clones covering the whole genome can be obtained from the BACPAC Resources Center at CHORI (bacpac. chori.org).

To design a microarray targeted to specific loci, there is a larger choice of clones which could be used, depending on the size of the genomic segments to cover and on the resolution which is required. While BAC clones are usually selected for the construction of whole-genome microarray, fosmid clones

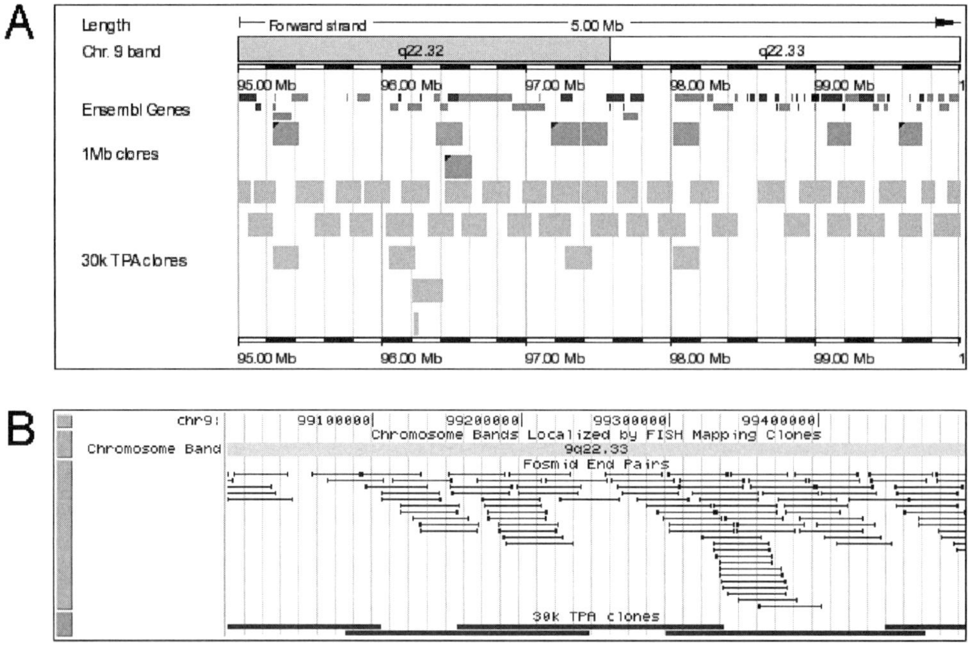

Fig. 3.1. Selection of large-insert clones for array-CGH using Genome Browsers (**A**) The Ensembl browser (www.ensembl.org) enables the user to visualize many physical or biological annotations in the context of the genome sequence. The box displays the respective positions of genes (Ensembl annotation, *top panel*), clones from the Sanger 1Mb set (*middle panel*) and clones from the 30k TPA set (*bottom panel*) between coordinates 95–100 Mb on human chromosome 9. Lists of clones from these two sets can be downloaded as delimited tables from the same website (select option "Graphical overview"). (**B**) Part of the same interval (99–100 Mb), displayed on the UCSC Genome Browser (genome.ucsc.edu), one alternative to Ensembl. *The bottom panel* shows positions of clones from the 30k TPA set. *The top panel* displays the positions of many fosmids mapped by pair-end sequencing. Some of the fosmid clones can be selected by their chromosomal locations for high-resolution coverage of the locus by array-CGH.

represent a good alternative for custom arrays. Overlapping fosmids provide better resolution than overlapping BACs (down to 10 kb in case of high redundancy in coverage versus approximately 50 kb) but can be prepared for spotting using the same protocols (*see* **Chapter 16**). The fosmid library WIBR-2 is particularly useful as it has been extensively characterized by end-sequencing: most clones from this library are precisely mapped on the human genome assembly and all read-pair positions can be visualized on the UCSC genome browser (genome.ucsc.edu, *see* **Fig. 3.1B**). Read-pair coordinates can be downloaded from the UCSC browser for further selection of the clones required to cover the regions of interest. All fosmids can be purchased at the BACPAC Resources Center (bacpac.chori.org).

For example, after selecting fosmids for the construction of a small custom microarray, we applied array-CGH for high-resolution breakpoint mapping of two deletions at 9q22.3, responsible for a syndrome involving mental retardation and overgrowth in two unrelated children *(9)*. The result obtained for one child is

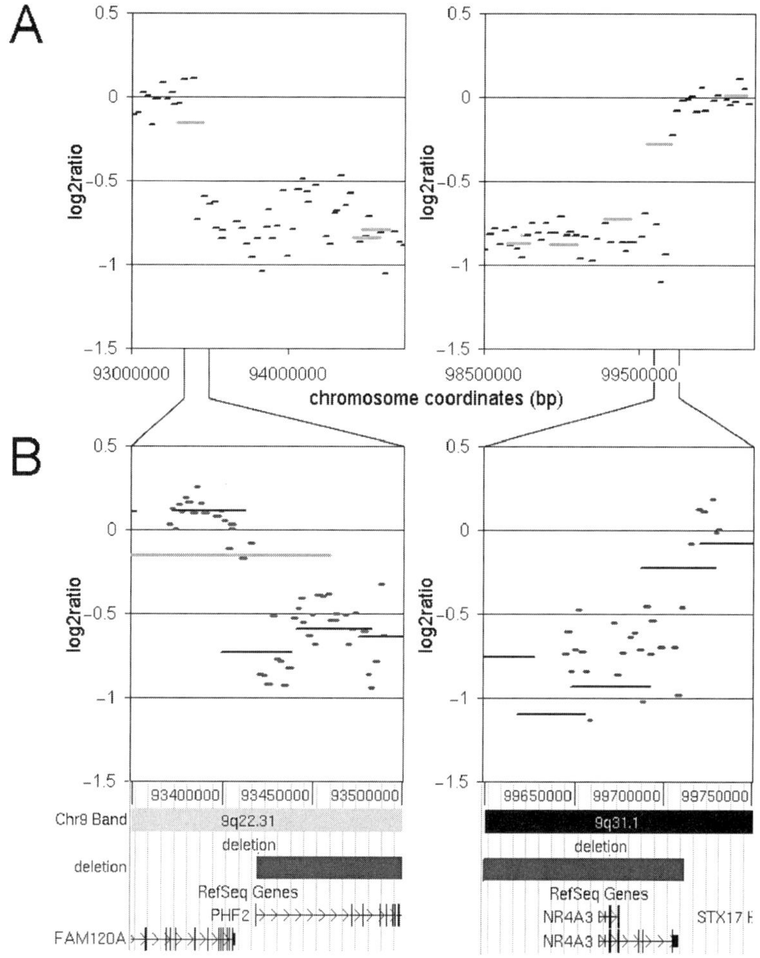

Fig. 3.2. High resolution breakpoint mapping by array-CGH (**A**) Array CGH profiles at the proximal (*left*) and distal (*right*) breakpoints of a 9q22.3 deletion detected in a patient with overgrowth syndrome *(9)*. The deletion was first detected with a microarray covering the whole genome at 1 Mb resolution (positions of 1 Mb clones are represented as *large grey bars*). One custom microarray comprising fosmids (represented as *short black bars*) was then constructed to cover the two breakpoint regions at tiling path resolution. CGH with the custom array refined the deletion breakpoints to intervals of less than 50 kb. Note that the 1 Mb array profile was normalized by a block median method, while the custom array was normalized by the median of log2ratios from 26 fosmids located on chromosome 18 and used as controls *(5)*. (**B**) Detailed views of the same deletion breakpoint intervals. Using a small custom microarray comprising small-insert clones (1.5 to 4 kb in length, represented as *small grey bars*), it was possible to map each deletion breakpoint at a resolution of less than 5 kb. Long-range PCR amplification and sequencing confirmed that array-CGH applied with increasing resolution enables accurate mapping of deletion breakpoints. The actual breakpoints are shown below the profiles on the UCSC browser: the proximal breakpoint disrupts the first intron of the PHF2 gene while the distal breakpoint is distal to the NR4A3 gene.

shown in **Fig. 3.2A**. Further increase in array-CGH resolution can be achieved by selecting small-insert clones (1.5–4 kb, *see* **Fig. 3.2B**) or PCR products (less than 1 kb), which can be used to cover all exons of any gene of interest *(7)*. Today, synthetic oligonucleotides have largely replaced these approaches to custom array construction. Several companies – such as Agilent Technologies, Inc. and NimbleGen Systems, Inc. – are now commercializing microarray platforms with custom oligonucleotide synthesis, which provides virtually unrestricted flexibility in the design of CGH.

2.2. Controls

The microarray design should always include a selection of control target sequences, which will be used to estimate the performance of the microarray as well as the quality of array-CGH hybridizations.

Some negative controls should be included to estimate the intensity of fluorescence resulting from the non-specific hybridization of genomic probes on the target DNA. For printed arrays, negative control spot positions commonly contain bacterial genomic DNA or DNA sequences from other species, such as *Drosophila*. After image acquisition and spot intensity quantification, the intensity of fluorescence on these negative controls should always be monitored and be extremely low when compared to the test intensities along the microarray.

It is also valuable if possible within the array design to include controls for the estimation of the dosage response on the array. For example, adding clones representing sequences on chromosome X can be used to estimate the ratio deviation due to the presence of one copy in a male test DNA compared to 2 copies in a reference female DNA. This strategy has been widely used to validate the performance of new microarray platforms *(6, 7)*.

In addition, it may be useful to include some normalization probes particularly for microarrays covering only small regions. Selecting a number of clones that are located in one or several regions of the genome unlikely to be variable in copy number in test and reference DNA samples can be critical for normalization steps (*see* **Fig. 3.2**). The control clones can either be located on a chromosome which is known to contain no gross anomaly or can cover genes which are known to be present in normal copy number in the test and the reference DNA. When working on copy number variations (CNV) in humans, one common strategy consists in selecting only clones located at chromosomal loci not reported to show variation in the literature (data available in the Database of Genomic Variants, projects.tcag.ca/variation).

At last, using one or a small group of clones that will be printed in replicate distributed regularly on the surface of the microarray can help in detecting problems of signal heterogeneity after hybridization and imaging. Furthermore, a control DNA sequence spotted in replicate along the array can be used to estimate and correct the spatial heterogeneity of log2ratio values (*see* **Section 3.1.3**).

3. Array-CGH Data Analysis

This section describes different strategies for the analysis of CGH profiles on large-insert clone microarrays that have been constructed and hybridized as described in **Chapters 16 and 17**.

3.1. Post-Processing

After hybridization and washes, microarray images are acquired on an array scanner. Test and reference fluorescence intensities are measured for each spot position, and test versus reference intensity ratios are calculated, usually after subtraction of local background fluorescence in each channel (*see* **Chapter 17, Note 6**).

3.1.1. Exclusion of Poor Quality Data Points

After ratio calculation, any data point which does not fulfill a series of quality criteria should be excluded from further analysis. The usual criteria are:

a. The signal fluorescence intensity should be significantly higher than the background intensity, at least in one channel. The exclusion threshold varies depending on image acquisition and quantification systems. Commonly, every data point with signal intensity lower than twice the background intensity is rejected. The background intensity can be either the local background intensity on the slide or the median signal intensity calculated from all negative controls (bacterial DNA or cloned DNA from an unrelated species).

b. If each target DNA sequence is printed in several copies (several spots), any clone with discordant replicate ratio values should be excluded from analysis.

c. Some microarray analysis programs, such as BlueFuse (BlueGnome, Ltd), calculate scores estimating quality criteria for each spot on the array (shape, regularity, concordance between the 2 color channels and/or signal vs. background ratio). These quality scores can be used to exclude poor quality spots.

3.1.2. Global Normalization Methods

After the exclusion step, intensity ratios are normalized to generate the final log2ratio profiles. One usual way to normalize whole genome log2ratio profiles is to subtract the median log2ratio value for all clones (or all clones located on autosomes) from each individual log2ratio (*as shown in* **Chapter 17 – Fig. 3.2B**). This method – called global median normalization – is suitable for experiments comparing the genomes of healthy individuals or in constitutional genetics (where there are only limited numbers of rearrangements along the genome). However, it may not perform as well with test samples showing many gross chromosomal rearrangements, such as DNA from cancer cells. In this particular situation, it may be preferable to normalize the log2ratio profiles using the modal rather than the median value: the modal value can be estimated for example using the Kernel method, which is available through the R software environment (www.r-project.org). Finally, array-CGH profiles that cover only one or several chromosomal loci of interest should be normalized by the median log2ratio value from control clones previously selected for their location in other non-variable regions of the genome (*see* **Section 2.2** and **Fig. 3.2**).

3.1.3. Other Normalization Methods

Apart from the global methods, additional normalization steps can be implemented to correct some technical biases that can occur during array-CGH.

A spatial normalization can be required when hybridizing large DNA microarrays using manual procedures: the hybridization process can result in uneven hybridization across the slide and lead to a gradient of log2ratio values along and/or across the array. A simple method for spatial normalization involves dividing the microarray into sub-arrays (or blocks) – each of them containing a sufficient number of spots – and normalizing each block individually. Contact printers using pins for spotting arrays often generate arrays already segmented into a number of blocks that can be convenient for this approach. Alternatively, control spots replicated across the slide that have been included during the array design (*as described in* **Section 2.2**) can be used for local normalization of ratios. The log2ratio gradient is then corrected by normalizing each block separately by the median value of all spots (or by the median value of all control replicates) from this same block. Note that the block median normalization using all spots is valid only if all the clones have been randomly distributed along the array with no use of their genome or chromosomal position to order them on the surface of the slide.

When analyzing DNA samples from various sources by array-CGH, we have observed that log2ratio values can sometimes show strong variations, resulting in a "wavy" profile (or "auto-correlation"). Although the experimental origin of this phenomenon is still unclear, the observed variations correlate with the GC content of the corresponding clone sequences. In consequence, to overcome this problem, we have introduced a GC correction, which consists of normalizing the log2ratios of each clone using the content in GC percent of that clone (*see* **Fig. 3.3**). This last step has enabled us to generate useful microarray data from some DNAs with poor quality by eliminating the wavy patterns often visible in array-CGH with these types of samples.

3.2. Automatic Detection of DNA Copy Number Changes

The last step of data analysis is the application of statistical methods for the automatic detection of significant copy number changes on log2ratio profiles. Many strategies can be followed

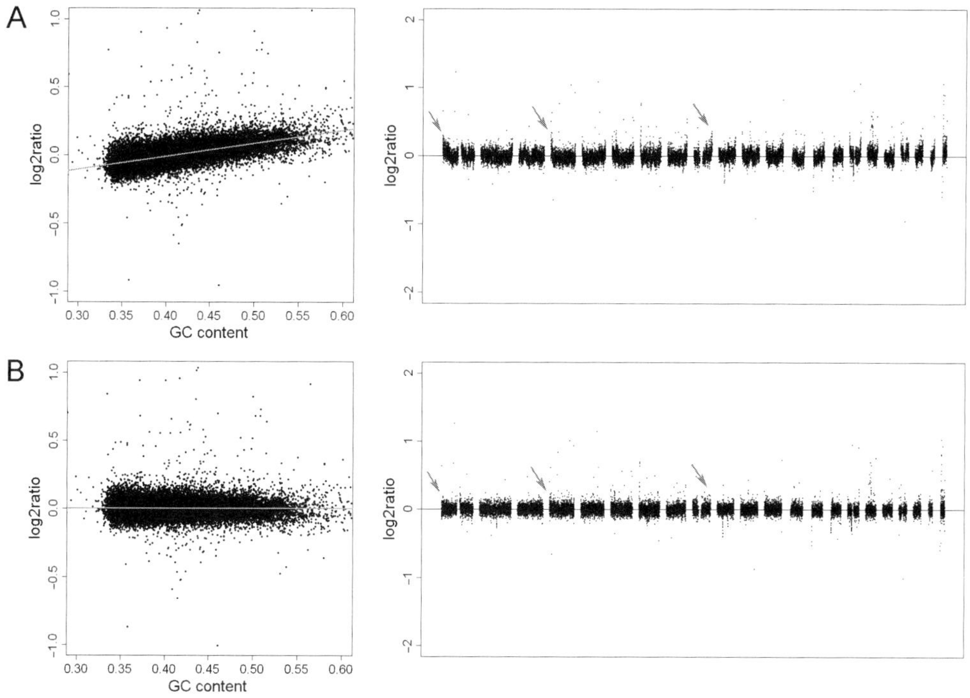

Fig. 3.3. Normalization of array-CGH log2ratio profile using clone GC content (**A**) After image quantification, log2ratio calculation and block median normalization, log2 ratios are plotted against the GC contents for all clones from autosomes: there is an apparent linear correlation between GC content and log2ratio (*left panel*: the fitted linear model is represented by the *grey line*). The influence of clone GC content on log2ratios results in local variations – called "waves" or "auto-correlation" – on the genome profile (*right panel*: only chromosomes 1 to 22 are displayed): waves are obvious for example on chromosomal arms 1p, 4p and 9q (*grey arrows*). (**B**) By normalizing the log2ratios using clone GC content, the linear correlation is eliminated (*left panel*: the fitted linear modal, in *grey*, gives a perfectly horizontal line at y=0). As a result, no wave is visible anymore on the log2ratio profile (*right panel*). Note that the GC correction in this example was performed by applying the linear model function "lm(y~x)" in the R language environment (the output corrected values are the "residuals").

to detect copy number changes objectively. The choice of the best method depends on the type of DNA samples that needs to be analyzed.

The simplest method is the arbitrary definition of one fixed threshold to determine which log2ratio values correspond to DNA gains or losses. Fixed thresholds were initially applied for the analysis of DNA samples from patients with constitutional anomalies because only few data points were expected to be variable by array-CGH *(10)*. Fixed thresholds can also be applied on profiles focused on particular loci and particularly for breakpoint mapping of previously identified chromosomal imbalances *(9)*. However, because the same fixed threshold is arbitrarily defined for every array-CGH profile, independently from differences in experimental log2ratio variability ("noise"), this simple method results in higher false discovery rates in "noisier" profiles.

One other simple strategy to avoid this problem consists in applying a threshold of significance which is proportional to the experimental variability. The variability can be roughly estimated with the standard deviation (SD) of all normalized log2ratios (considering that the experimental variability should be distributed normally). This method has been commonly applied by array-CGH studies in constitutional genetics with thresholds equal to three or four times the SD of all (autosomal) clones *(11)*.

With the development of larger microarrays covering the whole genome with up to 32,000 clones, the use of single thresholds has become inappropriate. In a normal distribution of 32,000 log2ratio values, 96 will be expected to be above a 3x SD threshold just by chance (and 3 or 4 above a 4x SD threshold). To address this problem, we have developed a more elaborate algorithm, CNVfinder, which enables the automatic detection and delineation of copy number changes using whole genome BAC/PAC array-CGH profiles *(12)*. CNVfinder is based on a robust estimation of the inherent log2ratio variance and applies different thresholds for single-clone and multi-clone copy number changes. It also incorporates a post-processing step to obtain the most likely bounds of each copy number change *(12)*. This algorithm, written in the Perl language, is freely available and can be downloaded at www.sanger.ac.uk/humgen/cnv/software.

CNVfinder is an efficient method to detect copy number variants on array-CGH profiles in constitutional genetics (**Fig. 3.4A,B**). However, this algorithm is not suitable for detecting gross chromosomal anomalies or to detect copy number changes in highly rearranged DNA samples, such as tumor genomes. Many other statistical methods have been developed recently and can be used for the detection of copy number alterations in tumor cells. Three of them, all freely available, are listed below.

- SW-array *(13)* applies the Smith-Waterman algorithm *(14)* to identify segments with deviating values within a log2ratio profile. SW-array is available as a package in the R Project (cran.r-project.org/doc/packages/cgh.pdf).
- aCGH-smooth *(15)* is a heuristic method, which identifies potential breakpoints and smoothes the observed array CGH values between consecutive breakpoints to a suitable common value. aCGH-smooth can be downloaded at www.few. vu.nl/~vumarray.
- DNA copy is a package in Bioconductor (www.bioconductor.org), which detects segments with abnormal copy number on array-CGH profiles using a method called circular binary segmentation *(16)*.

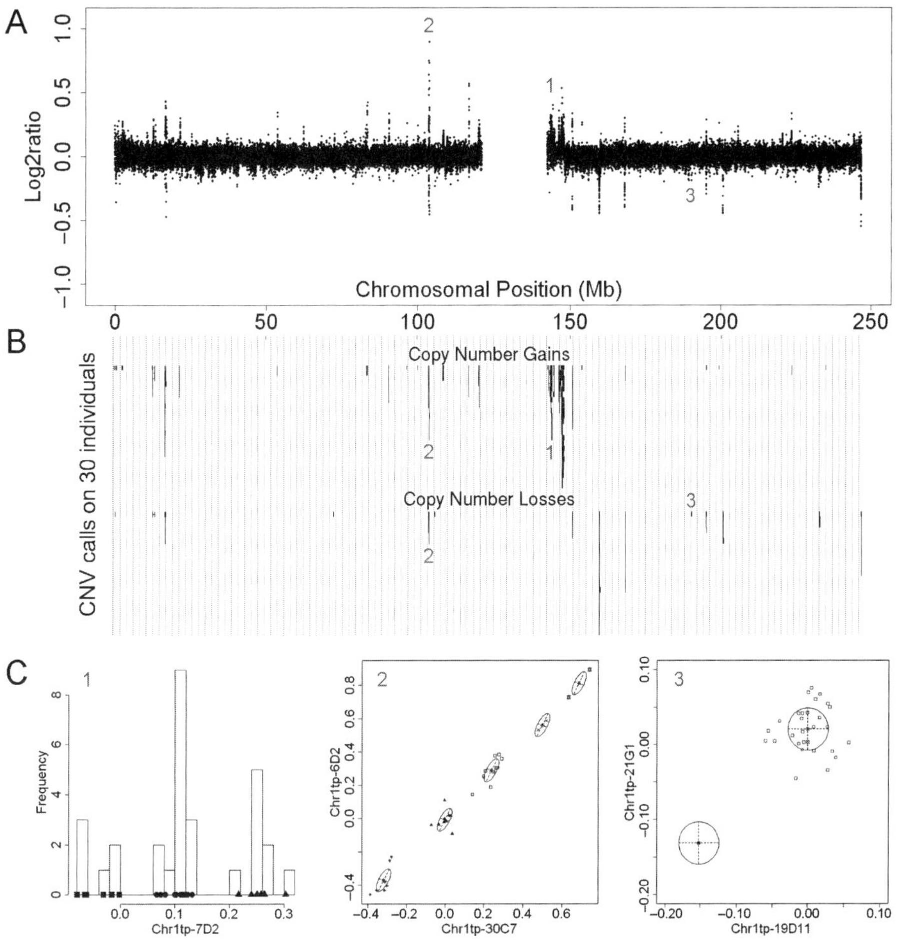

Fig. 3.4. **A**lternative strategies to investigate copy number variation by array-CGH (**A**) Superposition of 30 normalized array-CGH profiles on chromosome 1, resulting from the comparison of 30 human individuals from the general population with one single reference individual. The profile superposition enables to visualize regions showing copy number variation (CNV) in one or many individuals. (**B**) UCSC Browser display for the entire chromosome 1, showing all copy number changes detected automatically on each individual profile with the CNVfinder algorithm. The length of each bar represents the frequency of gains (*top panel*) and losses (*bottom panel*) at the corresponding clone positions. CNVfinder reports that the region 3 is lost in one individual, the region 1 is gained in 15 individuals and the region 2 is gained or lost in 15 and 5 samples, respectively. (**C**) Analysis of the distributions of log2ratio values in all 30 individuals for regions number 1 (clone Chr1tp-7D2), 2 (clones Chr1tp-30C7 and Chr1tp6D2), and 3 (clones Chr1tp-19D11 and Chr1tp-21G1). Applying an univariate model-based clustering method for Chr1tp-7D2 shows that log2ratio values can be classified in three different states: 6 samples are reported as deleted (*black squares*) and 9 as carrying duplicated (*black triangles*) in comparison to the 15 samples falling in the intermediary state (*black circles*). Bivariate model-based clustering using log2ratios for clones Chr1tp-30C7 and Chr1tp6D2 shows high levels of variation between individuals for the region 2: Five distinct clusters are detected using results from both clones. They correspond to five distinct copy numbers of region 2. This locus, which contains the AMY1 gene family, shows a highly unusual extent of copy number differentiation between human populations *(19)*. Bivariate model-based clustering for clones Chr1tp-19D11 and Chr1tp-21G confirms the deletion of region 3 for one single individual, as detected by CNVfinder. Normal mixture modeling and model-based clustering were performed using the MCLUST package in the R language environment (www.stat.washington.edu/mclust/).

3.3. Classification of CNV by Cross-Sample Analysis

Automatic detection methods have been designed to report, from each array-CGH profile, a list of chromosomal segments showing CNV in the test DNA compared to the reference using the rest of the genome (or some control chromosomal regions) as the baseline. Projects involving array-CGH analysis for larger cohorts of individuals may require additional information, such as the exact frequency of each CNV in the test population or the presence of any differences between samples showing the same gain or the same loss compared to the single reference sample. Cross-sample comparison can provide such additional information by examining for each clone – or each group of consecutive clones – the distribution of log2ratio values from all individuals (**Fig. 3.4C**). Cross-sample analysis has been successfully applied for a global description of CNV in the human genome *(17)* and will be more widely used for CNV association studies in a near future *(18)*.

4. Conclusion

Since its development in the late 1990s, array-CGH has been widely used by research laboratories for the detection of copy number changes in DNA samples from individuals with cancer and constitutional disease. It has quickly become a reference method for the diagnosis of patients with severe developmental defects. This method has also been instrumental for the discovery of an unexpected level of DNA CNV in the human genome, between individuals from the general population. New array-CGH platforms dedicated to variable regions of the genome are now in development for CNV association studies on common diseases. Array-CGH will certainly become a standard method in human genetics in years to come: new improved strategies will be required for better statistical interpretation of array-CGH results.

Acknowledgment

This work was supported by the Wellcome Trust.

References

1. Kallioniemi, A., Kallioniemi, O.P., Sudar, D., Rutovitz, D., Gray, J.W., Waldman, F. and Pinkel, D. (1992) Comparative genomic hybridization for molecular cytogenetic analysis of solid tumors. *Science*, **258**, 818–821.
2. du Manoir, S., Speicher, M.R., Joos, S., Schrock, E., Popp, S., Dohner, H., Kovacs, G., Robert-Nicoud, M., Lichter, P. and Cremer, T. (1993) Detection of complete and partial chromosome gains

and losses by comparative genomic in situ hybridization. *Hum genet*, **90**, 590–610.
3. Gebhart, E. and Liehr, T. (2000) Patterns of genomic imbalances in human solid tumors (Review). *Int J Oncol*, **16**, 383–399.
4. Cheung, V.G., Nowak, N., Jang, W., Kirsch, I.R., Zhao, S., Chen, X.N., Furey, T.S., Kim, U.J., Kuo, W.L., Olivier, M. et al. (2001) Integration of cytogenetic landmarks into the draft sequence of the human genome. *Nature*, **409**, 953–958.
5. Solinas-Toldo, S., Lampel, S., Stilgenbauer, S., Nickolenko, J., Benner, A., Dohner, H., Cremer, T. and Lichter, P. (1997) Matrix-based comparative genomic hybridization: biochips to screen for genomic imbalances. *Genes Chromosomes Cancer*, **20**, 399–407.
6. Pinkel, D., Segraves, R., Sudar, D., Clark, S., Poole, I., Kowbel, D., Collins, C., Kuo, W.L., Chen, C., Zhai, Y. et al. (1998) High resolution analysis of DNA copy number variation using comparative genomic hybridization to microarrays. *Nat Genet*, **20**, 207–211.
7. Dhami, P., Coffey, A.J., Abbs, S., Vermeesch, J.R., Dumanski, J.P., Woodward, K.J., Andrews, R.M., Langford, C. and Vetrie, D. (2005) Exon array CGH: detection of copy-number changes at the resolution of individual exons in the human genome. *Am J Hum Genet*, **76**, 750–762.
8. Selzer, R.R., Richmond, T.A., Pofahl, N.J., Green, R.D., Eis, P.S., Nair, P., Brothman, A.R. and Stallings, R.L. (2005) Analysis of chromosome breakpoints in neuroblastoma at sub-kilobase resolution using fine-tiling oligonucleotide array CGH. *Genes Chromosomes Cancer*, **44**, 305–319.
9. Redon, R., Baujat, G., Sanlaville, D., Le Merrer, M., Vekemans, M., Munnich, A., Carter, N.P., Cormier-Daire, V. and Colleaux, L. (2006) Interstitial 9q22.3 microdeletion: clinical and molecular characterisation of a newly recognised overgrowth syndrome. *Eur J Hum Genet*, **14**, 759–767.
10. Veltman, J.A., Schoenmakers, E.F., Eussen, B.H., Janssen, I., Merkx, G., van Cleef, B., van Ravenswaaij, C.M., Brunner, H.G., Smeets, D. and van Kessel, A.G. (2002) High-throughput analysis of subtelomeric chromosome rearrangements by use of array-based comparative genomic hybridization. *Am J Hum Genet*, **70**, 1269–1276.
11. Shaw-Smith, C., Redon, R., Rickman, L., Rio, M., Willatt, L., Fiegler, H., Firth, H., Sanlaville, D., Winter, R., Colleaux, L. et al. (2004) Microarray based comparative genomic hybridisation (array-CGH) detects submicroscopic chromosomal deletions and duplications in patients with learning disability/mental retardation and dysmorphic features. *J Med Genet*, **41**, 241–248.
12. Fiegler, H., Redon, R., Andrews, D., Scott, C., Andrews, R., Carder, C., Clark, R., Dovey, O., Ellis, P., Feuk, L. et al. (2006) Accurate and reliable high-throughput detection of copy number variation in the human genome. *Genome research*, **16**, 1566–1574.
13. Price, T.S., Regan, R., Mott, R., Hedman, A., Honey, B., Daniels, R.J., Smith, L., Greenfield, A., Tiganescu, A., Buckle, V. et al. (2005) SW-ARRAY: a dynamic programming solution for the identification of copy-number changes in genomic DNA using array comparative genome hybridization data. *Nucl Acids Res*, **33**, 3455–3464.
14. Smith, T.F. and Waterman, M.S. (1981) Identification of common molecular subsequences. *J Mol Biol*, **147**, 195–197.
15. Jong, K., Marchiori, E., Meijer, G., Vaart, A.V. and Ylstra, B. (2004) Breakpoint identification and smoothing of array comparative genomic hybridization data. *Bioinformatics (Oxford, England)*, **20**, 3636–3637.
16. Olshen, A.B., Venkatraman, E.S., Lucito, R. and Wigler, M. (2004) Circular binary segmentation for the analysis of array-based DNA copy number data. *Biostatistics (Oxford, England)*, **5**, 557–572.
17. Redon, R., Ishikawa, S., Fitch, K.R., Feuk, L., Perry, G.H., Andrews, T.D., Fiegler, H., Shapero, M.H., Carson, A.R., Chen, W. et al. (2006) Global variation in copy number in the human genome. *Nature*, **444**, 444–454.
18. McCarroll, S.A. and Altshuler, D.M. (2007) Copy-number variation and association studies of human disease. *Nat Genet*, **39**, S37–42.
19. Perry, G.H., Dominy, N.J., Claw, K.G., Lee, A.S., Fiegler, H., Redon, R., Werner, J., Villanea, F.A., Mountain, J.L., Misra, R. et al. (2007) Diet and the evolution of human amylase gene copy number variation. *Nat Genet*, **39**, 1256–1260.

Chapter 4

Design of Tag SNP Whole Genome Genotyping Arrays

Daniel A. Peiffer and Kevin L. Gunderson

Abstract

Whole genome association studies have recently been enabled by combining tag SNP information derived from the International HapMap project with novel whole genome genotyping array technologies. In particular, Infinium® whole genome genotyping (WGG) technology now has the power to genotype over 1 million SNPs on a single array. Additionally, this assay provides access to virtually any SNP in the genome enabling selection of optimized SNP content . In this chapter, we provide an overview of the tag SNP-based selection strategy for Infinium whole-genome genotyping BeadChips, including the Human 1 M BeadChip. These advances in both SNP content and technology have enabled both large-scale whole-genome disease association (WGAS) and copy number variation (CNV) studies with the ultimate goal of identifying common genetic variants, disease-associated loci, proteins, and biomarkers.

Key words: Tag SNP, HapMap, genotyping, copy number variation, BeadChip, BeadArray.

1. Introduction

Recent advances in the field of genomic technologies coupled with information from the International HapMap project are enabling an information explosion in genetic analysis. This is evidenced by the flood of research publications in the past year (2007) in which whole genome genotyping (WGG) technologies have enabled large-scale (thousands of individuals) genome wide association studies (GWAS). These efforts have yielded over 100 disease-associated loci in over 40 common diseases including Crohn's disease, heart disease, asthma, type 1 and 2 diabetes, prostate, colorectal and breast cancer, rheumatoid arthritis, and glaucoma with many more disease studies still in progress *(1–17)*.

Martin Dufva (ed.), *DNA Microarrays for Biomedical Research: Methods and Protocols, vol. 529*

DOI 10.1007/978-1-59745-538-1_4 Springerprotocols.com

The single nucleotide polymorphism (SNP) markers on these WGG arrays were largely derived from the efforts of the International HapMap project (www.hapmap.org/) that characterized the haplotype block structure of the human genome, and enabled selection of tag SNPs for several major populations *(18)*. Tag SNPs utilizing the linkage disequilibrium (LD) present in the human genome can serve as proxies for a much larger set of genetically redundant SNPs and can be used to capture a major fraction of the "variation" present within a population. The haplotype map and corresponding tag SNPs provide a framework for discovering associations between genes and disease. Importantly, incorporation of tag SNPs reduces genotyping demands by providing an equivalent power as threefold more randomly chosen SNPs *(19)*. The use of fewer SNPs also minimizes data handling and computation time and reduces the type I errors (false positives) from multiple hypothesis testing.

Based on the HapMap data, it was learned that the key technology needs for genome-wide association studies are: (1) the ability to accurately and economically genotype hundreds of thousands of loci across thousands of samples *(19)*, (2) a robust means of processing many samples easily and efficiently, (3) a technician-friendly automatable process that reduces sample tracking errors, and (4) a genotyping platform enabling unconstrained SNP selection allowing access to tag SNPs throughout the genome. This chapter describes how this SNP content has been incorporated into several BeadChip products with application to WGAS and DNA copy number studies. **Chapter 20** in this book describes the actual mechanics of performing the Infinium WGG assay.

2. Methods

2.1. Probe Design for Infinium SNP Assays

Probes in the Infinium assay are designed to be 50 bases in length immediately adjacent to the SNP site. Infinium II probes terminate just 3′ to the SNP allowing a single base extension read out of the SNP. The bioinformatics selection criteria for the probes are minimal. Probes are designed to either strand depending on which strand has a higher design score. A probe design score is assigned to each probe sequence based upon self-complementarity, runs of G's, multiple genomic hits, extremes in AT or GC content, and several other criteria. Only a few percent of SNP designs are rejected based upon these design rules.

2.2. Content is Derived from the International HapMap Project

As part of the Human Genome Project and SNP Consortium, we learned that genetic variation is characterized, at least in part, by the presence of over 10 million common, SNPs in the human genome. It was further learned that the LD patterns observed in human populations reveal a block-like structure; SNPs within a high LD region (termed haplotype block) tend to be transmitted together from generation to generation *(20, 21)*. Because of this structure, a set of tag SNPs (which serve as proxies for other SNPs) can be chosen based on levels of correlation between their alleles *(22, 23)*. The use of tag SNPs greatly reduces the genotyping burden in whole-genome disease association studies since a fewer number of tag SNPs need to be genotyped while maintaining the same information and power as if one had genotyped a much larger number of random SNPs *(19)*.

To examine LD and associated haplotypes in the genome, the International HapMap Consortium genotyped over 3.8 million SNPs (phase I and II) in four populations (CEU: Utah residents with ancestry from Northern and Western Europe; CHB: Asian populations Han Chinese in Beijing, China, and JPT: Japanese in Tokyo, Japan; YRI: Yoruba in Ibadan, Nigeria). Their results indicated that characterization of around 500,000 tag SNPs (representative SNPs) is adequate to provide sufficient genomic coverage for LD-based disease association studies in many populations, including several populations not directly typed in the HapMap project *(24)*.

Illumina has developed a WGG assay that enables unconstrained selection of SNPs in the genome and effectively unlimited multiplexing from a single sample preparation. Because of the ability to select any tag SNP in the genome, Illumina has created several different high-density tag SNP genotyping products. In early 2006, the world's first WGG BeadChip based upon tag SNPs – the HumanHap300 BeadChip containing over 317,000 tag SNPs was introduced. More recent introductions include the HumanHap300-Duo, the HumanHap550-Duo Beadchip containing over 560,000 tag SNPs, the HumanHap650Y BeadChip containing the Hap550 content plus 100 k YRI SNPs, and the Human 1 M BeadChip containing over one million SNPs for genotyping and genome-wide copy number variation (CNV) analysis (**Fig. 4.1**).

2.3. Intelligent Selection of SNP Content for BeadChip Products

The tag SNP concept and data from the HapMap project was used to intelligently select particular SNPs that would provide maximal coverage of the human genome, while at the same time still provide relatively even spacing for copy number profiling *(25)*.

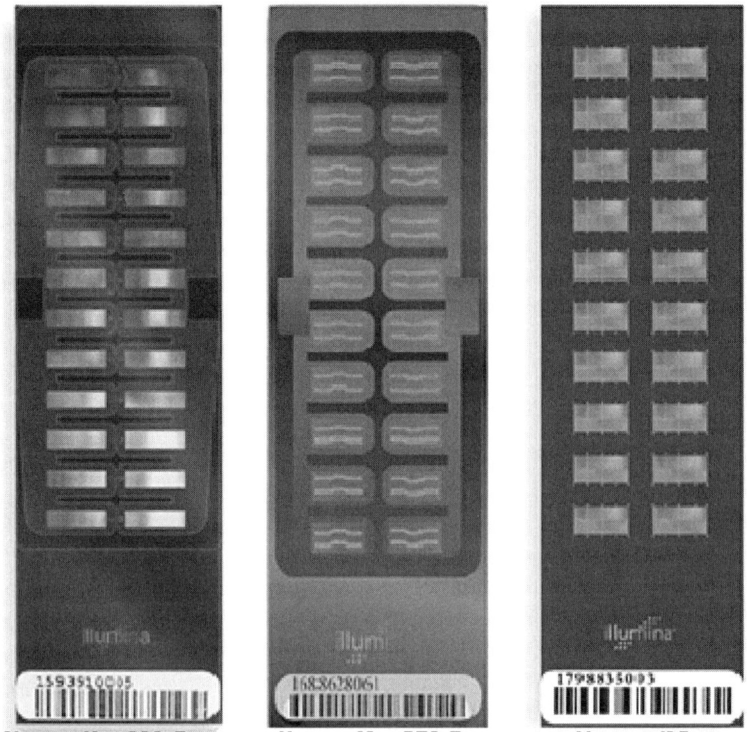

Fig. 4.1. **The Infinium family of BeadChips.** From left to right, examples of the HumanCNV300-Duo, Hap550-Duo, and Human 1 M BeadChips.

2.3.1. Content Selection for the HumanHap300 BeadChip

To select tag SNPs for the HumanHap300-Duo product, the genome was divided into 1 Mb non-overlapping segments, and pair wise r^2 values (measure of LD between two SNPs, 0 => no linkage and 1 => complete linkage between SNP alleles) were calculated for all SNP loci within 200 kb of each other. Approximately 314,000 SNPs were selected from a total of 775,000 SNPs (minor allele frequency (MAF) > 0.05 in CEU) in the Phase I HapMap data by employing an r^2 threshold. Priority was given to SNPs within 10 kb of a RefSeq gene or evolutionary conserved region (ECR). After this selection, gaps in coverage were filled with SNPs of r^2 value greater than 0.7. The HumanCNV370-Duo BeadChip is based upon the Hap300 BeadChip content, but includes an additional ~52,000 markers designed to target highly polymorphic CNV regions.

2.3.2. Content Selection for the HumanHap550 BeadChip

The HumanHap550-Duo product consists of the HumanHap300-Duo SNPs described above plus an additional ~240,000 tag SNPs selected from Phase II of the HapMap data (2.1 million SNPs with MAF greater than 0.5 in CEU population). These additional SNPs were selected using the LdSelect algorithm *(26)* in conjunction with the existing HumanHap300

SNPs. Those SNPs that were highly polymorphic across all populations were given priority, as well as SNPs within 10 kb of RefSeq genes and ECRs (r^2 = 0.8 in gene regions/ECRs; 0.7 for rest of genome). After the core tag SNPs for the CEU population were selected, additional tag SNPs were included from the Han Chinese/Japanese (CHB+JPT; all bins > 2 SNPs at r^2 = 0.8) and Yoruba populations (YRI; all bins > 4 SNPs at r^2 = 0.7). Apart from tag SNPs, 7,779 non-synonymous SNPs (nsSNPs), 96,098 SNPs in 2714 CNV regions, 177 mitochondrial SNPs, 1,800 SNPs in the MHC region, and SNPs in gap regions (no SNPs in regions > 100 kb) were included. The final product provides a mean SNP spacing of roughly one marker every 5 kb.

2.3.3. Content Selection for the Human 1 M BeadChip

For the Human 1 M BeadChip, we decided to expand on the content provided by the HumanHap550 BeadChip with an additional 510,000 SNPs from various sources. As discussed in the previous section, the HapMap project provided the scientific community with both validated SNPs and additional knowledge of population haplotype structure. However, it is now known that some genes are poorly covered by HapMap SNPs. As such, for the Human 1 M BeadChip, additional SNPs in gene regions outside of the HapMap project were chosen because many known variants associated with both simple Mendelian and complex disease traits have been found to reside either near (i.e., within 10 kb) or within coding regions. Due to the fact that both genes and ECRs can be so important to disease phenotypes, we targeted a high density of SNPs across all RefSeq transcript regions (>99%) and in putative ECRs of the genome. Since many diseases, such as Crohn's disease, have been linked to amino acid changing SNPs (nsSNPs), we also added nearly 25,000 nsSNPs *(27),* which may directly affect protein structure and/or function, and are therefore expected to have a major impact on phenotype.

Even though the Human 1 M content was focused primarily in genes, other classes of content were added to the Human 1 M BeadChip. We have chosen additional tag SNPs in all HapMap populations to fill the small number of bins that were not originally covered. We created a higher density of SNPs in ~200 genes involved with drug absorption, distribution, metabolism, and excretion (ADME), since they are an important class of genes for pharmacogenetic studies. We also targeted genes and other important regions showing CNV in the major histocompatibility complex (MHC), a region of the genome implicated in a large number of autoimmune and inflammatory diseases. The MHC region is gene-dense with a high proportion of genes involved in the immune system, including the human leukocyte antigen (HLA) membrane glycoproteins that mediate T-lymphocyte signalling.

Also, in collaboration with deCODE Genetics (www.decode.com) over 50,000 markers were added to specifically target a set of highly polymorphic CNV regions such as megasatellites, segmental duplications, and the "unSNPable" genome (defined as regions where genotyping failed in the HapMap project or regions lacking SNPs). Where possible, a SNP was used as this allows for the measurement of copy number changes apart from fluctuations in intensity, but with higher signal-to-noise ratios. In some cases, a SNP could not be found, so a non-polymorphic probe was utilized (essentially, a "fake" SNP). Overall, this product was designed to minimize the number of genome-wide gaps while providing even spacing of markers for profiling of copy number changes. The final product provides one marker every 1.5 kb.

2.4. Genomic Coverage of HapMap Infinium Products

Genomic coverage is defined as the number of SNPs that are in LD with a reference SNP set. To calculate the genomic coverage provided by each of the Infinium BeadChips, we used all common SNPs typed by the HapMap Project as a reference. Genomic coverage estimates were performed as previously described *(25)*. A high r^2 between two SNPs indicates high correlation, making these SNPs good proxies for each other. At a maximum r^2 of 1, two SNPs are in perfect LD and can serve as pure proxies; thus, only one SNP needs to be genotyped to know the genotype of the other. At any given r^2, different genotyping products have a certain genomic coverage, and therefore a certain power to detect association at a given sample size and odds ratio of the disease under study.

Using a strict r^2 threshold of 0.8, we find that the HumanHap300 captures 81% and 68% of the common HapMap SNPs in the CEU and CHB+JPT samples, respectively, whereas the HumanHap550 captures 90 and 87% of the common HapMap SNPs in the CEU and CHB+JPT samples, respectively. The HumanHap650Y captures 67% of common HapMap SNPs in YRI samples. Lastly, the Human 1 M captures 94, 92, and 73% of the common HapMap SNPs in the CEU, CHB+JPT, and YRI samples, respectively (**Fig. 4.2**). Genomic coverage provided by each Infinium BeadChip, as well as other details on the content of each array is shown in **Table 4.1**.

The genomic coverage provided by an array directly correlates to the ability to find an association to a disease under study. Since current whole-genome association technologies rely on genotyping SNPs near a disease locus, the power to detect association relies on the LD of the genotyped markers with the adjacent disease-causing SNP. Any reduction in power can be overcome by increasing the sample size. However, using a tag SNP approach optimizes the power at any given sample size, reducing the overall number of samples that must be run to detect a significant association *(25)*

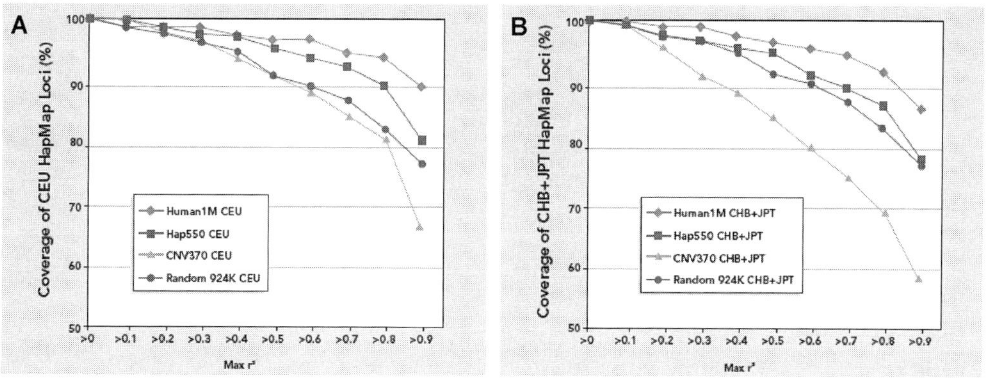

Fig. 4.2. **The Infinium products provide optimal genomic coverage for whole-genome association studies.** Max r^2 indicates the minimum LD with a tag SNP when calculating the fractions. Corresponding genomic coverage is shown for the Illumina Infinium HumanCNV370, HumanHap550, and Human 1 M BeadChips versus a microarray designed with 924,000 randomly selected SNPs across either (**A**) CEU populations, or (**B**) CHB + JPT populations. Note that these genomic coverage values are averaged over the chromosome and may vary based on the specific genomic location.

Table 4. 1
Specifications for BeadChips

	Human Hap300-DUOv2	Human CNV370-DUO	Human Hap550-DUO	Human Hap650yv3	Human 1 M
Number of Markers	318,237	370,404	561,466	660,918	1,072,820
Number of Samples	2	2	1	1	1
Genomic Coverage					
CEU (Mean/Median/ r^2>0.8)	0.87/0.97/ 0.81	0.87/1.0/ 0.81	0.93/1.0/ 0.90	0.94/1.0/ 0.91	0.95/1.0/ 0.94
CHB+JPT	0.81/0.94/ 0.68	0.82/0.95/ 0.68	0.91/1.0/ 0.86	0.92/1.0/ 0.88	0.95/1.0/ 0.92
YRI	0.57/0.55/ 0.34	0.59/0.58/ 0.34	0.74/ 0.86/ 0.57	0.81/0.93/ 0.67	0.85/1.0/ 0.73
Minor Allele Frequency*					
CEU (Mean/Median)	0.26/0.25	0.25/0.25	0.23/0.23	0.22/0.21	0.20/.019
CHB+JPT	0.23/0.23	0.23/0.23	0.21/0.20	0.20/0.19	0.19/0.17
YRI	0.23/0.22	0.22/0.22	0.22/0.21	0.22/0.20	0.20/0.18
Spacing (kb) (Mean/ Median)	9.25/5.5	7.9/5.0	5.3/2.9	4.5/2.4	2.7/1.7

(continued)

Table 4. 1 (continued)

	Human Hap300-DUOv2	Human CNV370-DUO	Human Hap550-DUO	Human Hap650yv3	Human 1 M
Marker Categories					
Markers within 10 kb of a known RefSeq Gene	143,317	164,485	252,793	292,543	565,718
Non-Synonymous SNPs**	6,663	7,181	7,107	7,350	23,288
MHC†/ADME‡/Indel SNPs	1,451/ 1,671/0	5,058/ 2,022/0	2,190/ 2,949/0	2,374/ 3,431/0	10,073/ 15,468/ 501
Sex Chromosome Content (X/Y/PAR loci)	9,035/0/2	12,556/ 1,412/ 361	13,820/ 10/15	16,472/ 10/15	40,097/ 2,283/ 686
Mitochondrial SNPs	na	na	163	163	163
		CNV Coverage			
Number of DGV§ Regions Represented	2,735	3,034	2,991	3,026	3,298
Number of Markers in DGV Regions	54,480	79,631	98,656	114,837	206,665
Average Markers per Region	19.9	26.2	33	38	62.6
Targets highly polymorphic CNV Regions (~9 K)	No	Yes	No	No	Yes

*Based on HapMap release 22 data
**Based on RefSeq and Ensembl databases
†MHC region as defined by de Bakker and colleagues[5]
‡Markers within 10 kb of a known ADME related gene
§Toronto Database of Genomic Variants (projects.tcag.ca/variation) containing 3,644 CNV regions as of March 2007

3. DNA Copy Number Applications

The BeadChip design for the Infinium assay is particularly optimal for copy number measurements as the signal intensity of each SNP (or probe) is the average of 15–20 individual beads. This built-in redundancy adds an additional level of statistical significance to each measurement, enabling the robust detection of single-copy changes. Also, the same assay with a simple normalization procedure is employed across all Infinium BeadChips, enabling copy number analysis on all products with marker densities ranging from ~7 kb to ~1 kb *(28)*. Numerous studies have shown the effectiveness of this approach for examining genome-wide copy number changes from tumors to various types of constitutional samples *(29, 30)*.

4. Summary

Illumina has developed a scalable WGG technology that provides high-quality tag SNP genotyping across more than one million loci on an accurate, efficient, cost-effective, and automated platform enabling large-scale whole-genome disease association and CNV studies. The relatively unlimited multiplex capabilities, the flexibility to choose SNPs or non-polymorphic probes based on scientific interest, and assay automation are key characteristics of the WGG technology. These characteristics enable products ranging from genome-wide tag SNP-based, custom, and multi-sample BeadChips with a common workflow from whole genome to fine-mapping. The Human 1 M BeadChip currently offers the most comprehensive genomic coverage across populations of any currently available genetic analysis product. In the future, integrated whole genome analysis of the genome will likely include a suite of methods including whole-genome SNP genotyping, whole genome epigenetic analysis, DNA copy and LOH analysis, gene expression, and focused large-scale genome sequencing.

Acknowledgments

The technology development summarized here would not have been possible without the efforts of many dedicated individuals. We would like to thank the people at Illumina for their valuable contributions in molecular biology, automation, oligo synthesis, chemistry, engineering, bioinformatics, software, manufacturing, and process development. Special thanks go to John Stuelpnagel, Cynthia Allred, and Rose Espejo for careful reading of the manuscript. The whole genome genotyping research was funded, in part, by grants from the NIH/NCI.

References

1. Thorleifsson, G., Magnusson, K. P., Sulem, P., et al. (2007) Common sequence variants in the LOXL1 gene confer susceptibility to exfoliation glaucoma. *Science* **317,** 1397–400.
2. Libioulle, C., Louis, E., Hansoul, S., et al. (2007) Novel Crohn disease locus identified by genome-wide association maps to a gene desert on 5p13.1 and modulates expression of PTGER4. *PLoS Genet* **3,** e58.
3. Moffatt, M. F., Kabesch, M., Liang, L., et al. (2007) Genetic variants regulating ORMDL3 expression contribute to the risk of childhood asthma. *Nature* **448,** 470–3.
4. Stacey, S. N., Manolescu, A., Sulem, P., et al. (2007) Common variants on chromosomes 2q35 and 16q12 confer susceptibility to estrogen receptor-positive breast cancer. *Nat Genet* **39,** 865–9.
5. Gudmundsson, J., Sulem, P., Steinthorsdottir, V., et al. (2007) Two variants on chromosome 17 confer prostate cancer risk, and the one in TCF2 protects against type 2 diabetes. *Nat Genet* **39,** 977–83.
6. Helgadottir, A., Thorleifsson, G., Manolescu, A., et al. (2007) A common variant on chromosome 9p21 affects the risk of myocardial infarction. *Science* **316,** 1491–3.
7. Gudbjartsson, D. F., Arnar, D. O., Helgadottir, A., et al. (2007) Variants conferring risk of atrial fibrillation on chromosome 4q25. *Nature* **448,** 353–7.
8. Saxena, R., Voight, B. F., Lyssenko, V., et al. (2007) Genome-wide association analysis identifies loci for type 2 diabetes and triglyceride levels. *Science* **316,** 1331–6.
9. Scott, L. J., Mohlke, K. L., Bonnycastle, L. L., et al. (2007) A genome-wide association study of type 2 diabetes in Finns detects multiple susceptibility variants. *Science* **316,** 1341–5.
10. Consortium, T. W. T. C. C. (2007) Genome-wide association study of 14,000 cases of seven common diseases and 3,000 shared controls. *Nature* **447,** 661–78.
11. Hunter, D. J., Kraft, P., Jacobs, K. B., et al. (2007) A genome-wide association study identifies alleles in FGFR2 associated with risk of sporadic postmenopausal breast cancer. *Nat Genet* **39,** 870–4.
12. McKinney, C., Merriman, M. E., Chapman, P. T., et al. (2008) Evidence for an influence of chemokine ligand 3-like 1 (CCL3L1) gene copy number on susceptibility to rheumatoid arthritis. *Ann Rheum Dis* **67**(3), 409–13.
13. Yeager, M., Orr, N., Hayes, R. B., et al. (2007) Genome-wide association study of prostate cancer identifies a second risk locus at 8q24. *Nat Genet* **39**(5), 645–9.
14. Sladek, R., Rocheleau, G., Rung, J., et al. (2007) A genome-wide association study identifies novel risk loci for type 2 diabetes. *Nature* **445,** 881–5.
15. Libioulle, C., Louis, E., Hansoul, S., et al. (2007) Novel Crohn Disease Locus Identified by Genome-Wide Association Maps to a Gene Desert on 5p13.1 and Modulates Expression of PTGER4. *PLoS Genet* **3,** e58.
16. Gudmundsson, J., Sulem, P., Manolescu, A., et al. (2007) Genome-wide association study identifies a second prostate cancer susceptibility variant at 8q24. *Nat Genet* **39**(5), 631–7.
17. Kasperaviciute, D., Weale, M. E., Shianna, K. V., et al. (2007) Large-scale pathways-based association study in amyotrophic lateral sclerosis. *Brain* **130**(9), 2292–2301.
18. Couzin, J. (2005) Genomics. New haplotype map may overhaul gene hunting. *Science* **310,** 601.
19. Hinds, D. A., Stuve, L. L., Nilsen, G. B., Halperin, E., Eskin, E., Ballinger, D. G., Frazer, K. A., and Cox, D. R. (2005) Whole-genome patterns of common DNA variation in three human populations. *Science* **307,** 1072–9.
20. Gabriel, S. B., Schaffner, S. F., Nguyen, H., et al. (2002) The structure of haplotype blocks in the human genome. *Science* **296,** 2225–9.
21. Salisbury, B. A., Pungliya, M., Choi, J. Y., Jiang, R., Sun, X. J., and Stephens, J. C. (2003) SNP and haplotype variation in the human genome. *Mutat Res* **526,** 53–61.
22. Cardon, L. R., and Abecasis, G. R. (2003) Using haplotype blocks to map human complex trait loci. *Trends Genet* **19,** 135–40.
23. Johnson, G. C., Esposito, L., Barratt, B. J., et al. (2001) Haplotype tagging for the identification of common disease genes. *Nat Genet* **29,** 233–7.

24. Altshuler, D., Brooks, L. D., Chakravarti, A., Collins, F. S., Daly, M. J., Donnelly, P., and Consortium, T. I. H. (2005) A haplotype map of the human genome. *Nature* **437,** 1299–320.
25. Eberle, M. A., Ng, P. C., Kuhn, K., et al. (2007) Power to Detect Risk Alleles Using Genome-Wide Tag SNP Panels. *PLoS Genet* **3,** e170.
26. Carlson, C. S., Eberle, M. A., Rieder, M. J., Yi, Q., Kruglyak, L., and Nickerson, D. A. (2004) Selecting a maximally informative set of single-nucleotide polymorphisms for association analyses using linkage disequilibrium. *Am J Hum Genet* **74,** 106–20.
27. Duerr, R. H., Taylor, K. D., Brant, S. R., et al. (2006) A genome-wide association study identifies IL23R as an inflammatory bowel disease gene. *Science* **314,** 1461–3.
28. Peiffer, D. A., Le, J. M., Steemers, F. J., et al. (2006) High-resolution genomic profiling of chromosomal aberrations using Infinium whole-genome genotyping. *Genome Res* **16,** 1136–48.
29. Jackson, E. M., Shaikh, T. H., Zhang, F., Wainwright, L. M., Storm, P. B., Hakonarson, H., Zackai, E. H., and Biegel, J. A. (2007) Atypical teratoid/rhabdoid tumor in a patient with Beckwith-Wiedemann syndrome. *Am J Med Genet A* **143,** 1767–70.
30. Poot, M., Eleveld, M. J., van 't Slot, R., van Genderen, M. M., Verrijn Stuart, A. A., Hochstenbach, R., and Beemer, F. A. (2007) Proportional growth failure and oculocutaneous albinism in a girl with a 6.87 Mb deletion of region 15q26.2–>qter. *Eur J Med Genet* **50**(6), 432–40.

Chapter 5

Fabrication of DNA Microarray

Martin Dufva

Abstract

There are many ways to fabricate DNA microarrays. The four main types of arrays are spotted arrays of pre-made DNA probes, in situ synthesis of DNA arrays, random bead arrays and suspension arrays. The different types of array can address different biological assays. Spotted arrays are suitable for application using small to medium number of probes, e.g. focused genotyping, bacterial diagnostics and gene expression analysis. In situ synthesized arrays and random bead arrays are suitable for applications where medium to large number of probes are needed such as genome wide screens for single nucleotide polymorphisms. Suspension arrays are suitable for applications requiring small number of probes. The chapter will include details of fabrication methods and a comparison of the strengths, weaknesses and future use of each method.

Key words: Spotted arrays, in situ synthesized arrays, random bead arrays.

1. Introduction

DNA microarrays expose thousands to millions of different probes to the target simultaneously. It is a challenge to produce arrays of millions of probes with sufficient quality to function in subsequent assay. There are many different ways to fabricate arrays and different methods result in different numbers of probes and probe characteristics (e.g. length, quality). All arrays, whether random, optical, or planar, the most popular, require the location of every probe, even when there are millions of them recorded. This recording of probe location is what allows the identification of probes that have located complementary sequences/targets in the sample. Recording probe locations on planar arrays are simply achieved by positioning the different probes in an ordered fashion so that probe 1 is placed at position (X1, Y1), probe 2 at position

Martin Dufva (ed.), *DNA Microarrays for Biomedical Research: Methods and Protocols, vol. 529*

DOI 10.1007/978-1-59745-538-1_5 Springerprotocols.com

(X2, Y1) etc. Another approach is to use DNA barcodes (*see* **Section 4.1** for details) or to link probes to differently fluorescing beads (*see* **Section 4.2** for details).

A few key characteristics describe microarrays:

(i) spot density: defined as the number of spots per unit area or the distance between spot centres,

(ii) total number of probes on the array: defined as the spot density × the area of the substrate,

(iii) spot morphology: defined as the shape of the spots and is dependent on the substrate and the fabrication technology and

(iv) number of sub-arrays on the slides: determined by size of the sub-arrays and the size of the slide. Sub-arrays can be identical allowing multiple samples to be hybridized to the same slide.

2. Spotted Arrays

2.1. Basic Operations

Spotted arrays are fabricated using a robotically controlled dispensing device (**Fig. 5.1**) called an arrayer, a spotter or a printer. Different spotters typically contain the same basic features but differ in the methodology used to deposit liquid to the surface. Spotters typically have a print head that is mounted on a robotically controlled arm. The arm moves the print head with the dispensing devices, pins, between the sample plate (96- or 384-well microtitre plates) to aspirate probe solution, the microarray

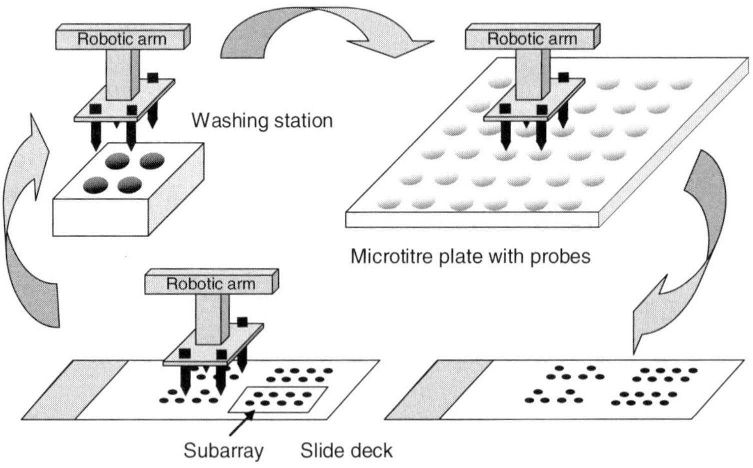

Fig. 5.1 Fabrication of spotted arrays. The robotic arm with the pins moves from the washing station to the probe source plate where the tips are filled with probes. The arm then moves to the slide deck to deposit the probes on the surface of microscope slides.

slides mounted on a slide deck to dispense the probe solution on slides and the washing station (**Fig. 5.1**). The washing station removes any traces of material from the pins before the pins take up new material for the next round of spotting. The number of rounds needed to spot an array is dependent on (*i*) the size of the array (i.e. the number of different probes), (*ii*) the number of pins that can operate in parallel during spotting and (*iii*) the amount of slides that can be spotted from one refill of the pins. Normally, the number of rounds is equal to the size of the probe set divided by the number of pins that are operating in tandem. This assumes that a quill pin is used where one aspiration of probe solution can support spotting of hundreds of slides consecutively before re-aspiration (also denoted 'refill'). If solid pins are used for spotting, the pins must be loaded, but not washed, between processing of each slide which is time consuming.

2.2. Time-Dependent Issues

Spotting arrays of 40,000 different probes (i.e. a complete transcriptome array) on hundreds of slides takes considerable time. It is highly desirable to minimize spotting time because evaporation of probe solutions from the microtitre plates during the spotting can destroy a batch of slides or yield different spot quality over the slides. This results from changes in the viscosity and probe concentration of the spotting buffer over time. Using many pins decreases the time it takes to fabricate arrays considerably because the probe set is basically divided into smaller pieces where each piece of the probe set is fabricated by a pin. The time earnings scales in essence linearly with the number of pins used where using one pin would result in printing times of about a month for 40,000 probes where 48 pins would require less than a day for the same set. Similarly, the size of the probe set to be printed scales essentially linearly with the time it takes to fabricate the arrays. The number of slides to be fabricated has in general less impact on the spotting times because washing and filling of pins with probe solution takes considerable time and is equal whether many or few slides are fabricated simultaneously. Evaporation from the microtitre wells can be minimized during long spotting runs by using cooled microtitre plates or adding compounds such as betaine to the spotting buffer *(1)*.

2.3. Precision

Robotics need to be very precise in order to place spots with 1–5 μm precision on all slides on a deck. The reason for this precision is that computer programmes are subsequently used to evaluate the spots. In such programmes, the spots are identified using grids that assume that the spots are located at equal distances from each other. Offsets in the fabrication method can cause troubles in the spot quantification process because the array cannot be fitted into pre-defined grids. If the deviation is too large, the arrays cannot be quantified without extensive human interventions.

2.4. Pin or Contact Spotters

Pin spotters are probably the most popular spotter types for fabrication of in-house microarrays. A pin spotter or contact spotter, as they are also called, is based on dispensing probe solution using small metal pins (usually made from stainless steel). The spotting mechanism is simple in principle; when the pin hits the solid support a droplet is deposited onto the surface. In practice however contact printing can be difficult. The volume spotted upon contact between the pin and the substrate is usually >0.5 nL depending on pin construction. The spotted *volume* is also dependent on the hydrophobicity of the surface, the spotting solution viscosity, the amount of time the pin rests on the slide surface, temperature (affects the viscosity of the spotting solution) and humidity (which affect evaporation rate and therefore the concentration and viscosity of the spotting solution). Careful optimization of spotting procedures is required to yield reproducible results and it is important to keep as many parameters during spotting and between spotting runs as similar as possible such as temperature and humidity. Humidity during spotting can often be controlled with built-in spotter features while the temperature is usually regulated in the room where the spotter is located. A large drawback of pin spotter is that the volume deposited is irreproducible even if the above parameters are carefully controlled. Typically, quill pins deposit large volumes of probe solution onto the slides just after aspiration. The volume subsequently decreases steadily before reaching a steady state where the variance of spotted volume is acceptable (±20%). To avoid the highly variable spots obtained right after filling up the pin, some spotters print dummy spots prior to printing the real spots on the slides. For gene expression analysis using the co-hybridization strategy, minor spot-to-spot and even batch-to-batch variance is not so critical because the control and the test sample are hybridized to the same slide during hybridization. For other applications where signal from different spots need to be compared, a >20% variation is unacceptable. Pin spotters are fast because many pins can be inserted into the print head and each of these pins works in tandem during spotting.

2.5. Non-contact Spotter

Non-contact spotters deliver probe solution to the slide surface using inkjet, bubble jet or electro-pietzo technology *(2)*. All of these technologies shoot droplets of probe solution onto the slides. The nuzzle ejecting the probe solution is *not* in contact with the slide surface and therefore the volume dispensed is not dependent on the chemistry of the surface. This enables spotting onto hydrophobic surfaces and results in increased spot density because the spots are smaller. Furthermore, non-contact spotters can deliver smaller volumes to the surface than contact spotters which also results in smaller spot size, higher spot density and a larger number of total spots on the array. Spotted volumes can be as low as 1 pL. This is similar to the volumes ejected using inkjet printers meant for traditional printing applications. However, even though as little as 1 pL can be dispensed using non-contact printing, volumes utilized for microarraying are usually >50 pL. The spot-to-spot variance of non-contact spotters can be less than 5% (personal observations) which is acceptable for all microarray applications. Another advantage is that it is possible to detect if droplets has been ejected during a run. The missing spots can then be backfilled into the array after the initial spotting run is completed. This error proofing feature is not available on contact spotters.

The drawback of many non-contact spotters is that they are relatively slow because not more than eight nuzzles can be used in the print head. Furthermore, most non-contact printers have the same kind of robotics as contact printers meaning that the arm is stopped above the target area during ejection of the droplets which is time consuming. ArrayJet has developed non-contact spotters that have 16 to 32 nuzzle that deliver droplets while the print head is moving. This technology makes non-contact spotters as fast as contact spotters for printing arrays.

3. On-Chip Synthesized Arrays

On-chip synthesis refers to the process of synthesizing DNA probes directly on the surface of a chip. The chip material can be made of silicon *(3)*, glass *(4)* or polymeric material *(5)*. The chip can however be of any size from a round wafer to a microscope slide.

3.1. DNA Synthesis Using Traditional Photolithography Process (Affymetrix Technology)

On-chip synthesized arrays were developed in the beginning of the 1990s and laid the foundation for Affymetrix, the largest microarray manufacturer, technology. The technology is based on guiding oligonucleotide synthesis on-chip surfaces using photolithography. The approach is borrowed from the semiconductor industry where computer chips are fabricated using similar photolithographic processes. In the Affymetrix approach, a special chemistry is utilized where each incorporated nucleotide has a photolabile blocking group attached to it. Upon light exposure, the photolabile group is destroyed resulting in exposure of sites for incorporation of the next nucleotide. Affymetrix uses a large number of masks to create arrays of oligonucleotides on silicon wafers where each mask defines which nucleotide is going to be incorporated in each synthesis cycle *(3)*. To create

Table 5.1
Characteristics of fabrication methods

Fabrication Method	Spot size (μm)	Probes/ Chip	Probe length	In house
Contact printing of using a robotic spotter with 1–48 pins	100–200	40,000	>100	Yes
Non-contact printing of probes using 1–32 nuzzels see above	100	50,000	>100	Yes
Fluorescent barcoded beads (Luminex, QDOTS) resolved on flow cytometer	N/A	<500	>100	Yes
Random bead arrays (Illumina)	N/A	500,000	>100	No
Photolithographic synthesis using masks (Affymetrix)	5	4.0 million	25	No
Spotting phosphoramidites (Agilent)	<100	244,000	60	No
Spotting phosphoramidites (Open source) *(11)*	100	98,000	N/A	Yes
Photolithographic synthesis without masks (Nimblegene)	16	2.1 million	85	No
Photolithographic synthesis without masks (FEBITT)	N/A	48,000	60	Yes

arrays of 25 nt probes 80–100 different masks as well as 80–100 process steps are needed. For further details on this fabrication process please *see* reference *(6)*. Microarray fabrication using masks for defining exposed areas can result in very small spot size and very small spot-to-spot distances (*see* **Table 5.2**). This allows arrays of millions of DNA probes per cm^2 to be manufactured (**Table 5.1**). Such large probe numbers are required for detailed SNP analysis of complex genomes *(7)*. The limitation of this technology is that only 25 nucleotide long probes are efficiently fabricated and these probe lengths are less suitable for applications that require longer probe lengths to obtain sufficient signals such as CGH.

3.2. DNA Synthesis Using Mask Less Photolithography (Nimblegene/FEBIT Technology)

A drawback of Affymetrix technology is that it requires 80–100 different masks for the fabrication of a single array. If the sequences of the probes on the arrays need to be changed, a new set of masks must be created and used. An alternative method is to use small mirrors where each mirror guides the synthesis of each spot on the array. As in the photolithographic approach using masks, light is determining the location where the respective nucleotides are going to be coupled in each synthesis step. The mirrors in digital micromirror devices (DMDs) can be individually angled to either reflect or deflect light. The mirrors are controlled by a computer programme that defines which spots (areas) of the array should be illuminated to be able to incorporate a nucleotide. Since the technology does not require any physical masks it is referred to as 'maskless' DNA synthesis *(4, 8)*. Since the control of light is obtained using a programmable device, it is a very flexible and cost efficient method to create small series of high-density arrays. FEBIT has for instance produced a complete microarray analysis machine that synthesizes the array using Nimlegen technology prior to each hybridization. Thus the microarray that is created can be changed between each run if desired just by programming the machine with the desired nucleotide sequences. This is a tremendous tool for array developers since many probes can be tested rapidly in the FEBIT Geniome one system. In contrast a set of masks can easily cost 100,000 $, assuming that each mask cost 1000$. Nimblegene (now a part of Roche diagnostics) has commercialized the mask less synthesis technology for producing DNA arrays and now produces arrays consisting of 384,000 or 2.1 million spots (**Table 5.1**). Another advantage of this approach is that it is possible to synthesize probes up to 85 nucleotides in length on-chip with sufficient yield of the resulting oligonucleotides. An interesting advantage of Nimblegene's synthesis method is the possibility to produce arrays with probes having free 3′ends and having the 5′ end anchored to the surface. A free 3′ end allows for using polymerization assay for genotyping such as mini-sequencing and primer extension *(6)*. Arrays of probes with free 3′ ends can have slightly lower incorporated nucleotide yields than arrays where nucleotides are added in the 3′ to 5′ direction *(9)*.

3.3. In-Situ Synthesis by Spotting Nucleotides

In situ synthesis of arrays can also be achieved by spotting phosphoramides in the order corresponding to the DNA sequences desired in the respective spot using a non-contact spotter *(10, 11)*. The chemistry in this method is efficient as it resembles the chemistry used for synthesis of oligonucleotides using column which is traditionally used for synthesis of oligonucleotides for PCR etc. It is quite common to synthesize oligonucleotides 60 nucleotides long using this approach. However, because array synthesis is based on dispensing liquids, which are more difficult to control spatially than light, the density of the arrays will not be high for light guided synthesis. Using this method, Agilent currently provides arrays with 244,000 spots per slide. Agilent also provides 2, 4 or 8 identical sub arrays on one slide to get cost and sample handling efficient solutions.

4. Bead Based Arrays

Bead arrays consist of optically or DNA bar coded beads onto which probes are linked. In contrast to planar array where probe identity is encoded by a fixed position in the array, bead array are random in terms of position in a plane or volume. The identity of the probes can be tracked by a tagging system where the beads are encoded by unique fluorescent properties or alternatively using DNA barcodes. Since each unique bead type contain only one probe type, it is easy to identify to which probes a target has been hybridizing with by identifying the identity of the respective beads. The beads can either be immobilized in an arrayed fashion as with Illumina's random bead arrays or they can be kept in suspension as with Luminex bead arrays.

4.1. Illumina Random Bead Arrays

Illumina's random bead arrays are fabricated and used according to the steps in **Fig. 5.2**. The first step is to link pre-synthesized oligonucleotides to beads. The oligonucleotide consists of two parts; one specific for the intended target and one bar DNA barcode *(12)*. Hence by immobilizing the oligonucleotide to

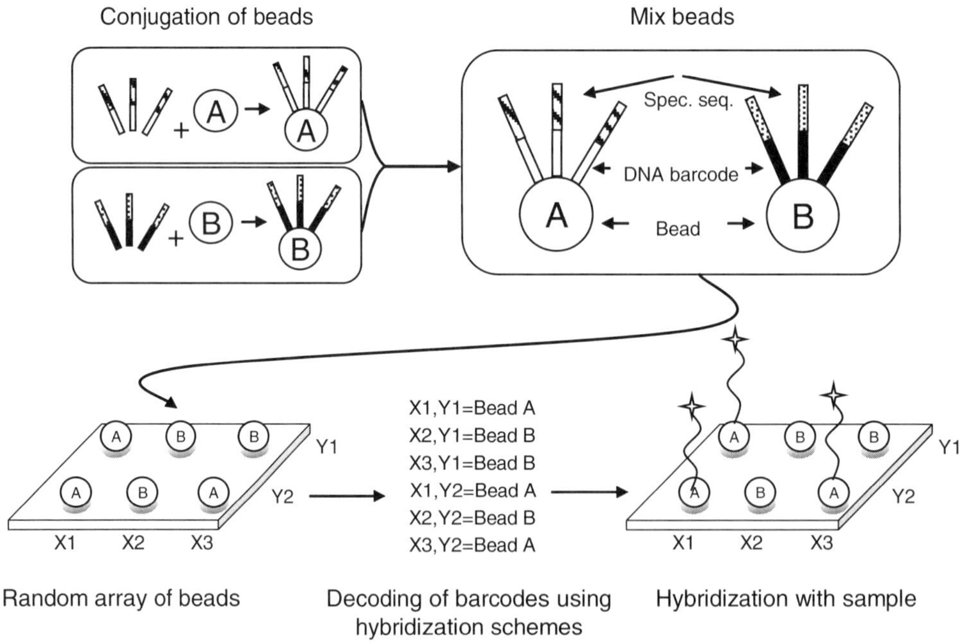

Fig. 5.2 Processing of Illumina random bead arrays. Random bead arrays are fabricated by first conjugating beads with respective oligonucleotide carrying a specific and bar code sequence. The beads carrying different oligonucleotides are then mixed and subsequently spread over a surface with small depression into it. The beads are immobilized in the small wells and will not move during hybridization and washing.

beads, the bead will be automatically encoded by the bar code. As **Fig. 5.2** illustrates, different probes are linked to unique bar codes. Each unique oligonucleotide, carrying a specific probe sequence and a unique bar code, is linked to beads in individual reactions *(12)* For arrays requiring large numbers of probes such as those used for gene expression studies, 40,000 probes, an independent bead reaction must be made for each probe. This requires robotic liquid handling based on the industry standard microtitre plate platform. Following beads modification, the modified beads are pooled and spread over a surface with micro-structured wells matching the size of the beads (**Fig. 5.2**). This self-assembly of the array does not include advanced, expensive and time consuming microfabrication steps as opposed to fabrication of arrays by spotting of photolithography.

For statistical integrity and to avoid having missing probes, the random arrays consist of 20–30 beads for each probe. As gene expression arrays require 40,000 probes, a total of about 1 million beads are needed. Since the beads are small (3–5 μm) and densely packed together, the actual surface of a 1 million bead array is only about 25 mm^2 with each unique probe taking up circle with a diameter of about 10–15 μm. This is significantly less than a corresponding spotted array and close to the size of the spots used for each probe in photolithography fabricated arrays (*see* **Table 5.2**).

The array of beads is decoded using a hybridization scheme that identifies each unique barcode. After this decoding step, the identity of each bead in the now planar array is known i.e. each probe has been assigned an X and Y coordinates in the plane (**Fig. 5.2**). This decoding process is usually performed prior to the hybridization of the sample *(13)*. Hybridization signals coming from the target can then be linked from the location of the signals on the array to the identity of the probes.

4.2. Optically Encoded Arrays (QDOTS and Luminex)

Beads can also be encoded by optical means and resolved on flow cytometer like instruments. Luminex encodes their beads by incorporating each of two different fluorochromes in 10 different concentrations. This results in beads with 100 unique fluorescent characteristics. Each unique bead colour is conjugated to a specific probe. Since there are 100 different beads with unique fluorescent characteristics, 100 different probes can be used in these arrays. Such arrays are also referred to as suspension arrays because the beads are kept in suspension during the experiments.

Although suspension arrays consist of small number of probes as compared to planar arrays, they still offer numerous advantages. Firstly, suspension arrays are easily integrated with robotic handling. This results in large throughput ability for

Table 5.2
Effects of miniaturization of DNA microarray spots

	Spot size						Unit
	1,000	200	100	15	10	5	µm
Technology	Spotting			Photolithography			
Spot centre distances	2,000	300	150	16	11	6	µm
Spot/cm^2	25	1,000	4,500	$4 \cdot 10^5$	$8.3 \cdot 10^5$	$2.8 \cdot 10^6$	Spots/cm^2
Molecules in spot[a]	$4.7 \cdot 10^{10}$	$1.9 \cdot 10^9$	$4.7 \cdot 10^8$	$1.1 \cdot 10^7$	$4.7 \cdot 10^6$	$1.2 \cdot 10^6$	Molecules
Pixels/spot (5 µm resolution	31,400	1,300	300	7	3	1	Pixels
GE array area (40 k spots)	1,600	36	9	0.1	0.048	0.014	cm^2
Diffusion time over 40 k array	Century	Decade	Decade	Days	Days	Hours	
Target volume	80,000	1,800	450	5	2.4	0.72	µL
Target amount	3,200	72	18	0.2	0.1	0.03	µg
Amount of cells	$3.2 \cdot 10^8$	$7.2 \cdot 10^6$	$1.8 \cdot 10^6$	20,000	10,000	3,000	Cells

[a] assuming 100 pmol/mm^2

sample preparation and hybridization to the arrays. Robotic processing has been further facilitated by the introduction of magnetic fluorescently encoded beads. Secondly, hybridization kinetics are excellent for suspension arrays as sample–probe mixing is much easier than on planar arrays. Thirdly, detection is accomplished via automated flow cytometers with automatic microtitre plate sample handling for high throughput analysis of many samples.

100-plex assays are sufficient for many applications however the main drawback of suspension arrays is that they cannot handle larger array experiments. Quantum dots or nanocrystals have better fluorescent characteristics than organic fluorochromes utilized by Luminex. Using six different quantum dots at 10 different intensity levels, about 1 million differently encoded beads can be produced *(14)*, which is sufficient for genome wide analysis. Proof of concept for the use of quantum dot encoded beads in SNP assays resolved on a flow cytometer has been demonstrated *(15)*.

5. Quality Control and Data Tracking

A common requirement for all fabrication methods described is to keep track of the probe contained in each spot or bead. For smaller arrays (50–100 probes) this is not an issue but larger arrays require data tracking systems that subsequently can be read by analysis software. Many spotters produce a 'GAL' file during spotting. The GAL file can be imported into detection/analysis software and create a link between an element in the array and probe identity.

All planar arrays, spotted or in situ synthesized, need to be fabricated in particle free environment. Particles will destroy spots and result in the inability to collect data from such spots. Particles are usually coming from air in the form of dust. A standard molecular biology laboratory is sufficient for spotting small arrays containing sufficient repeating spots. Such arrays are less vulnerable to particle because if one spot is lost due to a particle, the data from a replicate spot can be collected. Large arrays with no repeating spots are very vulnerable because particles on the array will result in the destruction of unique data points that need to be sorted out before analysis. One quality control possibility is to use fluorescent tags on all the probes, e.g. incorporating FITC fluorochrome *(16, 17)*. Microarrays can therefore be easily scanned before hybridization to sort out low quality arrays such as those with missing spots or spots with low amount of probes as determined by the fluorescent intensity. Despite the fact that slide to slide quality control as well as post hybridization quality control can be made, most microarrays are synthesized without a flourochrome on all the probes which is mainly due to increased costs of the probes and that few laboratories have support for detecting fluorochromes other than Cy3 and Cy5 which is used in the assay itself. Therefore, errors in the fabrication procedure will usually be detected when valuable samples have been consumed. Therefore, large arrays without repeating spots need to be fabricated in 'clean rooms'. Clean rooms are fabrication areas that have very low particle/particles counts. Clean rooms were developed several decades ago by the semiconductor industry because, just like with larger arrays, their devices can also be ruined by the presence of particles. It should be noted that particles during the fabrication process causes irregular spot morphology or altogether missing spots due to a particle lying in the surface of the slide to which DNA cannot bind after delivery from the dispensing device. Another common problem is that the dispensing device is not delivering a droplet each time it should. The latter is not necessaryly dependent on the particle count in the room but can also be caused by particles in the spotting solution. Spots can also be destroyed during the hybridization, washing, drying process where particles from the liquids/air can attach to the slides. Such particles will usually fluoresce brightly and should be sorted out during analysis of the slides.

6. Comparison of Technologies and Custom Made Arrays

6.1. Array Size and Quality

The type of assays that can be performed is in many cases limited to the number of spots that can be fitted into a microscope slide or the chip surface area. Some applications such as gene expression arrays can be fabricated using almost any technology except fluorescently encoded beads. Other applications such as tiling arrays, high-resolution CGH and SNP arrays demand more than an order of magnitude more spots on the chip surface if the aim is to get meaningful assays on the genomic scale.

Photolithography is an ideal technology for fabricating arrays with many different spots (**Table 5.1**). Affymetrix technology typically provides highest spot density (**Table 5.2**), however Nimblegene provides almost as many spots as Affymetrix within the array because of the larger array area used by Nimblegene. It should be noted however that Affymetrix is using many replicate spots, 8–20, to measure one type of transcript, gene segment or mutation. This means that for Affymetrix chips the effective number of analytes detected on the arrays is considerably lower. The many repetitive probes utilized give the advantage of having many types of controls for hybridization.

Despite lower spot densities, spotted array have several advantages over photolithographically fabricated arrays. Firstly, spotted arrays use probes synthesized off-chip which allows greater purification and quality control of the probes by using high pressure liquid chromatography or mass spectroscopy prior to spotting. This is a significant advantage of spotted arrays and it is easy to argue that probes of expected length are spotted. All on-chip synthesized arrays have yield issues when probe synthesis steps have yields of less than 100%. Even at 99% coupling efficiency *(4)* during synthesis of 25 nt probes, only 80% of the probes will be full length and correct sequence. For a 60 nt probe only 50% of the probes will be full length and have the proper nucleotide sequence. Pre-synthesized probes typically have yields of about 99.4% for each coupling step. For probes of shorter length there is little impact on overall yield compared to on-chip synthesized probes. However pre-synthesized 60 nt probes will be synthesized with 70 % correct sequence as opposed to only 50% by photolithographic methods (see above). Furthermore, pre-synthesized oligonucleotides can be subjected to purification to remove shorter species. Spotted array therefore utilizes probes that are almost 100% correct in terms of sequence and length. Secondly, spotted arrays are more economical to produce/purchase. On-chip synthesized arrays typically cost >200$ each whereas spotted arrays can be made far less expensively. The cost of a spotted slide will vary but associated costs include: (a) the slide itself, between $1–20 depending on the material it is printed on, (b) spotter costs at about $5/slide assuming 20,000 slides are fabricated during the life time of the spotter and, (c) costs for a probe set which can vary depending on the size of the arrays. In many cases though, the cost for the probe set per slides is quite small because of the minute amount of probes spotted on each slide.

6.2. Accessibility of Technologies

Contact and non-contact spotters are designed to be used 'in-house' and there are numerous installations of these around the worlds. The only commercially available on-chip synthesis machine is the Geniome ONE from FEBIT. The Geniome ONE machine is a complete microarray fabrication and analysis system and includes features such as light guided synthesis of arrays, automatic hybridization of samples and detection of hybridized target within the same instrument. The Geniome ONE platform is based on a small microfluidics device where arrays are synthesized using the same maskless in situ synthesis technology as Nimblegene. This allows the creation of 48,000 spots per array. In the future, more spots will be fit onto the chip since the higher density DMD will be used (see above). Since it contains all the functionalities for making, processing and reading arrays the Geniome ONE is expensive. The drawback of this machine is that the throughput is as low as only one chip per working day.

Agilent, Nimblegene and Affymetrix provide custom array production services that allow researchers to get access to in situ synthesized array technology. Agilent offers the eArray service where users can buy low volumes of custom made arrays/slides. Their 8 by 15 k array slide format allows for testing 15,000 different probes with eight samples per slide. Each such slide costs a little over $1000. Nimblegene offers a similar service.

7. Effects of Miniaturization

Is further miniaturization of arrays to obtain arrays with 10–100 million of probes required? What could such large and dense arrays be used for? There are several biologically relevant applications and question that cannot be addressed using current microarray technology. For example, gene expression arrays only measure the expression of known or predicted genes. One future application would be to have an array that covers the entire human genome without any gaps. A tiling array with 50 million probes each 60 nt long would cover almost the entire human genome. Such an array could be used to get a complete transcription map over human genomes with ~60 bp resolution. Tiling arrays for *Arabidopsis thaliana* have been completed, however 13 tiling arrays each consisting of ~400,000 probes and a 10 bp gap between 35 nt probes are required to cover the entire genome. Such arrays were used to map every expressed gene from *Arabidopsis thaliana (18)*. High-resolution transcription maps also enables location of novel microRNAs that easily could be missed using arrays with less sequence coverage of the genome. The probes and these arrays were fabricated by maskless on-chip synthesis. With current probe density offered by Nimblegene of 2.1 million probes per slide, the Arabidopsis array could be fit into 3 slides instead of 13. The real advantage comes when all the 5 million probes can be synthesized on one array. Single slide hybridization is advantageous compared to splitting the sample on many slides because slide to slide variation of hybridization can decrease the reliability of the results. Furthermore, it is advantageous to hybridize to one slide compared to many slides if many different samples are investigated, i.e. obtaining transcription maps from *Arabidopsis thaliana* grown at ten different experimental conditions would require 130 hybridization using low density arrays but only ten hybridization using slides with the complete probe set. It takes two to three weeks for 130 hybridizations, while 10 hybridizations take only a day. Hence increasing array density to be able to use a single slide is critical for throughput.

Other application where increased resolution is needed is CGH. A tiling array consisting of about 32,000 bacterial artificial chromosome (BAC) clones has been reported that covers the entire human genome with an average resolution of 100,000 bp *(19)*. This resolution is sufficient for identifying deletion or amplifications on the gene or cluster of genes depending on the density of genes within the deleted or amplified segment. The resolution of CGH can easily be increased by using oligonucleotide arrays i.e. 2.1 million probe arrays provided by Nimblegene would have a resolution of about 1,500 bp assuming that every probe is functional within the array. Using such array, deletions and amplification on the gene level can be studied. In contrast a super high-resolution CGH with 50 million tiling 60 nt probes would have a 60 bp resolution which allows for detection of changes within genes, i.e. deletion of exons and amplification or deletion of promoter regions.

As demonstrated above, it is possible to make new types of array to fine map genomes. Apart from increasing parallelism and throughput, miniaturization gives several more advantages and also disadvantages as described in **Table 5.2**. One disadvantage of super high-density microarrays are the large demands on the detection system. Most currently offered scanners have a maximum resolution of 5 μm per pixel. Spots of 2–4 μm in size cannot be detected using such scanners as it is desirable to have at least 25 pixels per spots for accurate quantification. Scanners with resolution down to 200 nm per pixel have been demonstrated *(20, 21)* and could provide enough resolution for reliable quantification of even micrometre-sized spots (**Table 5.2**). In fact, the resolution and sensitivity of such scanners is high enough to count individual molecules hybridized to the spots *(22)*. The sensitivity of the described scanner system is excellent (1.3 fM) and is enough for detecting low abundant transcripts from only 10,000 cells or 200 ng RNA *(22)*. Another disadvantage of miniaturizing the spots is that smaller spots contain fewer capture molecules. A 5 μm spot has about 240,000 active molecules on it assuming a probe density of 100 fmol/mm^2 and 20% of the probes being active (**Table 5.2**). In order to ensure a large dynamic range for target binding, spots cannot get much smaller without loosing the ability to capture and distinguish targets at large concentration ranges.

An important effect when miniaturizing arrays or assay is that less material in needed. There was a dramatic decrease in target required when going from dot blots to microarrays (**Table 5.2**). Instead of mL of hybridization solution, the target is dissolved in hundreds of microlitres or less (**Table 5.2**). In practice, microarrays are hybridized with higher target concentration than dot blot assays leading to higher sensitivity of microarrays compared with dot blots. Despite using higher target concentrations, a microarray assay usually requires less starting material compared with a dot blot assay. This enabled global gene expression to be investigated even in relatively small samples (10^7 cells). Further miniaturization of microrarrays is possible and will further lessen the demand of the sample size for expression analysis. However, if the size of the cellular sample is downscaled with the size of the array (*see* **Table 5.2**), problems with sample handling are likely to occur. For instance, a gene expression array with 5 μm spots would only require target from about 3,000 cells in a total volume of 2 μL assuming there is sufficient target molecule to elicit a signal in such a low cell count. Handling mRNA from 3,000 to 10,000 cells in a 2 μL hybridization solution is difficult because of problems with dispensing, evaporation and loss of material during purification procedures. Therefore, it would be desirable if sample preparation and hybridization to the microarrays where integrated onto a single Lab-on- chip system. Sample preparation has been demonstrated in a Lab-on-chip system in which mRNA from single cells can be extracted using poly-T coated magnetic beads *(23)*. It would be possible to insert a small foot print microarray of probes towards the transcriptome in such microfluidics devices to obtain a fully automatic user friendly miniaturized lab-on-chip system for gene expression analysis.

Acknowledgements

Thanks are due to David Sabourin, Lena Poulsen and Jesper Petersen for reading the manuscripts and helpful discussions.

References

1. Diehl, F., Grahlmann, S., Beier, M. and Hoheisel, J.D. (2001) Manufacturing DNA microarrays of high spot homogeneity and reduced background signal. *Nucleic Acids Res*, 29, E38.
2. Okamoto, T., Suzuki, T. and Yamamoto, N. (2000) Microarray fabrication with covalent attachment of DNA using bubble jet technology. *Nat Biotechnol*, 18, 438–441.
3. Fodor, S.P., Read, J.L., Pirrung, M.C., Stryer, L., Lu, A.T. and Solas, D. (1991) Light-directed, spatially addressable parallel chemical synthesis. *Science*, 251, 767–773.
4. Singh-Gasson, S., Green, R.D., Yue, Y., Nelson, C., Blattner, F., Sussman, M.R. and Cerrina, F. (1999) Maskless fabrication of light-directed oligonucleotide microarrays using a digital micromirror array. *Nat Biotechnol*, 17, 974–978.
5. Moorcroft, M.J., Meuleman, W.R., Latham, S.G., Nicholls, T.J., Egeland, R.D. and Southern, E.M. (2005) In situ oligonucleotide synthesis on poly(dimethylsiloxane): a flexible substrate for microarray fabrication. *Nucleic Acids Res*, 33, e75.
6. Dufva, M. (2005) Fabrication of high quality microarrays. *Biomol Eng*, 22, 173–184.
7. Samani, N.J., Erdmann, J., Hall, A.S., Hengstenberg, C., Mangino, M., Mayer, B., Dixon, R.J., Meitinger, T., Braund, P., Wichmann, H.E. et al. (2007) Genomewide association analysis of coronary artery disease. *N Engl J Med*, 357, 443–453.
8. Nuwaysir, E.F., Huang, W., Albert, T.J., Singh, J., Nuwaysir, K., Pitas, A., Richmond, T., Gorski, T., Berg, J.P., Ballin, J. et al. (2002) Gene expression analysis using oligonucleotide arrays produced by maskless photolithography. *Genome Res*, 12, 1749–1755.
9. Albert, T.J., Norton, J., Ott, M., Richmond, T., Nuwaysir, K., Nuwaysir, E.F., Stengele, K.P. and Green, R.D. (2003) Light-directed 5′->3′ synthesis of complex oligonucleotide microarrays. *Nucleic Acids Res*, 31, e35.
10. Hughes, T.R., Mao, M., Jones, A.R., Burchard, J., Marton, M.J., Shannon, K.W., Lefkowitz, S.M., Ziman, M., Schelter, J.M., Meyer, M.R. et al. (2001) Expression profiling using microarrays fabricated by an ink-jet oligonucleotide synthesizer. *Nat Biotechnol*, 19, 342–347.
11. Lausted, C., Dahl, T., Warren, C., King, K., Smith, K., Johnson, M., Saleem, R., Aitchison, J., Hood, L. and Lasky, S.R. (2004) POSaM: a fast, flexible, open-source, inkjet oligonucleotide synthesizer and microarrayer. *Genome Biol*, 5, R58.
12. Gunderson, K.L., Kruglyak, S., Graige, M.S., Garcia, F., Kermani, B.G., Zhao, C., Che, D., Dickinson, T., Wickham, E., Bierle, J. et al. (2004) Decoding randomly ordered DNA arrays. *Genome Res*, 14, 870–877.
13. Yeakley, J.M., Fan, J.B., Doucet, D., Luo, L., Wickham, E., Ye, Z., Chee, M.S. and Fu, X.D. (2002) Profiling alternative splicing on fiberoptic arrays. *Nat Biotechnol*, 20, 353–358.

14. Han, M., Gao, X., Su, J.Z. and Nie, S. (2001) Quantum-dot-tagged microbeads for multiplexed optical coding of biomolecules. *Nat Biotechnol*, 19, 631–635.
15. Xu, H., Sha, M.Y., Wong, E.Y., Uphoff, J., Xu, Y., Treadway, J.A., Truong, A., O'Brien, E., Asquith, S., Stubbins, M. et al. (2003) Multiplexed SNP genotyping using the Qbead system: a quantum dot-encoded microsphere-based assay. *Nucleic Acids Res*, 31, e43.
16. Hessner, M.J., Wang, X., Hulse, K., Meyer, L., Wu, Y., Nye, S., Guo, S.W. and Ghosh, S. (2003) Three color cDNA microarrays: quantitative assessment through the use of fluorescein-labeled probes. *Nucleic Acids Res*, 31, e14.
17. Hessner, M.J., Wang, X., Khan, S., Meyer, L., Schlicht, M., Tackes, J., Datta, M.W., Jacob, H.J. and Ghosh, S. (2003) Use of a three-color cDNA microarray platform to measure and control support-bound probe for improved data quality and reproducibility. *Nucleic Acids Res*, 31, e60.
18. Stolc, V., Samanta, M.P., Tongprasit, W., Sethi, H., Liang, S., Nelson, D.C., Hegeman, A., Nelson, C., Rancour, D., Bednarek, S. et al. (2005) Identification of transcribed sequences in Arabidopsis thaliana by using high-resolution genome tiling arrays. *Proc Natl Acad Sci U S A*, 102, 4453–4458.
19. Ishkanian, A.S., Malloff, C.A., Watson, S.K., DeLeeuw, R.J., Chi, B., Coe, B.P., Snijders, A., Albertson, D.G., Pinkel, D., Marra, M.A. et al. (2004) A tiling resolution DNA microarray with complete coverage of the human genome. *Nat Genet*, 36, 299–303.
20. Hesse, J., Sonnleitner, M., Sonnleitner, A., Freudenthaler, G., Jacak, J., Hoglinger, O., Schindler, H. and Schutz, G.J. (2004) Single-molecule reader for high-throughput bioanalysis. *Anal Chem*, 76, 5960–5964.
21. Sonnleitner, M., Freudenthaler, G., Hesse, J. and G.J., S. (2005) High-throughput scanning with single-molecule sensitivity. *Proc SPIE*, 5699, 202–210.
22. Hesse, J., Jacak, J., Kasper, M., Regl, G., Eichberger, T., Winklmayr, M., Aberger, F., Sonnleitner, M., Schlapak, R., Howorka, S. et al. (2006) RNA expression profiling at the single molecule level. *Genome Res*, 16, 1041–1045.
23. Hong, J.W., Studer, V., Hang, G., Anderson, W.F. and Quake, S.R. (2004) A nanoliter-scale nucleic acid processor with parallel architecture. *Nat Biotechnol*, 22, 435–439.

Chapter 6

Immobilization Chemistries

Sascha Todt and Dietmar H. Blohm

Abstract

Among the parameters which influence the success of a microarray experiment, the attachment of the nucleic acid captures to the support surface plays a decisive role.

This article attempts to review the main concepts and ideas of the multiple variants which exist in terms of the immobilization chemistries used in nucleic acid microarray technology. Starting from the attachment of unmodified nucleic acids to modified glass slides by adsorption, further strategies for the coupling of nucleic acid capture molecules to a variety of support materials are surveyed with a focus on the reactive groups involved in the respective process.

After a brief introduction, an overview is given about microarray substrates with special emphasis on the approaches used for the activation of these – usually chemically inert – materials. In the next sections strategies for the "undefined" and "defined" immobilization of captures on the substrates are described. While the latter approach tries to accomplish the coupling via a defined reactive moiety of the molecule to be immobilized, the former mentioned techniques involve multiply occurring reactive groups in the capture.

The article finishes with an example for microarray manufacture, the production of aminopropyltriethoxysilane (APTES) functionalized glass substrates to which PDITC homobifunctional linker molecules are coupled; on their part providing reactive functional groups for the covalent immobilization of pre-synthesized, amino-modified oligonucleotides.

This survey does not seek to be comprehensive rather it tries to present and provide key examples for the basic techniques, and to enable orientation if more detailed studies are needed. This review should not be considered as a guide to how to use the different chemistries described, but instead as a presentation of various principles and approaches applied in the still evolving field of nucleic acid microarray technology.

Key words: DNA microarray, immobilization chemistry, capture, survey.

Martin Dufva (ed.), *DNA Microarrays for Biomedical Research: Methods and Protocols, vol. 529*

DOI 10.1007/978-1-59745-538-1_6 Springerprotocols.com

[. . .] "with careful experimental design and appropriate data transformation and analysis, microarray data can indeed be reproducible and comparable among different formats and laboratories" [. . .]

Editorial (2006) Making the most of microarray data.
Nature Biotechnology **24**, 1039

1. Introduction

In 1989 Saiki and co-workers *(1)* were among the first to use the experimental set-up of the reverse dot blot, thereby creating the precursor of the nucleic acid microarray with oligonucleotide captures for detecting and genotyping any nucleic acid (in this case mutations of beta-thalassemia). Today, many parameters have been identified which influence the course of a microarray experiment, specifically the hybridization reaction. These parameters can be divided into sequence- and assay-specific ones. Among the latter the best known are the buffer composition *(2, 3)*, incubation time *(4)*, reaction temperature *(5, 6)*, the chemical properties of the microarray substrate (e.g., hydrophobicity, hydrophilicity) including its homogeneity *(7)*, as well as the spacing between and the conformation of the captures *(8, 9, 10)*. The influence of these parameters on the experimental outcome of microarray experiments has been extensively studied and is not the subject of this review.

Since Saiki's approach of UV-crosslinking polyT tailed oligonucleotides to nylon membranes to achieve spatially resolved immobilization of captures on a solid support, a large spectrum of alternative strategies has been developed (see for example *(7, 11, 12, 13, 14)*), reaching from adsorption of DNA strands to silanized glass slides *(5)* or polymers *(15)*, up to specific covalent bond formation between reactive moieties of the substrate and photoactivatable reactive groups attached to the ends of the capture molecules *(16)*

Due to their different components and compounds, the various substrate chemistries for nucleic acid microarrays differ immensely, not only regarding preparative effort and costs. However, their common aim is to guarantee a stable immobilization of the captures on the solid support, to maintain the molecules' inherent ability to interact sequence specifically with their targets, complementary nucleic acid strands, and to prevent at the same time the non-specific adsorption of nucleic acids in the sample to the microarray surface.

There are essentially two ways to fabricate microarrays, the spotwise synthesis of captures directly on the slide and the deposition of previously synthesized nucleic acid captures using suitable spotting devices (refer to chapters 5 and 7 of this book). As shortly described in **Section 4.6**, the spotwise synthesis at the surface is achieved e.g., by photolithographic methods, enabling densities of more than one million spots per array *(17)*. The alternative, the spotting of pre-synthesized nucleic acids, is based on chemical synthesis of captures, usually produced by the classical phosphoramidite chemistry *(18)*, or on biological reactions like cloning, PCR, or reverse transcription of mRNA. During these synthetic steps the captures are often provided with chemical modifications which are part of the respective immobilization chemistry.

Figure 6.1 depicts the building blocks for the immobilization of pre-synthesized captures to a solid support.

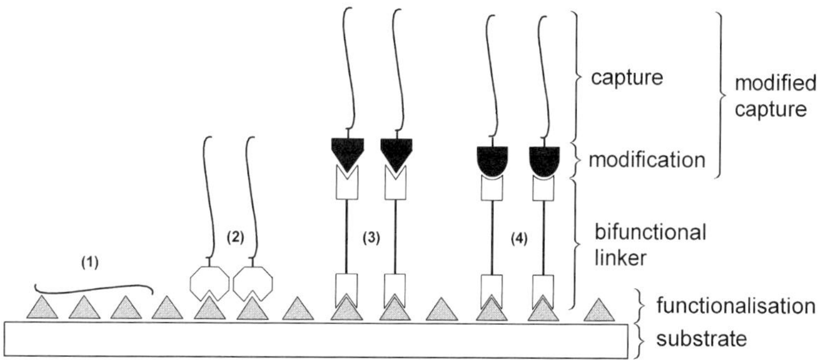

Fig. 6.1. The substrate provides the basis for the microarray architecture. Usually being chemically inert (e.g., a glass microscope slide), it has to be functionalized for the immobilization of the captures (elsewhere in the literature also referred to as (capture) probes). In **Section 2** some procedures for this functionalization process – leading to reactive groups on the support's surface (triangles) – are described in combination with different substrate materials. Unmodified captures (as illustrated in (**1**)) as well as capture molecules, which have been modified with one ((**2**)–(**4**)) or more reactive moieties, can bind or covalently attach to the reactive groups on the surface. With an adequate functionalization of the slide, the unmodified as well as the multiple modified captures can interact with the surface at more than one contact point ((**1**) for the unmodified capture), whereas this binding situation is supposed to be much better defined in case of singly modified captures. Examples for the settings of (**1**) and (**2**) are described in **Sections 3, 4.1, and 4.2**. A variation of capture immobilization is the use of a linking molecule that acts as a "connector" between the reactive groups generated at the support and the capture molecules. These linker molecules usually consist of a chemically inert spacer unit (e.g., CH_n) and reactive groups at both of their ends and can be conceptually subdivided into two groups: homobifunctional molecules (**3**), which carry two identical reactive groups (*see* **Section 4.3**), and heterobifunctional linker molecules (**4**), whose reactive groups target at different moieties (*see* **Section 4.4**).

2. Functionalization of Substrates

2.1. Two-Dimensional Surfaces

Glass has been conventionally used as a substrate for microarray production because of its ready availability, low price, its relatively homogeneous chemical surface as well as its inertness and low autofluorescence. For the attachment of capture probes to silica based surfaces like silicon *(19, 20, 21)*, oxidized silicon *(22)*, or fused silica *(22, 23, 24)* first the surface has to be activated, which is generally achieved by the use of cleaning procedures that result in a consistent surface abundance of ≡Si-OH groups. Common techniques are solvent cleaning in methylene chloride, isopropyl alcohol, RCA or piranha solution (H_2SO_4:H_2O_2, 2:1) and the use of UV–ozone. The following functionalization can be realized by the use of organosilanes, which react with the surface ≡Si–OH groups and whose functionalities (e.g., –OH, –NH2, –SH, –COOH) may be used either directly or subsequently converted or derivatized for capture immobilization.

The by far most common functionalization of glass substrates is the introduction of reactive amino groups: Zammatteo, Guo and Beier et al. *(25, 26, 27)* presented the use of 3-aminopropyltrimethoxysilane (APTMS, **Fig. 6.2(a)**) and Benters et al. *(28)* its triethoxy variant APTES (**Fig. 6.3**). Chrisey and co-workers *(24)* described the use of the carbon spacer-based groups Trimethoxysilylpropyldiethylenetriamine (DETA, **Fig. 6.2(c)**) and N-(2-aminoethyl)-3-aminopropyltrimethoxysilane (EDA, **Fig. 6.2(b)**), differing in their spacer length, and the phenyl containing m,p-(aminoethylaminomethyl)phenethyltrimethoxysilane (PEDA, **Fig. 6.2(d)**), of which Joos et al. *(29)* used the smaller variant *p*-aminophenyl-trimethoxysilane (PEDA, **Fig. 6.2(g)**).

By contrast, the heterobifunctional linker N-(2-trifluoroethanesulfonatoethyl)-N-(methyl)-triethoxysilylpropyl-3-amine (NTMTA), described by Kumar et al. *(30)*, carries at its silane core, instead of an amino functionality, a trifluoroethanesulfonyl-group for the coupling reaction with oligonucleotides, which have to be functionalized with an amino or mercapto moiety (*see also* **Section 4.4**).

Another opportunity to functionalize the glass surface is the use of epoxy groups by applying 3-glycidyloxypropyltrimethoxysilane (GOPTS, **Fig. 6.2(e)**) *(22, 31, 32, 33)* or a combination of GOPTS and the above mentioned APTMS *(34)*.

Microarray surfaces can also be coated with gold as required for surface plasmon resonance spectroscopy (SPRS) *(35)*. A gold layer can be realized for example by silane based (3-mercaptopropyl) trimethoxysilane (MPTS, **Fig. 6.2(f)**) *(19, 36, 37, 38)*, to which thiol-modified capture molecules can be covalently linked (*see* **Section 4.2**). With the silanization reagent maleimido-undecyltrichlorosilane (MUTS) Wang and co-workers *(39)* presented an additional way to introduce thiol moieties on glass substrates.

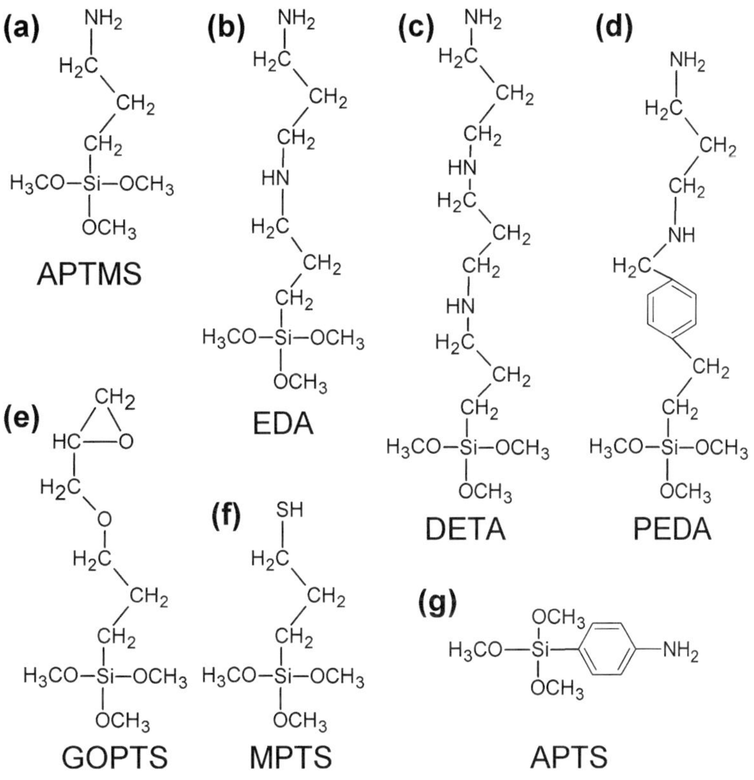

Fig. 6.2. Examples for amine-silanization reagents. (**a**) 3-aminopropyltrimethoxysilane, (APTMS). (**b**) N-(2-aminoethyl)-3-aminopropyltrimethoxysilane (EDA). (**c**) Trimethoxysilylpropyldiethylenetriamine (DETA). (**d**) m,p-(aminoethylaminomethyl)phenethyltrimethoxysilane (PEDA). (**e**) 3-glycidyloxypropyltrimethoxysilane (GOPTS). (**f**) (3-mercaptopropyl) trimethoxysilane (MPTS). (**g**) *p*-aminophenyl-trimethoxysilane (APTS).

Silanized glass has not only been used as a substrate for the direct immobilization of capture oligonucleotides but also as a basis in such microarray architectures which have a second layer of functional linker molecules (see below).

In case of the polyfunctional trichloro- or trimethoxysilanes, Si-O-Si linkages are not only formed between the glass surface and the silane moiety but also between the silane molecules themselves (**Fig. 6.3**) *(40, 41)*, generating a high stability of the so formed siloxane film. Unfortunately, the elegance and simplicity of this approach is compromised by the disadvantageous polymerization reaction which can easily lead to inhomogeneous and irreproducible surfaces when compared with monofunctional reagents like monochloro- or monomethoxysilanes *(42)*.

Apart from these and other silica-based substrates like silica optical fibers functionalized by DETA and GOPTS *(23, 43)*, synthetic polymers have also been used as microarray surfaces, as for example poly(methyl-methacrylate) (PMMA) *(16, 44, 45)*, poly(-dimethylsiloxane) (PDMS) *(46)*, polypropylen (PP) *(47, 48)*,

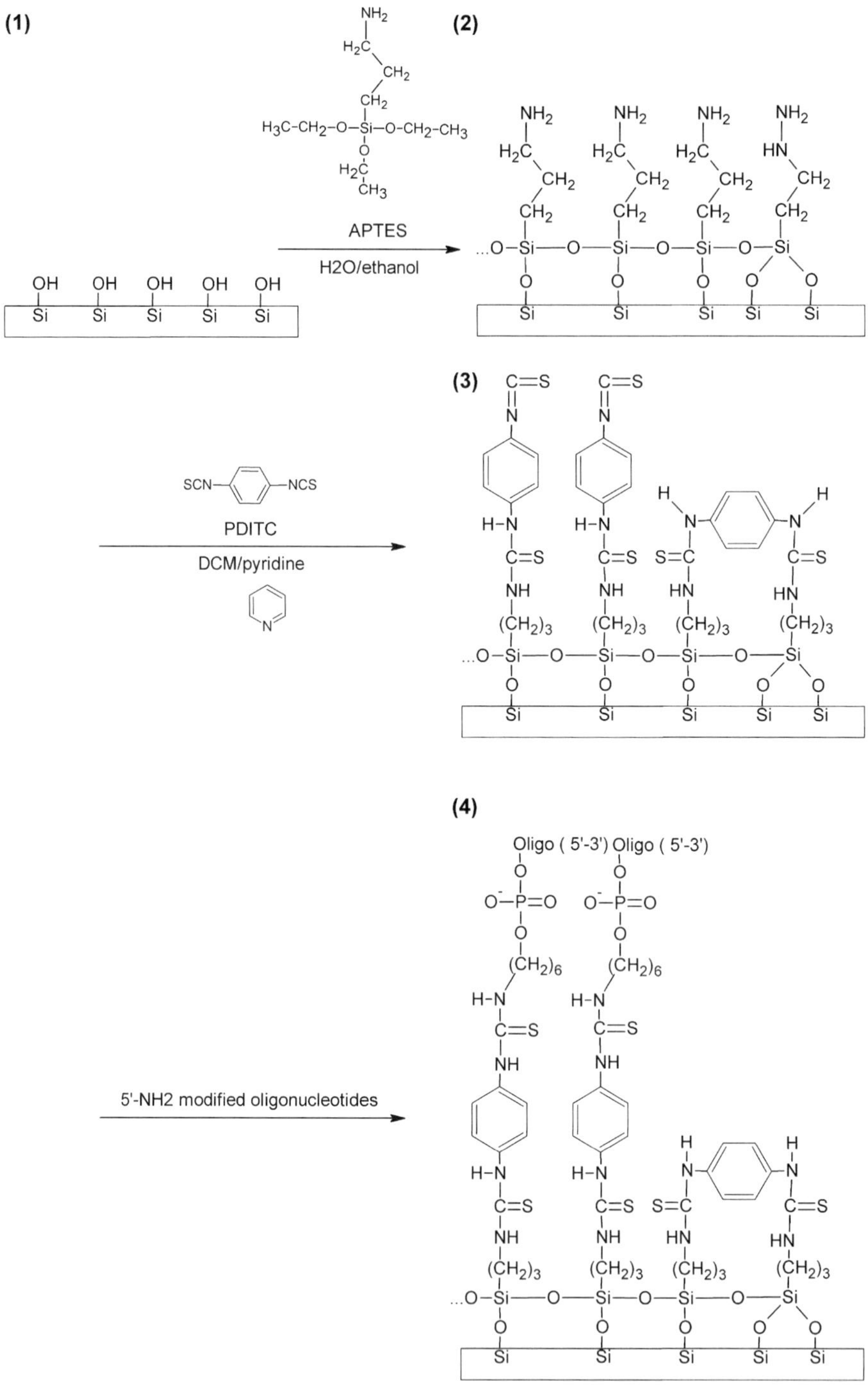

Fig. 6.3. Schematic representation of the immobilization of 5′-amino-modified capture molecules on an APTES-PDITC-activated glass surface: After a cleaning step the hydroxylated glass surface (**1**) is exposed to a mixture of ethanol/H_2O/APTES. By condensation, APTES reacts with one or more reactive groups of the SiO_2-surface, thereby generating an amino-modified support (**2**). The coupling of the amino reactive homobifunctional PDITC (**3**) provides the opportunity for immobilizing amino-modified capture molecules (**4**).

polycarbonate (PC) *(49, 50)*, polyethylene terephtalate (PET) *(45)*, polystyrene (PS) *(16, 50, 51)* and cyclic olefin copolymer *(16, 52)*. A more recent development is the use of diamond surfaces *(53)*; see also Krüger *(54)* and references therein for the use of single- and polycrystalline diamond films for DNA immobilization.

In some cases silanization has been combined with surfaces other than glass. Liu et al., *(46)* silanized a PDMS surface with the help of MPTS. Alternatively, the amino modification of a polymer has been accomplished by Fixe et al. using the heterobifunctional hexamethylene-diamine, which is able to react with the methyl ester of PMMA *(44)*. Amino-modified PP is also commercially available *(47)*.

2.2. Three-Dimensional Surfaces

Since Shchepinov et al., *(47)* showed that the use of spacer molecules between the surface and the capture oligonucleotides can lead to 150-fold increase in the hybridization efficiency (percentage of capture molecules having bound to their complementary targets at the end of a standardized hybridization reaction, Todt and Blohm in prep.), steric hindrance is considered to be a serious factor influencing the outcome of microarray experiments in terms of detectable signal intensity. The approach of Shchepinov et al., to use spacer molecules like polyT, has been extended by the development of substrates which exhibit a three-dimensional surface. To increase the sterical accessibility of the captures and to diminish their uneven distribution on a planar surface, porous or dendritic surface structures have been utilized.

To combine the advantages of glass surfaces with those of porous structures, agarose coated glass slides have been applied as microarray substrates by several different groups *(55, 56, 57)*. Also, the combination of polyacrylamide gels with glass substrates has been described by Guschin et al *(58)*, Rubina et al. *(59)* and others. The composition of the acrylamide mixtures has been optimized with regard to several parameters: Suriano et al. *(60)* even proposed a terpolymer, whose three components contribute different properties for microarray production; two of them optimize the attachment of the gel film to the glass substrate, the third one the covalent binding of amino-modified captures to the gel.

Liu and co-workers *(46)* turned up the order of the protocol presented e.g., by Guschin et al. by coupling acrylamides to the captures and immobilizing these constructs to a thiol-activated PDMS surface. The nylon membranes mentioned above *(1)* and the nitrocellulose based, commercially available FAST^{TM} slides *(61)* are other examples for three-dimensional surfaces to which the capture molecules may be immobilized.

In addition, dendrimeric structures have been explored to generate a three-dimensional construct on flat substrates, thereby increasing its surface and as a consequence the amount of attachable reactive groups. Dendrimeric structures can be built of a central core of hexachlorocyclo-triphosphazene ($N_3P_3Cl_6$) with diverse reactive groups at their periphery (e.g., aldehyde, thiol, epoxy); APTES or GOPTS activated glass surfaces can be used as a basis *(28, 62, 63, 64)* on which dendrimeric films are generated that are very stable and enable the production of re-usable microarrays.

3. Undefined Capture Immobilization

The interactions of different molecular constituents of the surface with certain moieties of the nucleic acid bases, e.g., primary amino groups of adenine, guanine and cytosine, hydrophobic elements of the base environment, and specifically the negative electrostatic charges of the backbone, can contribute to an interaction of the nucleic acid with the substrate which can be used for – although undefined – immobilization of the capture molecules.

3.1. Adsorption-Based Immobilization

Belosludtsev and co-workers *(5)* presented a method for the adsorptive, non-covalent attachment of oligonucleotides to an APTMS-modified glass slide. The captures are spotted and dried at 50°C, whereby the oligonucleotides retain their ability to hybridize to the complementary strand.

Call and co-workers *(65)* immobilized unmodified DNA on glass substrates pretreated with 3 M H_2SO_4 and 3 M HCl, by baking the array after spotting at 130°C for 30–60 min.

3.2. Immobilization by Covalent Coupling

As described by Saito et al. *(66)* light of the wavelength of 254 nm increases the reactivity of thymines with primary amines, which can be used to immobilize nucleic acid captures to amino-modified surfaces.

3.2.1. UV-Light-Mediated Coupling

In one of the earliest contributions, Church et al. *(67)* covalently immobilized DNA fragments generated by restriction enzymes onto nylon membranes by taking advantage of the UV-light-induced thymine reactivity with the amino groups of nylon. Wang et al. exploited the same chemistry to immobilize unmodified 70-mer oligonucleotides and cDNA onto amino-modified glass slides *(68)*, and Kimura and co-workers fixed polyT tagged captures on PET-, PC-, and PMMA-surfaces using a similar protocol *(45)*. Dufva et al., investigated the dose dependency of UV-crosslinking on the hybridization efficiency of capture oligonucleotides on agarose films *(57)*, confirming reports of Wang et al. about amino-modified glass substrates *(68)*.

In fact, UV-light of the wavelength of 254 nm cannot only be used for activating thymine bases but also for the activation of the –C=O groups of PC *(49)*. This reaction has also been applied to the coupling of unmodified PCR products to poly-L-lysine coated slides *(69, 70)*.

3.2.2. Epoxy and Aldehyde Functionalities

According to Chiu et al., *(34)* the electrostatic interaction between the anionic phosphate backbone with the cationic amino groups on the surface can be further augmented by covalent linkages between epoxy groups and primary amines of the DNA bases, if the surface has been covered by a mixture of APTMS and GOPTS.

Exploiting the reactivity of aldehydes with primary amines represents another widely used approach for immobilizing unmodified nucleic acid captures. Aldehyde groups on the surface can be generated e.g., by the treatment of agarose with sodium periodate ($NaIO_4$) *(55, 56)*. Proudnikov et al. *(71)* described an immobilization method which is based on partial depurination of the DNA. After incorporating amino groups into these sites, the immobilization of the so treated DNA on aldehyde-derivatized polyacrylamide gels is accomplished as above by the reduction of the resulting Schiff's base. A method for preparing aldehyde functionalized glass substrates was published by Zammatteo et al. *(25)*.

4. Defined Capture Immobilization

Different types of interactions between the nucleic acid capture molecules and the substrate surface are supposed to reduce the conformational freedom of the former, with consequences in terms of their affinity and specificity for their hybridization targets *(12)*. It is generally believed that the hybridization efficiency between capture and target can be fundamentally improved if the capture molecules are immobilized on the surface via exactly defined and exclusive single binding sites *(5)* – which furthermore react highly specifically under spotting conditions and are stable during hybridization.

4.1. Immobilization Via Biotin-Streptavidin Binding

The protein streptavidin has four highly affine biotin binding sites exhibiting a binding constant of about 10^{15} mol^{-1}. Spread out on a substrate surface it is suitable for very specific attachment of biotin labeled probes, e.g., nucleic acid captures. Miyachi and colleagues *(72)* developed a procedure to coat a substrate with strepavidin by adsorbing and embedding the protein into a layer of hexamethyldisiloxane (HMDS), which had first been affixed to the substrate. Subsequent application of 5′-biotin-modified oligonucleotides resulted in the preparation of a DNA-microarray.

4.2. Direct Covalent Coupling

An amino group attached to the 5′- or 3′-end of a nucleic acid capture can directly react with an aldehydes on a surface (*see* **Section 3.2.2**) *(25, 73, 74)*. The exposed position of the terminal primary amino group reduces its steric limitation and presumably supports its preferable reaction with the aldehydes. The above mentioned idea of an easier accessibility of end-tethered probes is supported by the observation of Dufva et al. who reported about a four- to fivefold higher amount of target molecules that were hybridized to terminally amino-modified captures compared to their unmodified variants immobilized on aldehyde activated glass slides *(57)*.

Whether this also holds true for aldehyde moities in three-dimensional surface structures as in activated agarose gels *(57)*, three-dimensional acrylamide gels *(58)* and dendrimeric structures *(63)* or whether and to what extent these aldehyde groups also react with the amino functions inside the bases remains unclear.

Another direct coupling method consists of the attachment of amino-modified capture molecules to epoxylated surfaces via a terminal amine linkage as described e.g., by Belosludtsev and co-workers *(5)* or Taylor et al. *(64)*.

Mahajan et al. *(31)* discovered that the reaction time for the coupling of phosphorylated oligos to epoxylated glass surfaces can be enormously shortened by microwave irradiation. Whereas 4 h is required under thermal conditions, microwave irradiation needs only about 10 min and results nevertheless in heat stable capture fixation, as indicated by the finding that the functionality of the captures decreased after 20 PCR like cycles by only 7%. This approach obviously allows the possibility of integrated PCR microarray systems.

The attachment of amino-modified captures to the epoxy based photoresist SU-8 has been described by Marie et al. *(75)*. Surprisingly, the authors found that the treatment of SU-8 with epoxy ring opening agents had no effect on binding or functionality of the captures; thus, it is not fully understood which functional groups are involved in the immobilization process.

To attach amino-modified capture molecules to carboxylated glass surfaces, whose preparation is described e.g., by Zammatteo et al. *(25)*, a mixture of *N*-hydroxysulfosuccinimide (NHS) and 1-ethyl-3(3-dimethylaminopropyl)-carbodiimide (EDC) can be used; the latter reagent promotes the formation of the amide bond between the nucleic acid capture and the surface *(15)*. In this case the reactive surface group is the carboxylic acid moiety, hence this chemistry is applicable to a wide range of materials such as gold *(76)*, cycloolefin polymer *(77)* and polystyrene beads *(51)*. It has also been suggested for the attachment of amino-modified captures to the commercially available CodeLink^{TM} slides, the surface of which consists of a crosslinked hydrophilic polymer network containing amine-reactive groups *(78)*.

Another possibility for directed immobilization is the chemisorption of thiol-modified captures on gold surfaces (*see also* **Section 2.1**). Since sulfur is not a component of nucleic acid, a high selectivity of the attachment should be guaranteed. In spite of this Herne and co-workers *(79)* deduced from measurements of the thickness of the layer of captures (38±2 Å vs. expected 160 Å for 25-mers) that the purine and pyrimidine bases of the capture molecules can interact with the gold surface and are probably not orientated perpendicularly to the surface. Wolf et al. *(80)* described sequence-dependent non-specific interactions between gold surfaces and nucleic acids, which however can be minimized with the addition of mercaptohexanol (MCH), whose thiol group competes with the unspecific adsorption of DNA to the surface *(20)*. A variant of this approach is the use of 5′-disulfide-modified captures, the binding of which is realized by a thiol disulfide exchange reaction between the sulfhydryl group of a mercaptosilane film on the glass slide and the disulfide-modified oligonucleotides *(38)*.

4.3. Homobifunctional Linker

The most simple example of a homobifunctional linker – a molecule that presents identical reactive groups at both of its ends and a spacer molecule in between that is supposed to be chemically inert (cf. **Fig. 6.1** *(3)*) – is an ion with a pair number of charges: in a conceptual paper, Bujoli and co-workers presented the use of zirconium (Zr^{4+}) ions in combination with gold or silica substrates that have been activated with a phosphonate monolayer film for the immobilization of captures carrying phosphate groups at their ends *(81)*.

A common example for a less reduced homobifunctional linker molecule is the acid anhydride of succinic acid (HOOC $(CH_2)_2$ COOH) which can be employed as a linker between APTES-modified substrates and amino-modified captures. Such a scheme has been presented by Joos et al. *(29)*, who effected the amide bond between the carboxyl and amino groups via a carbodiimide-mediated condensation. Similarly, Benters et al., used the hombifunctional glutaric acid (HOOC $(CH_2)_3$ COOH) for the attachment of amino functionalized dendrimers to APTES activated substrates *(62)*. These authors described also the use of 1,4-phenylenediisothiocynate (PDITC) for the same purpose, as well as the immobilization of NH_2-modified captures (*see also* **Fig. 6.3** and **Section 6**). The latter approach had been previously described by Guo and co-workers *(26)*. The use of glutaraldehyde for coupling of identically modified captures to amino-modified PMMA was shown by Fixe and colleagues *(44)*.

The polysaccharide chitosan, containing varying amounts of amino groups, offers a large number of identical reactive groups, similar to dendrimeric structures. Consolandi and co-workers described the preparation of microarray slides with chitosan of three different molecular weights (120, 400, and 650 kDa). PDITC is subsequently used to couple amino-modified capture molecules to the chitosan activated surface *(33)*.

Dendrimeric structures as well as macromolecules such as chitosan can actually be regarded as homo-multi-functional linkers.

4.4. Heterobifunctional Linker

Unlike their homobifunctional counterparts, heterobifunctional linkers present different reactive moieties at their ends (*see* **Fig. 6.1** *(4)*), each targeting different reactive groups and thus diminishing the possibility of unwanted crosslinking reactions (cf. **Fig. 6.3** *(3)*) for the homobifunctional linker (PDITC).

After covering a gold surface with a monolayer of the heterobifunctional 11-mercaptoundecylamine (MUAM), Brockman and colleagues employed a second heterobifunctional linker, sulfosuccinimidyl 4-(N-maleimidomethyl)cyclohexane-1-carboxylate (SSMCC), to attach thiol-modified captures. SSMCC contains both, *N*-hydroxysulfosuccinimide (NHS) ester- and maleimide functionality that are highly reactive towards amines and thiols, respectively *(82)*. Jordan and co-workers *(83)* used SSMCC for the attachment of captures on a gold slide surface that has been prepared with poly-L-lysine. In this construct the NHS-ester group of the SSMCC reacts with the pendant amino groups of the poly-L-lysine. To prepare the gold surface for the electrostatic adsorption of poly-L-lysine it was first covered with 11-mercaptoundecanoic acid (MUA), the carboxylic acid variant of MUAM.

Fixe and co-workers reported the use of the linker sulfo-EMCS, which carries the same reactive groups as SSMCC *(44)*. Another heterobifunctional linker targeting the same reactive groups was presented by Chrisey and co-workers who used 4-(maleimidophenyl) butyrate (SMPB), whose succinimide ester moiety is reactive towards primary amino groups and the maleimide towards thiol groups *(24)*.

Kumar et al., investigated the strategy of attaching reactive silanol groups to the captures to be immobilized. In a second step these silanized capture molecules could be immobilized to glass substrates. The heterobifunctional linkers they used were N-succinimidyl-3-(2-pyridyldithiol)-propionate (SPDP) or succinimidyl-6-(iodoacetyl-amino)-hexaonate (SIAX) *(37)*.

4.5. Photoactivatable Linker

Besides being heterobifunctional linkers, photoactivatable linker molecules contain a moiety that is sensitive to light activation.

Because of its reactivity with any polymer which exposes C-H groups, anthraquinone is a prominent component of photoactivatable coupling linkers. It is often used for the immobilization of captures on polymer substrates as described e.g., by Koch et al. *(50)* who immobilized capture molecules that carry an anthraquinone modification at their 5′-end to polystyrol and polycarbonate surfaces of microtiter plates.

Kumar and co-workers used Anthraquinone-2-carboxamide (NTPAC), an anthraquinone derivative containing *N*-(3-trifluoroethanesulfonyloxypropyl) for the immobilization of amino or mercapto-modified oligonucleotides *(48)*. Under ultraviolet irradiation (365 nm) anthraquinone reacts cleanly with polypropylene and other polymers.

Dankbar et al. *(16)* investigated the photolinker-mediated attachment of biomolecules to polymer surfaces and described in detail the influences of different parameters on these photoactivated coupling processes. Besides anthraquinone, other photolinkers like ketyl-reactive benzophenone, nitrene-reactive nitrophenyl azide and carbene-reactive phenyl-(trifluoromethyl)-diazirine have been used for the preparation of DNA microarrays by coupling them to the captures which were then immobilized on the polymeric substrates PMMA, PS or COC.

As an alternative to the photoactivatable covalent coupling, Sabanayagam and co-workers presented a photoactivatable biotin derivate (1-[4-azidosalicylicylamido]-6-[biotinamido]-hexane) covalently linked to an MPTS layer covering the surface as binding substrate *(84)*.

4.6. Synthesis on the Chip

Instead of attaching pre-synthesized captures to the surface of substrates, the oligonucleotides can be synthesized directly on the substrate's surface by photolithographic methods as mentioned above. The most prominent example is the procedure introduced by Fodor and colleagues *(85, 86)*, which is the basis of Affymetrix's gene chip^{TM} arrays (refer chapter 5). Fodor described the use of an amino-modified surface *(85)* for the attachment of the first nucleotide "building block" via the 3′-hydroxyl group. The next base is then coupled to the 5′-hydroxyl group via a light sensitive phosphoramidite chemistry. For the same strategy Pease et al. *(86)* used hydroxylated surfaces.

Various modifications and optimizations of this approach have been introduced. Examples are the use of photoresists to increase the number of spots on the microarrays to up to one million and more per cm^2 *(87)*, the renounce of masks for the photolithographic steps *(88)*, and the variation of the protecting groups by Hasan et al. *(89)* and Beier et al. *(90)* that have led to improved yields of 12% and more.

Shchepinov et al. *(47)* described the synthesis of spatially resolved captures via phosphoramidite chemistry on an amino-modified polypropylene support in both the 3′–5′ or the 5′–3′ orientation, and Moorcroft and co-workers *(91)* showed that the GOPTS activated polymer PDMS can also serve as an adequate substrate for the fabrication and synthesis of DNA microarrays.

5. Conclusions

Experimental results obtained by using different DNA microarray platforms and assay methods have suffered for a long time from the allegation of not being reproducible. This problem is surely caused by deficiencies concerning the many steps of a microarray experiment, including extraction and preparation of the nucleic acids for the analysis, technical solutions during microarray production, capture immobilization procedures to guarantee the highest possible hybridization efficiency, data processing, and others.

The immobilization methods summarized in this review cover only one of the many steps of the whole experimental set-up, but give an impression of how versatile the experimental settings and consequently the results can be. Nevertheless, the data of a large study published by the microarray-quality-control-consortium in the September's issue of Nature Biotechnology 2006 provoked the author of the editorial to state that "with careful experimental design and appropriate data transformation and analysis, microarray data can indeed be reproducible and comparable among different formats and laboratories" *(92)*.

As pointed out by Fixe et al. *(44)*, the highest possible capture density does not necessarily lead to the highest hybridization efficiency and Gong et al. as well as Peterson et al. *(8, 93)* showed that the highest hybridization efficiencies are reached at intermediate DNA capture densities. Since the density of functional groups on the microarray surface depends largely on the immobilization protocol, the quality of the raw materials (in case of aminosilanes see Chrisey et al. *(24)*) and on the experience of the experimenter, the focus in this field is not to reach a maximum of density of reactive groups at a microarray surface or to determine this parameter with different methods, but the concepts and the mechanisms of the different chemistries as contribution to maximize the hybridization efficiency, as far as it is influenced by the immobilization of the captures on the substrates. It is therefore difficult to provide a universal guideline on the usage of the single chemistries; in most of the cases an optimization of the protocol is inevitable and the ideal immobilization chemistry depends on several preparational parameters as summarized by Dufva *(11)*.

Neither the developments concerning new microarray surfaces *(22)* nor their preparation methods have yet achieved their fullness and new methods like plasma enhanced vapor deposition *(21, 53)* will probably find their way into the field. Also the surface activation reactions and other modifications, which target the sensitivity of the final fluorescence measurement, presently the most common read-out, are challenged by new variants as proposed by Sabanayagamn et al. *(94)*, showing that there is still space for optimization.

6. Example: Immobilization of 5′-NH_2 Modified Oligonucleotides on APTES – PDITC Activated Glass Surfaces

Standard microscope glass slides are used as microarray substrates. Prior to silanization of the substrates' surface with APTES (*see* **Fig. 6.3**), which provides the amine functionality for the attachment of the homobifunctional PDITC, the slides are cleaned and activated in freshly prepared piranha solution (H_2SO_4:H_2O_2, 2:1), rinsed in distilled water and dried under nitrogen; a process that is expected to lead to the uniform hydroxylation of the silicon atoms at the surface.

The subsequent modification of the glass surface with APTES has been described by Benters et al. *(28)*. In short, the pre-cleaned slides are treated with a mixture of ethanol/H_2O/APTES (95:3:2 v/v) for at least 1 hr. The slides are then thoroughly washed in pure ethanol and acetone. Under these conditions covalent bonds are formed between the alkoxysilanes and one or more hydroxy groups of the SiO_2-surface, as well as among the alkoxysilane molecules (*see* **Fig. 6.3** *(2)*), leading to a stable film-like surface structure *(40, 41)*.

If APTMS is used instead of APTES, Beier et al. *(27)* suggested the use of methanol and water for the last cleaning steps.

Scanning electron and atomic force microscopy, documented in the publication of Benters et al. *(28)*, indicates as a result of the protocol described that the flatness of the surface is excellent since the maximal roughness is less than 25 nm.

Alternatively, the silanization with APTES can be carried out under water free conditions immersing the substrates for at least one hour in a mixture of toluene/APTES/diisopropylethylamine (89:10:1 v/v) at 80°C in a nitrogen atmosphere, followed by a thorough wash with toluene and acetone.

The second step, the attachment of PDITC, which carries amino reactive thiocyanate groups at both ends of the molecule, can be accomplished by immersing the silanized substrates at room temperature for 2 h in a 10 mM solution of PDITC in dichloromethane (DCM) and 1% (v/v) of pyridine which catalytically activates the reaction. Subsequently, the slides are washed thrice in DCM and dried in a nitrogen stream. Ideally, the PDITC linkers are hoped to be arranged like cornstalks in a grain field, but obviously structures as depicted in **Fig. 6.3** *(3)* cannot be excluded because of the homobifunctionality of the linker.

For the third step, the covalent coupling of the oligonucleotides, the latter must contain a primary amino group which is usually attached to their 5′-end via a C_6-linker. Spotting of ~ 250 pL of a 10 μM oligonucleotide solution (in water) and incubating the microarrays for 2 h at room temperature in a humid chamber, followed by washing off the unbound oligonucleotides with TETBS (150 mM NaCl, 20 mM Tris, 5 mM EDTA, 0.05% (v/v) Tween 20, pH 7,35), and TBS (150 mM NaCl, 20 mM Tris, pH 7,35) results in a capture surface densities of ~ 70 fmol oligos/mm^2 as determined by autoradiography experiments *(28)*.

Acknowledgment

We are deeply grateful to A. Guiseppi-Elie, Clemson University, for reading the manuscript and for very helpful and important contributions in some details. S.Todt was supported by the project PTJ-BIO/0312892 of BMBF (German Federal Ministry of Education and Research).

References

1. Saiki, R., Walsh, P., Levenson, C. and Erlich, H. Genetic (1989) Analysis of amplified DNA with immobilized sequence-specific oligonucleotide probes. *Proc. Natl. Acad. Sci.* **86**, 6230–6234.
2. Nakano, S., Fujimoto, M., Hara, H. and Sugimoto, N. (1990) Nucleic acid duplex stability: influence of base composition and cation effects. *Nucleic Acids Res.* **27**, 2957–2965.
3. Rose, K., Mason, J. and R., L. (2002) Hybridization Parameters Revisited: Solutions Containing SDS. *BioTechniques.* **33**, 54–58.
4. Dai, H., Meyer, M., Stepaniants, S., Ziman, M. and Stoughton, R. (2002) Use of hybridization kinetics for differentiating specific from non-specific binding to oligonucleotide microarrays. *Nucleic Acids Res.* **30**, e86.
5. Belosludtsev, Y., Iverson, B., Lemeshko, S., Eggers, R., Wiese, R., Lee, S., Powdrill, T. and Hogan,M. (2001) DNA microarrays based on noncovalent oligonucleotide attachment and hybridization in two dimensions. *Anal. Biochem.* **292**, 250–256.
6. Peterson, A., Wolf,L.K. and Georgiadis, R. (2002) Hybridization of mismatches and partially matched DNA at surfaces. *J. Am. Chem. Soc.* **124**, 14601–14607.
7. Oh, S., Hong, B., Choi, K. and Park, J. (2006) Surface modification for DNA and protein microarrays. *OMICS.* **10**, 327–343.
8. Peterson, A., Heaton, R. and Georgiadis, R. (2001) The effect of surface probe density on DNA hybridisation. *Nucleic Acids Res.* **29**, 5163–5168.
9. Vainrub, A. and Pettitt, P. (2002) Coulomb blockage of hybridization in two-dimensional DNA-arrays. *Phys. Rev. E.* **66**, 041905.
10. Relógio, A., Schwager, C., Richter, A., Ansorge, W. and Valcarcel, J. (2002) Optimization of oligonucleotide-based DNA microarrays. *Nucleic Acids Res.* **30**, e51–.
11. Dufva, M. (2005) Fabrication of high quality microarrays. *Biomol. Eng.* **22**, 173–184.
12. Beaucage, S. (2001) Strategies in the preparation of DNA oligonucleotide arrays for diagnostic applications. *Curr. Med. Chem.* **8**, 1213–1244.
13. Pirrung, M. (2002) How to make a DNA chip. *Angew. Chem. Int. Ed.* **41**, 1276–1289.
14. Sobek, J., Bartscherer, K., Jacob, A., Hoheisel, J. and Angenendt, P. (2006) Microarray technology as a universal tool for high-throughput analysis of biological systems. *Comb. Chem. High Throughput Screen.* **9**, 365–380.

15. Nikiforov, T. and Rogers, Y. (1995) The use of 96-well polystyrene plates for DNA hybridization-based assays: an evaluation of different approaches to oligonucleotide immobilization. *Anal. Biochem.* **227**, 201–209.
16. Dankbar, D. and Gauglitz, G. (2006) A study on photolinkers used for biomolecule attachment to polymer surfaces. *Anal. Bioanal. Chem.* **386**, 1967–1974.
17. Clough, J. (2004) GeneChip expression microarrays part 2: industrial and clinical applications. *Int. Biotechnol. Lab.* **22**, 16–18.
18. Caruthers, M. (1991) Chemical synthesis of DNA and DNA analogues. *Acc. Chem. Res.* **24**, 278–284.
19. Jordan, C., Frey, B., Kornguth, S. and Corn, R. (1994) Characterization of poly-L-lysine adsorption onto alkanethiol-modified gold surfaces with polarization-modulation Fourier transform infrared spectroscopy and surface plasmon resonance measurements. *Langmuir.* **10**, 3642–3648.
20. Levicky, R., Herne, T., Tarlov, M. and Satija, S. (1998) Using self-assembly to control the structur of DNA monolayers on gold: a neutron reflectivity study. *J. Am. Chem. Soc.* **120**, 9787–9792.
21. Manning, M. and Redmond, G. (2005) Formation and characterization of DNA microarrays at silicon nitride substrates. *Langmuir.* **21**, 395–402.
22. Hong, B., Oh, S., Youn, T., Kwon, S. and Park, J. (2005) Nanoscale-Controlled Spacing Provides DNA Microarrays with the SNP Discrimination Efficiency in Solution Phase. *Langmuir.* **21**, 4257–4261.
23. Uddin, A., Piunno, P., Hudson, R., Damha, M. and Krull, U. (1997) A fiber optic biosensor for fluorimetric detection of triple-helical DNA. *Nucleic Acids Res.* **25**, 4139–4146.
24. Chrisey, L., Lee, G. and O'Ferrall, C. (1996) Covalent attachment of synthetic DNA to self-assembled monolayer films. *Nucleic Acids Res.* **24**, 3031–3039.
25. Zammatteo, N., Jeanmart, L., Hamels, S., Courtois, S., Louette, P., Hevesi, L. and Remacle, J. (2000) Comparison between different strategies of covalent attachment of DNA to glass surfaces to build DNA microarrays. *Anal. Biochem.* **280**, 143–150.
26. Guo,Z., Guilfoyle,R.A., Thiel, A., Wang, R. and Smith, L. (1994) Direct fluorescence analysis of genetic polymorphisms by hybridization with oligonucleotide arrays on glass supports. *Nucleic Acids Res.* **22**, 5456–5465.
27. Beier, M. and Hoheisel, J. (1999) Versatile derivatisation of solid support media for covalent bonding on DNA-microchips. *Nucleic Acids Res.* **27**, 1970–1977.
28. Benters, R., Niemeyer, C.M. and Wöhrle, D. (2001) Dendrimer-activated solid supports for nucleic acid and protein microarrays. *Chem. Bio. Chem.* **2**, 686–694.
29. Joos, B., Kuster, H. and Cone, R. (1997) Covalent attachment of hybridizable oligonucleotides to glass supports. *Anal. Biochem.* **247**, 96–101.
30. Kumar, P., Choithani, J. and Gupta, K. (2004) Construction of oligonucleotide arrays on a glass surface using a heterobifunctional reagent, N-(2-trifluoroethanesulfonatoethyl)-N-(methyl)-triethoxysilylpropyl-3-amine (NTMTA). *Nucleic Acids Res.* **32**, e80.
31. Mahajan, S., Kumar, P. and Gupta, K. (2006) Oligonucleotide microarrays: immobilization of phosphorylated oligonucleotides on epoxylated surface. *Bioconjug. Chem.* **17**, 1184–1189.
32. Maskos, U. and Southern, E. (1992) Oligonucleotide hybridizations on glass supports: a novel linker for oligonucleotide synthesis and hybridization properties of oligonucleotides synthesised in situ. *Nucleic Acids Res.* **20**, 1679–1684.
33. Consolandi, C., Severgnini, M., Castiglioni, B., Bordoni, R., Frosini, A., Battaglia, C., Bernardi, L. and De Bellis, G. (2006) A structured chitosan-based platform for biomolecule attachment to solid surfaces: application to DNA microarray preparation. *Bioconjug. Chem.* **2006** ,*17*, 371–377
34. Chiu, S., Hsu, M., Ku, W., Tu, C., Tseng, Y., Lau, W., Yan, R., Ma, J. and Tzeng, C. Synergistic effects of epoxy- and amine-silanes on microarray DNA immobilization and hybridization. *Biochem. J.* **374**, 625–632.
35. Georgiadis, R., Peterlinz, K. and Peterson, A. (2000) Quantitative measurements and modeling of kinetics in nucleic acid monolayer films using SPR spectroscopy. *J. Am. Chem. Soc.* **122**, 3166–3173.
36. Halliwell, C. and Cass, A. (2001) A Factorial Analysis of Silanization Conditions for the Immobilization of Oligonucleotides on Glass Surfaces. *Anal. Chem.* **73**, 2476–2483.

37. Kumar, A., Larsson, O., Parodi, D. and Liang, Z. (2000) Silanized nucleic acids: a general platform for DNA immobilization. *Nucleic Acids Res.* **28**, E71.
38. Rogers, Y., Jiang-Baucom, P., Huang, Z., Bogdanov, V., Anderson, S. and Boyce-Jacino, M. (1999) Immobilization of oligonucleotides onto a glass support via disulfide bonds: A method for preparation of DNA microarrays. *Anal. Biochem.* **266**, 23–30.
39. Wang, Y., Prokein, T., Hinz, M., Seliger, H. and Goedel, W. (2005) Immobilization and hybridization of oligonucleotides on maleimido-terminated self-assembled monolayers. *Anal. Biochem.* **344**, 216–223.
40. Strother, T., Hamers, R. and Smith, L. (2000) Covalent attachment of oligodeoxyribonucleotides to amine-modified Si (001) surfaces. *Nucleic Acids Res.* **28**, 3535–3541.
41. Waddell, T., Leyden, D. and DeBello, M. (1981) The nature of organosilane to silica-surface bonding. *J. Am. Chem. Soc.* **103**, 5303–5307.
42. Ulman, A. (1996) Formation and Structure of Self-Assembled Monolayers. *Chem. Rev.* **96**, 1533–1554.
43. Liu, X. and Tan, W. (1999) A fiber-optic evanescent wave DNA biosensor based on novel molecular beacons. *Anal. Chem.* **71**, 5054–5049.
44. Fixe, F., Dufva, M., Telleman, P. and Christensen, C. (2004) Functionalization of poly(methyl methacrylate) (PMMA) as a substrate for DNA microarrays. *Nucleic Acids Res.* **32**, e9.
45. Kimura, N. (2006) One-step immobilization of poly(dT)-modified DNA onto non-modified plastic substrates by UV irradiation for microarrays. *Biochem. Biophys. Res. Commun.* **347**, 477–484.
46. Liu, D., Perdue, R., Sun, L. and Crooks, R. (2004) Immobilization of DNA onto Poly(-dimethylsiloxane) Surfaces and Application to a Microelectrochemical Enzyme-Amplified DNA Hybridization Assay. *Langmuir.* **20**, 5905–5910.
47S chepinov, M., Case-Green, S. and Southern, E. (1997) Steric factors influencing hybridisation of nucleic acids to oligonucleotide arrays. *Nucleic Acids Res.* **25**, 1155–1161.
48. Kumar, P., Agarwal, S. and Gupta, K. (2004) N-(3-Trifluoroethanesulfonyloxypropyl) an-thraquinone- 2-carboxamide: a new heterobifunctional reagent for immobilization of biomolecules on a variety of polymer surfaces. *Bioconjug. Chem.* **15**, 7–11.
49. Li, Y., Wang, Z., Ou, L. and Yu, H. (2007) DNA detection on plastic: surface activation protocol to convert polycarbonate substrates to biochip platforms. *Anal. Chem.* **79**, 426–433.
50. Koch, T., Jacobsen, N., Fensholdt, J., Boas, U., Fenger, M. and Jakobsen, M. (2000) Photochemical immobilisation of anthraquinone conjugated oligonucleotides and PCR amplicons on solid surfaces. *Bioconjug. Chem.* **11**, 474–483.
51. Lund, V., Schmid, R., Rickwood, D. and Hornes, E. (1988) Assessment of methods for covalent binding of nucleic acids to magnetic beads, Dynabeads, and the characteristics of the bound nucleic acids in hybridization reactions. *Nucleic Acids Res.* **16**, 10861–10880.
52. Kinoshita, K., Fujimoto, K., Yakabe, T., Saito, S., Hamaguchi, Y., Kikuchi, T., Nonaka, K., Murata, S., Masuda, D., Takada, W., Funaoka, S., Arai, S., Nakanishi, H., Yokoyama, K., Fujiwara, K. and Matsubara, K. (2007) Multiple primer extension by DNA polymerase on a novel plastic DNA array coated with a biocompatible polymer. *Nucleic Acids Res.* **35**, e3.
53. Yang, J., Song, K., Zhang, G., Degawa, M., Sasaki, Y., Ohdomari, I. and Kawarada, H. (2006) Characterization of DNA hybridization on partially aminated diamond by aromatic compounds. *Langmuir.* **22**, 11245–11250.
54. Krüger, A. (2006) Hard and soft: biofunctionalized diamond. *Angew. Chem. Int. Ed. Engl.* **45**, 6426–6427.
55. Wang, H., Li, J., Liu, H., Liu, Q., Mei, Q., Wang, Y., Zhu, J., He, N. and Lu, Z. (2002) Label-free hybridization detection of a single nucleotide mismatch by immobilization of molecular beacons on an agarose film. *Nucleic Acids Res.* **30**, e61.
56. Afanassiev, V., Hanemann, V. and Wölfl, S. (2000) Preparation of DNA and protein micro arrays on glass slides coated with an agarose film. *Nucleic Acids Res.* **28**, E66.
57. Dufva, M., Petronis, S., Bjerremann Jensen, L., Krag, C. and Christensen, C. (2004) Characterization of an inexpensive, nontoxic, and highly sensitive microarray substrate. *Biotechniques.* **37**, 286–292, 294, 296.

58. Guschin, D., Yershov, G., Zaslavsky, A., Gemmel, A., Shick, V., Proudnikov, D., Arenkov, P. and Mirzabekov, A. (1997) Manual manufacturing of oligonucleotide, DNA, and protein microchips. *Anal. Biochem.* **250**, 203–211.
59. Rubina, A., Pan'kov, S., Dementieva, E., Pen'kov, D., Butygin, A., Vasiliskov, V., Chudinov, A., Mikheikin, A., Mikhailovich, V. and Mirzabekov, A. (2004) Hydrogel drop microchips with immobilized DNA: properties and methods for large-scale production. *Anal. Biochem.* **325**, 92–106.
60. Suriano, R., Levi, M., Pirri, G., Damin, F., Chiari, M. and Turri, S. (2006) Surface behavior and molecular recognition in DNA microarrays from N,N-dimethylacrylamide terpolymers with activated esters as linking groups. *Macromol. Biosci.* **6**, 719–729.
61. Stillman, B. and Tonkinson, J. (2000) FAST slides: a novel surface for microarrays. *Biotechniques.* **29**, 630–635.
62. Benters, R., Niemeyer, C., Drutschmann, D., Blohm, D. and Wöhrle, D. (2002) DNA microarrays with PAMAM dendritic linker systems. *Nucleic Acids Res.* **30**, E10.
63. Le Berre, V., Trévisiol, E., Dagkessamanskaia, A., Sokol, S., Caminade, A., Majoral, J., Meunier, B. and Françoise, J. (2003) Dendrimeric coating of glass slides for sensitive DNA microarrays analysis. *Nucleic Acids Res.* **31**, e88.
64. Taylor, S., Smith, S., Windle, B., and Guiseppi-Elie, A. (2003) Impact of surface chemistry and blocking strategies on DNA microarrays. *Nucleic Acids Res.* **31**, e87.
65. Call, D., Chandler, D. and Brockman, F. (2001) Fabrication of DNA microarrays using unmodified oligonucleotide probes. *Biotechniques.* **30**, 368–72, 374, 376 passim.
66. Saito, I., Sugiyama, H., Furukawa, N. and Matsuura, T. (1981) Photochemical ring opening of thymidine and thymine in the presence of primary amines. *Tetrahedron Lett.* **22**, 3265–3268.
67. Church, G. and Gilbert, W. (1984) Genomic sequencing. *Biochem.* **81**, 1991–1995.
68. Wang, H., Malek, R., Kwitek, A., Greene, A., Luu, T., Behbahani, B., Frank, B., Quackenbush, J. and Lee, N. (2003) Assessing unmodified 70-mer oligonucleotide probe performance on glass-slide microarrays. *Genome Biol.* **4**, R5.
69. Diehl, F., Grahlmann, S., Beier, M. and Hoheisel, J. (2001) Manufacturing DNA microarrays of high spot homogeneity and reduced background signal. *Nucleic Acids Res.* **29**, E38.
70. Shalon, D., Smith, S. and Brown, P. (1996) A DNA microarray system for analyzing complex DNA samples using two-color fluorescent probe hybridization. *Genome Res.* **6**, 639–645.
71. Proudnikov, D., Timofeev, E. and Mirzabekov, A. (1998) Immobilization of DNA in polyacrylamide gel for the manufacture of DNA and DNA-oligonucleotide microchips. *Anal. Biochem.* **259**, 34–41.
72. Miyachi, H., Hiratsuka, A., Ikebukuro, K., Yano, K., Muguruma, H. and Karube, I. (2000) Application of polymer-embedded proteins to fabrication of DNA array. *Biotechnol. Bioeng.* **69**, 323–329.
73. Schena, M., Shalon, D., Davis, R. and Brown, P. (1996) Quantitative monitoring of gene expression patterns with a complementary DNA microarray. *Science.* **270**, 467–470.
74. Lindroos, K., Liljedahl, U., Raitio, M. and Syvänen, A. (2001) Minisequencing on oligonucleotide microarrays: comparison of immobilisation chemistries. *Nucleic Acids Res.* **29**, E69–9
75. Marie, R., Schmid, S., Johansson, A., Ejsing, L., Nordström, M., Häfliger, D., Christensen, C., Boisen, A. and Dufva, M. (2006) Immobilisation of DNA to polymerised SU-8 photoresist. *Biosens. Bioelectron.* **21**, 1327–1332.
76. Frey, B. and Corn, R. (1996) Covalent Attachment and Derivatization of Poly(L-lysine) Monolayers on Gold Surfaces As Characterized by Polarization-Modulation FT-IR Spectroscopy. *Anal. Chem.* **68**, 3187–3193.
77. Laib, S. and Maccraith, B. (2007) Immobilization of Biomolecules on Cycloolefin Polymer Supports. *Anal. Chem.* 79, 6264–6270.
78. Gong, P. and Grainger, D. (2004) Comparison of DNA immobilization efficiency on new and regenerated commercial amine-reactive polymer microarray surfaces. *BIOSURF V: Funct. Polym. Surf. Biotechnol.* **570**, 67–77.
79. Herne, T. and Tarlov, M. (1997) Characterization of DNA Probes Immobilized on Gold Surfaces. *J. Am. Chem. Soc.* **119**, 8916–8920.
80. Wolf, K., Gao, Y. and Georgiadis, R. (2004) Sequence-dependent DNA immobilization: specific versus nonspecific contributions. *Langmuir.* **20**, 3357–3361.
81. Bujoli, B., Lane, S., Nonglaton, G., Pipelier, M., Léger, J., Talham, D. and Tellier, C. (2005) Metal phosphonates applied to biotechnologies: a novel approach to oligonucleotide microarrays. *Chemistry.* **11**, 1980–1988.

82. Brockman, J., Frutos, A. and Corn, R. (1999) A Multistep Chemical Modification Procedure To Create DNA Arrays on Gold Surfaces for the Study of Protein-DNA Interactions with Surface Plasmon Resonance Imaging. *J. Am. Chem. Soc.* **121**, 8044–8051.
83. Jordan, C., Frutos, A., Thiel, A. and Corn, R. (1997) Surface Plasmon Resonance Imaging Measurements of DNA Hybridization Adsorption and Streptavidin/DNA Multilayer Formation at Chemically Modified Gold Surfaces. *Anal. Chem.* **69**, 4939–4947.
84. Sabanayagam, C., Smith, C. and Cantor, C. (2000) Oligonucleotide immobilization on micropatterned streptavidin surfaces. *Nucleic Acids Res.* **28**, E33.
85. Fodor, S., Read, J., Pirrung, M., Stryer, L., Lu, A. and Solas, D. (1991) Light-directed, spatially addressable parallel chemical synthesis. *Science.* **251**, 767–773.
86. Pease, A., Solas, D., Sullivan, E., Cronin, M., Holmes, C. and Fodor, S. (1994) Light-generated oligonucleotide arrays for rapid DNA sequence analysis. *Proc. Natl. Acad. Sci.* **91**, 5022–5026.
87. McGall, G., Labadie, J., Brock, P., Wallraff, G., Nguyen, T. and Hinsberg, W. (1996) Light-directed synthesis of high-density oligonucleotide arrays using semiconductor photoresists. *Proc. Nat. Acad. Sci.* **93**, 13555–13560.
88. Singh-Gasson, S., Green, R., Yue, Y., Nelson, C., Blattner, F., Sussman, M. and Cerrina, F. (1999) Maskless fabrication of light-directed oligonucleotide microarrays using a digital micromirror array. *Nat. Biotechnol.* **17**, 974–978.
89. Hasan, A., Stengele, K., Giegrich, H., Cornwell, P., Isham, K., Sachleben, R., Pfleiderer, W. and Foote, R. (1997) Photolabile protecting groups for nucleosides: Synthesis and photodeprotection rates. *Tetrahedron.* **53**, 4247–4264.
90. Beier, M. and Hoheisel, J. (2000) Production by quantitative photolithographic synthesis of individually quality checked DNA microarrays. *Nucleic Acids Res.* **28**, E11.
91. Moorcroft, M., Meuleman, W., Latham, S., Nicholls, T., Egeland, R. and Southern, E. (2005) In situ oligonucleotide synthesis on poly(dimethylsiloxane): a flexible substrate for microarray fabrication. *Nucleic Acids Res.* **33**, e75.
92. Editorial (no author) (2006) Making the most of microarray data *Nat. Biotechnol.*, **24**, 1039.
93. Gong, P., Harbers, G. and Grainger, D. (2006) Multi-technique comparison of immobilized and hybridized oligonucleotide surface density on commercial amine-reactive microarray slides. *Anal. Chem.* **78**, 2342–51.
94. Sabanayagam, C. and Lakowicz, J. (2007) Increasing the sensitivity of DNA microarrays by metal-enhanced fluorescence using surface-bound silver nanoparticles. *Nucleic Acids Res.* **35**, e13.

Chapter 7

Fabrication Using Contact Spotter

Annelie Waldén and Peter Nilsson

Abstract

Many steps of optimization are needed to achieve large-scale fabrication of high-quality DNA microarrays. These steps involve the printing instrument, the probes to be printed, microarray slides, and spotting buffer together with the surrounding environment, such as humidity and temperature. Robust microarray production requires not only appropriate reagents, equipments, and established procedures, but also devoted and experienced personnel. It is a challenging and craftsman like activity, but at the same time highly rewarding in terms of flexibility and cost efficiency. Outlined here is the workflow of a high-throughput microarray production line.

Key words: Contact spotter, microarrays, printing, spotting, fabrication.

1. Introduction

Spotted DNA Microarrays are fabricated by robotics that transfer pre-synthesized DNA probes with high precision to the microarray slide surfaces in strictly ordered patterns. Either a direct contact between the microarray spotting pins and the slide surface as described here or a non-contact piezo electric principle is utilized. The slides are generally chemically or membrane coated glass slides enabling the attachment of the DNA probes to the surface.

After the pioneering work at Stanford University *(1–4)* in the mid 1990s, the development of microarray production became relatively rapid and wide spread. The amount of microarray facilities increased steadily, but that trend changed around 2003–2004 when the microarray production became more and more centralized. The main reason is that it is a time consuming process and to be able to produce high-quality microarrays it

Martin Dufva (ed.), *DNA Microarrays for Biomedical Research: Methods and Protocols, vol. 529*

DOI 10.1007/978-1-59745-538-1_7 Springerprotocols.com

requires experience and expertise. It is important to be aware of the effort needed, but at the same time, it should also clearly be stated the enormous possibilities that opens up if you have direct access and possibility to customize your own microarray slides. The two main advantages are flexibility and cost. Commercial microarrays have become more accessible and prices have been reduced but still large-scale microarray-based projects usually require large amount of slides, and commercial arrays is thereby not an alternative for many research groups. The access to the content, i.e., the collections of suitable DNA probes to be spotted on the arrays can be a limiting factor for in-house production, but there are now many commercial sources for e.g., whole genome oligonucleotide collections for a wide variety of species.

2. Materials

2.1. Contact Spotter

1. Technology: There are a number of contact printers or spotters available on the market with various amounts of pins and slide and plate positions and thereby various capacity. The range of instruments cover the whole spectra from small and completely manual instruments to fully automated instruments for high-throughput and large-scale production. In the following, various parts of an array instrument, which are important to consider is described.
2. With or without plate stacker: If producing high volumes of microarrays containing several thousands probes, the choice between a contact spotter with or without a plate stacker is easy. Many hours can be saved during one print run by placing plates in the stacker for walk away automation. It is important to ensure that the well plates and lids are compatible with the stacker unit. Problems that could occur are related to plates, lids, and stacker unit. Plates and/or lids may be too small, big, high or bent and also vary between different lots of the same plate type. Fine-tuning adjustments in the stacker unit may be a tricky issue. A system without stacker may be more suitable when printing smaller probe collections, i.e., up to approximately 2000 probes. In spotters without stacker system you may place between 3–5 plates at a time and a walk away automation is not feasible and therefore this system demands more hands on work.
3. Number of slides per print run: The maximum number of slides within a print run differs between instruments, from a few slides up to several hundred slides. In general, the extra time it takes to print larger amount of slides is relatively small, since the main time consuming step is the wash cycles and not the actual printing.

4. Plate Chiller: To have a controlled temperature and minimize evaporation of the spotting solution during a print run, a chilled and humidified source plate holder may be an option. Many chilled source plate holders works best when all the positions are covered with plates.
5. Microarray spotting pins: The sample is loaded onto the surface of the slide with a combination of direct contact between the microarray spotting pins and the microarray slide surface together with the capillary force. There is a variety of microarray spotting pins available. The choice lies between a solid pin, split pin, and quill pin. A solid pin is solid through the whole pin. With these pins a new sample pick up between every spot is required and a higher sample volume is needed due to this. A split pin has a notch were it collect sample and therefore you will be able to print several hundred spots with only one sample pick up. The difference between a split pin and a quill pin is the reservoir in the quill pin that holds a certain volume apart from the volume in the notch. The material the pins are made of also differs. Stainless steel and tungsten are two examples of material which are widely used for producing microarray spotting pins. The spot size varies due to the tip size of the pin and when a higher amount of spots on the microarray is desirable a smaller tip size has to be chosen.

   The number of microarray spotting pins that fits in to the head will also determine the different possibilities you have to place replicated spots on the slides. The print run time also depends on the number or pins within the microarray head, higher number of pin will shorten your print run time due to fewer wash cycles between sample pick ups.
6. Printing with single or multiple hits on the microarray: If it is desirable to increase the spot intensities or decrease doughnut effect it is advisable to try multiple hits per spot. When printing with multiple hits you have the opportunity, in some instrument, to try both cyclic (print all slides and then go back for another round in the same position) and immediate (prints the consecutive hits immediately after printing the first one). Printing with several hits per spot will usually result in an increase of the spot diameter. The result may be that the desirable number of spots on the array may not fit into the printing area. A higher consumption of sample will also be an effect of many multiple hits.
7. Wash system: A thorough optimization of the wash system is highly recommended to ensure there is no carry over between spots. The optimum wash cycle for the contact spotter depends on the wash station, number of pins in the microarray head, the spotting buffer, and the printed probe. Depending on the design of the wash station, which in many cases unfortunately is relatively non-optimized, an increase in the number of wash

cycles may be needed due to how many pins you are using, higher number of pins could mean more or longer wash cycles. A more viscous spotting buffer may need a longer wash or more cycles to prevent carry over between sample pick up.

The most common wash solutions used in contact spotters is water and ethanol when printing cDNA or oligonucleotides. Ethanol concentration may vary, but 70–80% is what many manufacturers recommend. To set up a wash program that fulfills your criterion you have to test several different wash cycles. A good starting point is 4–6 water steps for 2–3 s each followed by 2–3 ethanol steps for 3–5 s. A quick dry between every wash step may also be included but the most important dry step is after the last ethanol step. Start with a dry time of 10–15 s and see if this is enough to dry the spotting pins. If the split/quill pins are still wet after this step they will not be able to pick up any sample and only ethanol will be printed on the slides. An increase of the dry time may be needed depending on the time of the year. During warm weather conditions with a high humidity in the surroundings a longer dry time may be needed to achieve microarray pins that are completely dry.

8. Print adjustments: Due to variation in thickness of slides an adjustment of the print depth of the pins may be needed. The depth of how far down the pins must go to obtain a good contact between the microarray spotting pins and the glass surface needs to be empirically adjusted. A thinner slide needs an increase of the print depth. The recommendation when trying out new different types of slides is to do several print runs with the different types of slides separately.
9. Velocity: The velocity of the microarray head movement can be adjusted in most contact spotters. To minimize the doughnut effect, a decreased velocity could have a positive effect.

2.2. Source Plates

The microtiter plates have to be compatible with the stacker system. They have to be stable over time when storing them in freezer or fridge. When sealing the well plates between print runs make sure they are compatible with the material in the plate. For example the glue on the aluminum foil that is used may stick on the top of the well plate then the lid will get caught onto the plate. The glue can also slide into the wells and contaminate the probes.

Maintain a dust free environment when preparing the printing plates. Dust and/or fiber enter the wells easily and could cause a clogged pin during a print run. This will give smearing spots which leads to difficulties in distinguishing one spot from another.

2.3. Array Lab

To place the contact spotter in a separate room is highly recommended. It is not totally necessary to have a complete dust free environment as in a clean room but it is important to have a controlled surrounding environment, due to risk of clogged pins during a print run. A set temperature and humidity in both printing room and spotter is important to achieve high-quality microarrays. In a surrounding with a low temperature and humidity, the result will be none or very small spots. With opposite conditions, the slide will contain very big spots which merge into each other. A temperature between 18–22°C and humidity around 45–55% is a starting point. Different combinations of spotting buffer and slides are also dependent on the surrounding environment. The distributors usually have their own recommendations as to which criteria are best, but it is important to be aware that the temperature and humidity must be optimized for each individual location and application.

2.4. Pattern

When choosing a suitable print pattern, the goal is to fit in all probes on one slide and with as many replicates as possible. The density needs to be balanced against the spot size in order to generate clearly separated spots. The number of pins in the microarray head decides how many blocks (subarrays) and fields there will be on the microarray, **Fig. 7.1**. If a multi-well mask is used in the hybridization step an adjustment of the distance between the blocks may be needed. A mask enables multiple hybridizations on one glass slide. Each block or field can be hybridized separately by placing a silicon mask with wells where the blocks or fields will fit in.

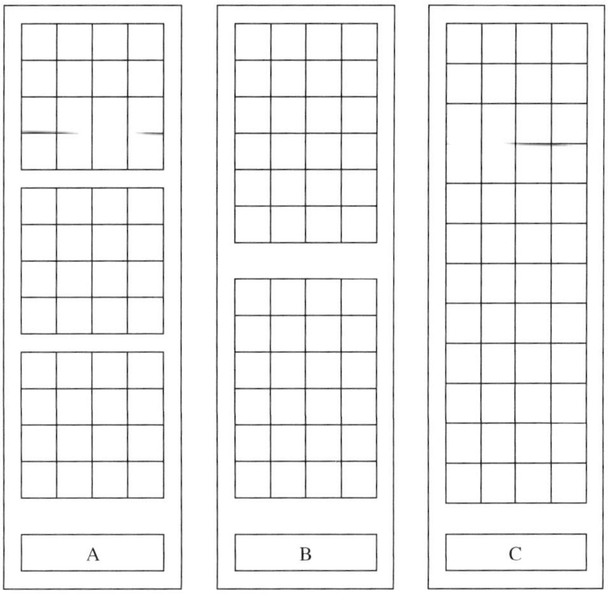

Fig. 7.1. Slide **A**: A 16-pins Microarray head gives three fields with 16 blocks in each field. Slide **B**: A 24-pins Microarray head gives two fields with 24 blocks in each field. Slide **C**: A 48-pins Microarray head gives one field with 48 blocks.

2.5. Spotting Solution

Only a certain combination of spotting buffer and slides is possible. A good spotting buffer should give spots of equal size and with a nice round morphology, no change over time, and be resistant to evaporation during long print runs. A certain auto fluorescence can be seen in some spotting buffers and it should be investigated if the chosen spotting buffer has this effect before starting your production.

Dimethylsulfoxide, DMSO, in different concentrations, usually 30–50%, is a widely used spotting buffer for both oligonucleotides and cDNA. Increasing DMSO concentration results in larger spots but also increased resistance to evaporation. Sodiumdodecylsulfate, SDS, betain, and sarcosyl are other examples of additives to the spotting buffer which also gives larger and more homogenous spots. A lot of commercial spotting buffers are also available and often in a combination together with a specific slide chemistry.

2.6. Slides

There are a number of commercial slides available to choose from when selecting suitable slides for each application. A low variation within slides, between slides and between batches of slides is an important issue to consider. The size and the print area of the slide must fit in to the contact spotter, hybridization chamber, and the microarray scanner that will be used. The standard slide size is $75 \times 25 \times 1$ mm, but be aware of slides that are produced with the very similar sizes of 1×3 inch, i.e., approximately 76×26 mm. The most widely used slide is glass slide with different types of coating, such as amino silane, epoxy, amine, and aldehyde coating which give different bonding between the surfaces and probe. Membrane slides, glass slide with a membrane attached to the glass surface, and mirror slides are also available. The application decides which slide and what coating to choose.

2.7. Placement of Controls and Replicates

The goal when placing the controls on the array is to spread them out as much as possible. There are a number of different controls suitable for different purpose such as specificity probes, spike-ins, dilution series, positive controls, and negative controls. A good thing when dealing with low intensities is to print fluorescent probes to help placing the grid in the correct way. These shall always give intensity in the hybridization and can be placed in each corner of the slide/block/field. Plate ID controls can also be placed in each plate to give an easy way to detect if any plate has been rotated when placed in the instrument. A change in the printed plate order will also be seen by using these controls.

2.8. Printing Parameters

Humidity and temperature are important printing parameters to optimize in order to achieve high-quality microarrays. Spotting buffer and slides can be sensitive to a variation in both temperature and humidity. Some slides have to be printed at a certain humidity not to loose any activity on the slide surface which will results in unusable slides. A controlled system with a humidifier where temperature and humidity can be displayed is a recommendation. If the plates and the slides share the same environment, there could be a conflict between the optimal humidity for avoiding evaporation from the plates and for obtaining the best spot morphology.

2.9. Log Files

To keep track of changes and deviations, log files for almost every instrument and step involved is needed. With a log file for the contact spotter it will be easier to track changes that may affect the print run. In this log file everything related to the instrument must be noted to increase the possibility for successful troubleshooting when something goes wrong. The printing plate log can contain information regarding spotting buffer, volume, concentration, if the volume within the plate has been leveled out or if the plate has been exchanged. If printing several batches and/or several different projects a slide log is a must to keep track of all different slides in the production.

2.10. Post-Processing of Slides

There are a number of different post-processing steps depending of what kind of slide you are using. This includes UV-crosslinking, baking, dehydration, humidity chamber, washing steps, and blocking to mention some of them. The goal with the post-processing step is to attach the probes to the surface of the slides, to minimize the background intensity by blocking, and to wash away unbound probes. All information regarding these steps will be provided by the distributor if using commercial slides.

2.11. Post-Printing Controls

A quick control of how the print run has progressed is to pre-scan some slides from the printed batch. These pre-scanned slides will show if all block is present on the slides and if the pins have been working properly during the print run. Syto61 labeling and random-nonamer hybridization is both fine for spot morphology control. If having doughnuts problem it will not be visible when using the random-nonamer hybridization which is a disadvantage. RNA hybridization for microarray quality control is recommended. You will be able to do comparisons between slides, within slides, and between print bathes. To set up criterions of what fail and pass depend on the application.

2.12. Standard Operating Procedures

Standard Operating Procedures for all steps makes it easier to track deviations. All steps may not be fixed to a certain number or time; these could be set as variable in the Standard Operation Procedure. If using a spotter with split pins you may have to change blotting parameter every print run, i.e., the amount of pre-spots that is used to remove excess of probe containing liquid from the print tips. The wash cycles may also be changed from time to time due to optimization of the wash system in the spotter, which has to be done several times a year to prevent carry over contamination.

3. Methods

This method describes the procedure for a routine print, exemplified with the printing of HUM 46 k cDNA microarray using a contact spotter.

3.1. Reagents and Equipments

1. Contact spotter with tungsten split pins
2. Laboratory Centrifuge: SIGMA
3. Plate Shaker
4. Microarray slides with barcode: Ultra GAPS, Corning
5. MilliQ water
6. Wash solution 1: MilliQ water
7. Wash solution 2: 80% EtOH
8. Clean room wipes
9. Microarray scanner: DNA Microarray scanner, Agilent Technologies
10. Liquid dispenser: QFill, Genetix
11. Magnifying glass
12. Ultrasonic Bath
13. Slide barcode reader: Gryphon™ M100
14. Powder free nitrile gloves
15. Vacuum cleaner with HEPA filter
16. Adhesive cover
17. Sterile water
18. Clone collections: HUM cDNA clones in 30%DMSO spotting solution in a total volume of 20 μl per well, Genetix 384-well plates
19. 12-Channel 384 Equalizer pipette, 1–30 μl (Techtum)
20. Syto61 (Invitrogen) and/or Random-nonamer-cy3(MWG)
21. UV-Crosslinker

3.2. Procedure

Dust or fiber can easily give problems during a print run. Always use laboratory coat, suitable for a dust free environment, during a print run.

1. Preparation of the microarray laboratory

   The microarray laboratory has a temperature of 18–20°C and humidity of 45–50%.

   Use clean room wipes, moist with 70% ethanol and wipe all clean areas. Vacuum clean the lab.

2. Preparations of the source plates.
   1. Remove the source plates from –20°C storage. Leave the lids and adhesive cover on.
   2. Thaw the source plates to room temperature; do not let the plates thaw in a pile.
   3. Spin down the source plates in the laboratory centrifuge to remove any sample on the inside of the adhesive cover.
   4. Volume adjustment within the source plates. Due to evaporation during a print run it is necessary to adjust the volume to achieve equal volume within the plates to get an even spot size.
   5. Remove the lids and adhesive cover from 3–6 plates.
   6. Check volume in 10 wells per plate. The well volume has to be 20 μl in each well
   7. Add appropriate volume of sterile water if needed by using liquid dispenser or 12-channel pipette.
   8. If the volume has been adjusted reseal the plates and spin them down once more, to spin down any liquid on the wall of the wells.
   9. Place the source plates on a shaker for 30 min at 80 rpm, to get an even mixture within the spotting solution.
   10. Remove the adhesive covers and lids.
   11. Manually inspect the wells within a source plate by using a magnifying glass and remove any dust and fibers. This is a very important step but unfortunately also very tedious (**Sections 4.1, 4.7**)
   12. Place the lids on the source plates.
   13. Place the plates in the in-stacker. Make sure that well A1 is in the bottom right corner and the first source plate to be printed in the bottom and the last one at the top.
3. Prepare the contact spotter
4. Cleaning the microarray spotting pins.
   1. Place the microarray spotting pins in a floatable pin cleaning rack.
   2. Place them in an ultrasonic bath containing MilliQ water and sonicate for 5 min.
   3. Dry the microarray spotting pins by air (**Note 14**).
   4. Check each pin in the microscope to ensure the pins are clean (**Sections 4.1, 4.7, 4.8**).
   5. Place the microarray spotting pins in the microarray head. Place each pin in their "home-position." By having a fixed position for each pin it is easier to track loss in performance.

5. Cleaning of the slide bed.
   1. The slide bed of the machine should be cleaned before each print run to prevent dust and fibers entering the pin tips during a print run.
   2. Wipe off the slide bed with a clean room wipe moist with MilliQ water.
   3. Close the door.
   4. Fill up the Wash bottles with appropriate wash solutions, MilliQ water, and 80% EtOH.
   5. Empty the Waste bottles.
6. Create the Print script "HUM cDNA Microarray."
   1. When this has been done once all information can be saved and when the next batch is to be printed the script can be re-opened and re-started with the same parameters.
   2. Number of source plates to be printed: 120
   3. Microarray Head: 48-Microarray Head
   4. Number of blots: 50
   5. Number of slides to be printed: 75
   6. 90 slides will fit onto the bed but the first 15 slides will be dedicated to blotting and will be discarded after the print run.
   7. Number of hits per spot: 1
   8. Dwell time: 3000 ms
   9. Pattern: 31 × 31
   10. Space between spots: 140 μm
   11. Inking time: 3000 ms
   12. Stamp time: 10 ms
   13. Print adjustment: 110 μm
   14. Set the wash cycle to:

       4 × 3000 ms water wash, dry time 500 ms between each water wash.

       1 × 3000 ms water wash, dry time 5000 ms.

       1 × 5000 ms 80% EtOH wash, dry time 5000 ms

       1 × 5000 ms 80% EtOH wash, dry time 15000 ms

       Wait time: 5000 ms
7. Data tracking
   1. To create the correct gal-file (linking the probe identity and spot position) for this print run: Import the text file "HUM cDNA Microarray," containing plate name, clone ID, and the well location.
   2. Check the box for Data tracking within the software.

3. Add the plates in the same order as they will be printed, 1–120.
4. Check the box for GAL 4.0.
5. Name the file: HUM cDNA Microarray Batch No
6. Place microarray slides onto the bed, the barcoded end to the right.
7. Start the print run.

8. During printing
   1. Empty waste bottles when needed. The instrument is equipped with a liquid sensor and will stop when the waste bottles are full and then inform you to empty the wash bottles.
   2. Fill up wash bottles when needed. The instrument is equipped with a liquid sensor and will stop when the wash bottles are empty and inform you to fill up the wash bottles.
   3. Remove printed source plates from the "out-stacker," reseal with adhesive cover, place the lid on and store the source plates in –20°C.
   4. Place source plates in the "in-stacker."
9. Post printing
   1. Remove printed source plates from "out-stacker."
   2. Remove printed source plates from the "out-stacker," reseal with adhesive cover, place the lid and store the source plates in –20°C.
   3. Log the slide barcodes in the Slide log using the barcode reader.
   4. Remove printed slides from bed.
10. Post-processing procedure.
    1. UV-Crosslink the slides in 150 mJ/cm^2.
    2. Place the slide in a slide box.
    3. Label the slide boxes with HUM cDNA Microarray, batch number, printing date, and slide position.
    4. Store slides in the slide box dark at room temperature.
11. Print run control
    1. Pre-scan every 5th slide in the microarray scanner within a print run for a quick printing control.
    2. Syto61 labeling and/or Random-nonamer hybridization for spot morphology control, using Standard Operation Procedure.
12. Microarray quality control
    1. RNA hybridization for an appropriate and complete Microarray Quality Control, using Standard Operation Procedure.

4. Notes

There are five key components to manufacture high-quality microarrays: the spotter, the spotting pins, probe preparation, slide surface, and printing environment. The quality of microarray printing is completely depending on that all of these five criteria are fully optimized. Almost all problems listed below can be related to one of the key components. It is although important to point out that many of the potential issues concerning the quality of the final scanned images are often related to later events, i.e., the sample preparation, hybridization, and washing.

1. One or a few pins are not printing at all: The pin is sticking in the head due to dirty pin or head. Clean the pin and/or head according to manufacture's instructions. The Pin is bent and has to be replaced with a new one.
2. Occasional spots are missing from the array: Print Depth needs to be increased to compensate for variations in slide thickness. Source plates are incorrectly filled or ink depth is insufficient. The wells in source plate may contain dust of fibers and need to be removed. In inadequate cleaning of pins during the print run, optimization of the wash cycle is a must.
3. Some slides are not printed: Slides of varied thickness are being used and to eliminate this always use slide of same thickness. If testing out new slide, do several print runs with the different slides separately. The slide holder has not been placed flat on the bed, contact distributor/manufacturer for service. One slide holder Datum Point (calibration point) is incorrect, calibrate the datum point.
4. Spacing of blocks is not even: Pin Tip/s is damaged and needs to be replaced. Microarray head has been incorrectly re-assembled.
5. Spacing within blocks is not even: Surface is not stable enough and robot is vibrating. Drives need servicing.
6. Doughnuts: The humidity is to low during the print run, rapid drying of slides. The velocity of the microarray head is set to high. Inappropriate combination between pin, slide chemistry, and/or spotting solution.
7. Spots are merging into each other: Too high volume in wells of source plate and the volume or ink depth needs to be decreased. The chosen pattern is too tight, choose a smaller pattern or number of blotting spots has to be increased. Combination between slide surface and spotting solution is incorrect; try another combination of slide and buffer that may fit your application. The humidity in surrounding environment is too high. Pin/s is slightly dirty or damaged, clean

the pin or change to a new one. The contact time between slide surface and pins is set to high. The dwell time during sample pick up in well plate is set to high.

8. Comet tails: Overloading the binding capacity of the slide due to high concentration of DNA in arraying material. Poor UV-crosslinking or inadequate baking gives inadequate fixation of DNA to slide surface, try different UV-crosslinking and baking.
9. Damaged or particle contamination of slide surface: Print adjustment is set to high. Use powder free nitrile gloves and handle the slides carefully. Always store slides in a dust free environment and use slides within expiry date.
10. High background fluorescence: High inherited background fluorescence of slide substrate, always use powder free nitrile gloves.
11. Irregular spot morphology: Poor pin design, bent pins or pin tip is damaged, change to new a new pin. The contact time between slide surface and pin/s is set to high. The dwell time during sample pick up in well plate is set to high. The volume in wells of source plate is too high.
12. Low signal intensity: Too low concentration of material arrayed, two or more hits per spot gives more arrayed material on the array but also bigger spots. Scanner defect, check if the scanner works properly.
13. Source plates stuck together in stacker: When thawing micro-titer plates in a pile, liquid between the plates can cause the stacker system to malfunction.
14. Spot to spot contamination during a print run: An Optimization of the wash cycles is needed.

References

1. Schena M, Shalon D, Davis RW, Brown PO. (1995) Quantitative monitoring of gene expression patterns with a complementary DNA microarray. Science. 270(5235): 467–70.
2. Shalon D, Smith SJ, Brown PO. (1996) A DNA microarray system for analyzing complex DNA samples using two-color fluorescent probe hybridization. Genome Res. 6(7):639–45.
3. Schena M, Shalon D, Heller R, Chai A, Brown PO, Davis RW. (1996) Parallel human genome analysis: microarray-based expression monitoring of 1000 genes. Proc Natl Acad Sci U S A. 93(20):10614–9.
4. DeRisi J, Penland L, Brown PO, Bittner ML, Meltzer PS, Ray M, Chen Y, Su YA, Trent JM. (1996) Use of a cDNA microarray to analyse gene expression patterns in human cancer. Nat Genet. 14(4):457–60.

Chapter 8

RNA Preparation and Characterization for Gene Expression Studies

Michael Stangegaard

Abstract

Much information can be obtained from knowledge of the relative expression level of each gene in the transcriptome. With the current advances in technology as little as a single cell is required as starting material for gene expression experiments. The mRNA from a single cell may be linearly amplified to an excess of 10^6-fold. Reverse transcription and fluorescent labeling of the amplified RNA yields a stable target for subsequent hybridization to DNA microarrays.

Key words: RNA, RNA amplification, reverse transcription, random priming, random pentadecamer, fluorescent labeling, aminoallyl-dUTP, aa-dUTP, gene expression.

1. Introduction

Critical to the correct interpretation of a gene expression experiment is that the starting material is representative of the sample, from which it was originally taken. Any subsequent analysis will not provide correct information if the starting material is not representative of the sample or culture from which it was taken.

1.1. RNA Isolation and Amplification

The amount of mRNA required for target preparation for gene expression analysis on DNA microarrays was in an introductory technology paper 5 μg *(1)*. Assuming 1–3 % of a cell's total RNA is mRNA, the amount of total RNA required was 165–500 μg. As a typical cell contains between 10 and 30 pg of total RNA, the number of cells required was 5.5×10^6–5.0×10^7, to obtain 5 μg mRNA per hybridization reaction. This significant requirement of cells severely limited the use of microarrays to

Martin Dufva (ed.), *DNA Microarrays for Biomedical Research: Methods and Protocols, vol. 529*

DOI 10.1007/978-1-59745-538-1_8 Springerprotocols.com

experiments where large quantities of cells were available. One solution to this limitation was in fact discovered 5 years before the introduction of microarray technology. Van Gelder et al. *(2)*, devised a protocol for producing amplified heterogeneous populations of RNA from limited quantities of cDNA by linear amplification. Their method is commonly referred to as the Eberwine method. The working principle of the method involves reverse transcription of mRNA directly from the total RNA pool with an oligo dT primer, bearing a T7 RNA polymerase promoter site. The resulting mRNA–cDNA hybrid is then converted to double stranded cDNA with RNase H, DNA polymerase I, and a T4 DNA polymerase. Antisense RNA is subsequently transcribed in vitro by T7 polymerase, resulting in a ~2,000-fold linear amplification of the original mRNA sequence *(3)*. By repeating the method using the amplified RNA (aRNA) product as input to a second round of amplification, an excess of 10^6-fold amplification may be achieved *(4)*. This level of amplification allows for analysis of the entire transcriptome from as little as a single cell *(4–6)*. The method is less likely to introduce a bias in transcript abundance due to the linearity of the method opposed to an exponential based method such as RT-PCR. It is however, extremely important that all samples to be tested, including controls are amplified using the same method *(7–9)*.

As microarray analysis compares transcript abundance between sample and control, conservation of transcript abundance during RNA amplification is vital to allow identification of significantly regulated genes. Generally RNA amplification has been found acceptable, although minor changes in transcript abundance can be introduced during amplification when using highly diluted samples *(10)*.

1.2. Reverse Transcription

Reverse transcription of RNA into cDNA is a core method for analysis of gene expression. The reverse transcription reaction should result in a cDNA population which reflects the original mRNA population in terms of transcript abundance and complexity.

The choice of reverse transcription polymerase has been found to influence the relative yield of cDNA product up to 100-fold *(11, 12)*. Furthermore, this effect is both gene-dependent *(12)* and priming strategy-dependent *(11, 13)*. Priming of the reverse transcription reaction with oligo-dT primers may induce a 3′ bias in the resulting transcript *(14)*. Gene specific primers have only practical relevance if relatively few transcripts are targeted *(15, 16)*. Random primers can and will anneal to all RNA molecules containing the complementary sequence. As the target mRNA population typically comprises less than 2% of the total RNA pool, cDNA synthesis from total RNA with random primers will produce a large quantity of fragments predominately originating from ribosomal RNA. These cDNA fragments will lack specificity to the arrayed probes on the DNA microarray and can give rise to elevated background signals. As a result, random priming is preferred when using poly(A) purified RNA like mRNA or aRNA. Traditionally, random priming has been performed using random hexamers *(17)*. Recently, it was demonstrated that increased coverage of the transcriptome could be achieved using random primers 12 or 15 nucleotides long *(13)*.

Detection of the cDNA product requires it to be labeled. For practical reasons, fluorescence is preferred over radioactivity. Labeling can be performed during the reverse transcription reaction in form of labeled primers. This method in theory ensures that each cDNA molecule contains one fluorescent label. To increase the number of labels per cDNA molecule, a small amount of aminoallyl-conjugated dUTP (aa-dUTP) can be added to the dNTP mix. During the reverse transcriptions some aa-dUTP will replace some dTTP in the cDNA molecules and serve as an anchor-point for subsequent attachment of e.g., fluorescent dyes *(18)*.

1.3. Characterization

Following reverse transcription, the amount of resulting cDNA is characterized usually using spectroscopic analysis. The amount of incorporated dye can be measured and the relationship between amount of cDNA and dye be used as a quality parameter to assess the success of the reverse transcription reaction.

As enzyme-dependent reaction will very between reactions, it is advantageous to normalize the amount of dye prior to combining the sample and control for the subsequent co-hybridization reaction on the DNA microarray. This reduces dye bias between sample and control as well as ensures only samples containing sufficient dye and cDNA are used on microarrays.

2. Materials

2.1. RNA Isolation and Amplification

1. 0.5 M solution of NaOH. Store in 1 L blue cap bottle or similar at room temperature (RT).
2. 60 μL quartz micro-cuvette (Hellma, Müllheim, Germany).
3. Agilent 2100 Bioanalyzer™ with the RNA 6000 nano kit (Agilent).
4. Riboamp™ RNA amplification kit (Arcturus Engineering, Mountain View, CA, USA).
5. Diethylpyrocarbonate (DEPC)-treated water (RNase-free water, Sigma).

2.2. Reverse Transcription

1. Random pentadecamer primers (Sigma-Genosys, Haverhill, UK or any other oligo vendor) in 3 mg/mL solution stored in aliquots at –20°C (*see* **Note 1**).
2. SuperScript^{TM} II reverse transcriptase, First-strand buffer and Dithiothreitol (DTT; Invitrogen). Such as product number 18064-014 or alternative. Store at –20°C.
3. 1 M phosphate buffer:
   a. Prepare a 100 mL of 1 M solution of K_2HPO_4 (Sigma) in MilliQ water by dissolving 1.742 g K_2HPO_4 in 100 mL MilliQ water.
   b. Prepare 100 mL of 1 M KH_2PO_4 (Sigma) in MilliQ water by dissolving 1.361 g KH_2PO_4 in 100 mL MilliQ water.
   c. To make 10 mL of 1 M phosphate buffer (KPO_4, pH 8.5–8.7): combine 9.5 mL of 1 M K_2HPO_4 with 0.5 mL 1 M KH_2PO_4.
4. 0.1 M phosphate buffer: Prepare 10 mL 0.1 M phosphate buffer, pH 8.0, by combining 1 mL 1 M phosphate buffer with 9 mL MilliQ water.
5. Aminoallyl-dUTP solution:
   a. Add 19.1 μL of 0.1 M phosphate buffer to the lyophilized 5-aminoallyl-dUTP (aa-dUTP; Sigma). Such as product number A0410 or alternative.
   b. Allow the content to dissolve completely by slowly pipetting up and down.
   c. Measure the concentration in the spectrophotometer by removing 0.5 μL of the dissolved aa-dUTP and adding to 2.5 mL 0.1 M phosphate buffer (1:5000 dilution).
   d. Measure the OD_{289} of the aa-dUTP sample. Use the 0.1 M phosphate buffer as background.
   e. The concentration of aa-dUTP in mM is calculated by multiplying the observed OD_{289} (corrected for background) by 704 (*see* **Note 2**).
6. 50X dNTPs solution containing 500 μM of dATP, dCTP and dGTP, 300 μM dTTP (Larova Biochemie GMBH, Teltow, Germany, product number NU-1005S) and 200 μM aa-dUTP. Store at –20°C. The 50X labeling mix is prepared with a 2:3 aa-dUTP:dTTP ratio by mixing the components listed below:

   dATP (100 mM) 5 μL (final concentration: 25 mM)
   dCTP (100 mM) 5 μL (final concentration: 25 mM)
   dGTP (100 mM) 5 μL (final concentration: 25 mM)
   dTTP (100 mM) 3 μL (final concentration: 15 mM)
   aa-dUTP (100 mM) 2 μL (final concentration: 10 mM)

Use the actual concentration determined for the aa-dUTP solution rather than the 100 mM indicated above. Calculate the volume of aa-dUTP required to have a final concentration of 10 mM (*see* **Note 2**) and substitute that volume for the 2 μL aa-dUTP above. Store at –20°C.

7. RNasin^{TM} (Promega, Mannheim, Germany). Such as product number N2511 or alternative. Store at –20°C.
8. Solution of ethylenediamine tetraacetic acid (EDTA; 0.5 M, Sigma). Store at RT.
9. Solutions of NaOH and HCl, both 1 M. Store in appropriate size blue cap bottles or similar at RT.
10. Phosphate wash buffer: 5 mM phosphate buffer, pH 8.0, 80% ethanol (*see* **Note 3**). 100 mL of the ready to use phosphate wash buffer is prepared in a 250 mL nuclease-free blue-cap bottle or similar by combining:

    1 M phosphate buffer, pH 8.5 0.5 mL
    MilliQ water 15.25 mL
    96% ethanol 84.25 mL

    Store at RT.
11. QIAquick^{TM} PCR Purification Kit (Qiagen).
12. Phosphate elution buffer: 4 mM phosphate buffer, pH 8.5 (*see* **Note 3**). To prepare 100 mL ready to use phosphate elution buffer combine 0.4 mL 1 M phosphate buffer with 99.6 mL MilliQ water. Since only 60 μL are used per sample processed, aliquot in eppendorf tubes and store at RT.
13. 0.1 M sodium carbonate buffer, pH 9.0 (*see* **Note 4**). Prepare 10 mL sodium carbonate buffer (0.1 M, pH 9.0) by:
    a. Dissolving 1.08 g of Na_2CO_3 (Sigma) in 8 mL MilliQ water.
    b. Adjust pH to 9.0 with 1 M HCl and 1 M NaOH.
    c. Bring the volume up to 10 mL with MilliQ water.
    d. Store at RT. Carbonate buffers are stable for about 1 month.
14. NHS ester of Cy-3 and Cy-5 (GE Healthcare such as product numbers PA23001 and PA25001 respectively or alternative), dissolved in dimethyl sulofoxide (DMSO *see* **Note 5 and 6**). Cy-3 and Cy-5 are light sensitive solutions should be kept in dark as much as possible to maximize reproducibility. Aliquot and store at –80°C.
15. 0.1 M sodium acetate solution, pH 5.2. To prepare a 100 mL solution of a 0.1 M sodium acetate solution, pH 5.2 in MilliQ water.
    a. Dissolving 0.8203 g sodium acetate (Sigma) in 90 mL MilliQ water.

b. Adjust the pH to 5.2 with 1 M HCl or 1 M NaOH.

c. Bring the volume up to 100 mL with MilliQ water.

Store at RT.

2.3. Characterization

1. 60 µL quartz micro-cuvette (Hellma, Müllheim, Germany).

3. Methods

The process of generating micrograms of labeled cDNA from as little starting material as a single cell is quite complex and contains numerous steps. All steps must be performed correct in order to obtain a reproducible result. An overview of the process is found in **Fig. 8.1**. As all further treatment of the samples and subsequent interpretation of the biological meaning of the overall hypothesis

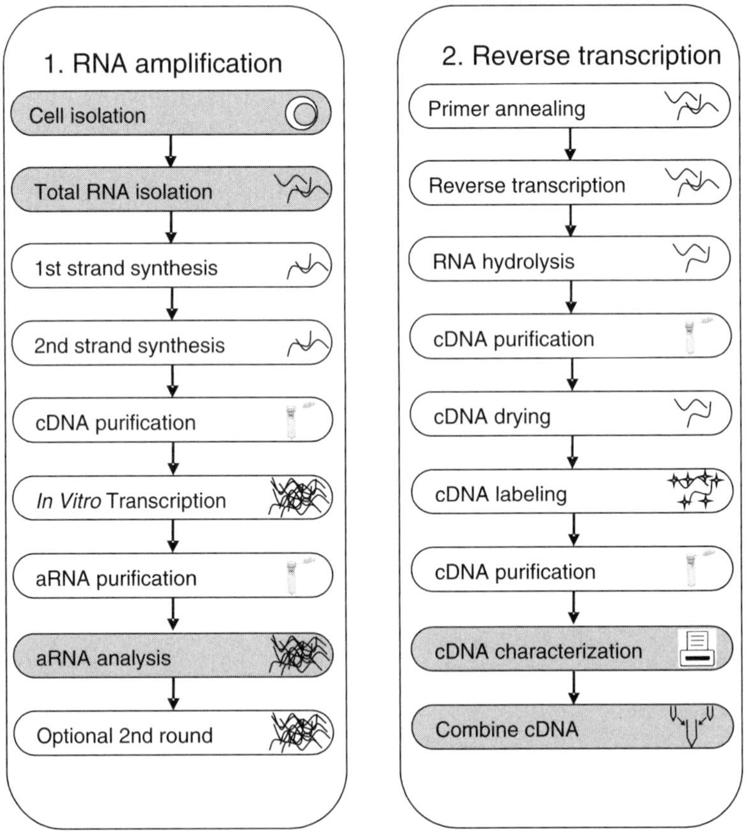

Fig. 8.1. Process overview. The major steps in the long process are highlighted. Steps at which critical information regarding the success of previous steps can be obtained are indicated with *greyed* backgrounds.

investigated rely on the method being carried out reproducibly. Hence, it is extremely important that the method provided in this section is performed in a standardized manner. Any variation, however small it may seem, may infer with the efficiency of the method and hence interfere with any subsequent interpretation of the results.

3.1. RNA Isolation and Amplification

The protocol in this section should be stopped at specific steps only. This is mentioned after the relevant steps.

3.1.1. Cell Isolation

1. Harvest cells according to laboratory standard operating procedure (SOP) for the specific cell line in question.

3.1.2. Total RNA Isolation

1. Isolate cellular total RNA according to laboratory SOP (*see* **Note 7**).
2. Measure the concentration of the isolated RNA using a standard spectrophotometer and a 60 μL quartz micro-cuvette. Wash the cuvette with DEPC-treated water and dry with compressed air. Use 1 μL of the isolated total RNA in 59 μL of the elution buffer used in step 1 above, in a 60 μL micro-cuvette (*see* **Note 8**).
3. Inspect the quality of the isolated RNA on the Agilent 2100 Bioanalyzer™ using the RNA 6000 nano kit (*see* **Note 9**).

3.1.3. 1st Round of Amplification; 1st Strand Synthesis

1. For each amplification reaction, 2 μg total RNA is sufficient as starting material. Depending upon cell line used around 35–40 μg of aRNA should be obtained following 1st round of amplification. Add DEPC-treated water to a total volume of 10.0 μL and place in a 0.2 mL RNase-free microcentrifuge tube. Add 1.0 μL of Primer A (Beige-labeled Vial – A) to each sample, mix well, and spin down.
2. Incubate at 65°C for 5 min then chill the sample(s) to 4°C for at least 1 min. Hold the sample at 4°C until ready to proceed. Spin the contents down before proceeding to the next step (*see* **Note 10**).
3. Thaw 1st Strand Synthesis components (Red-labeled Vials) and place on ice (1st Strand Enzyme Mix does not require thawing and can be placed directly on ice; *see* **Note 11**).
4. To each chilled sample vial add 7 μL 1st Stand Master Mix (Red Vial # 1) and 2 μL 1st Strand Enzyme Mix (Red Vial # 2). If multiple samples are processed in parallel it is more convenient to prepare a Complete 1st Strand Master Mix using the volumes mentioned above multiplied by the number of samples.
5. Mix thoroughly, spin down, and incubate at 42°C for 45 min.
6. Chill sample(s) to 4°C for at least 1 min, and then spin down briefly.

7. Add 2.0 μL of 1st Strand Nuclease Mix (Gold-labeled Vial) to each sample, mix thoroughly, spin down, and incubate at 37°C for 20 min. Followed by 5 min at 95°C.
8. Chill sample(s) to 4°C for at least 1 min and hold at 4°C until ready to proceed. Spin down briefly. The protocol can be stopped at this point. Sample(s) may be stored at –20°C overnight.

3.1.4. 2nd Strand Synthesis

1. Thaw sample(s) on ice if stored overnight at –20°C. Add 1.0 μL of Primer B (Pink-labeled Vial – B) at 4°C. Mix thoroughly and spin down. Incubate sample(s) at 95°C for 2 min, then chill and maintain the sample(s) at 4°C for at least 2 min.
2. Thaw 2nd Strand Synthesis components (White-labeled Vials) and place on ice. 2nd Strand Enzyme Mix does not require thawing and can be placed directly on ice.
3. To each sample vial add 29 μL 2nd Strand Master Mix (White Vial # 1), 1 μL 2nd Strand Enzyme Mix (White Vial # 2). Mix thoroughly and spin down. If multiple samples are processed in parallel, it is advantageous to prepare a Complete 2nd Strand Master Mix using the volumes mentioned above, multiplied by the number of samples. Add 30 μL to each sample.
4. Incubate the sample(s) as follows: 25°C for 5 min, 37°C for 10 min, 70°C for 5 min maintain at 4°C until ready to proceed. The samples should be left on hold for a maximum of 30 min.

3.1.5. cDNA Purification

1. Add 250 μL of DNA Binding Buffer (DB) to an empty Purification Column seated in the collection tube provided with the kit. Hold for 5 min at room temperature. During this time the sample(s) can carefully be placed in the centrifuge.
2. Centrifuge at 16,000 × *g* for 1 min. Leave the flow-through in the collection tube.
3. Add 200 μL of DB to the 2nd Strand Synthesis sample tube(s), mix well, and pipette the entire volume into the purification column(s).
4. To bind cDNA to column(s), centrifuge at 100 × *g* for 2 min (or lowest speed setting available), immediately followed by centrifugation at 10,000 × *g* for 30 s to remove flow-through.
5. Add 250 μL of DNA Wash Buffer (DW) to each column and centrifuge at 16,000 × *g* for 2 min. Check the purification column(s) for any residual wash buffer. If any wash buffer remains, re-centrifuge at 16,000 × *g* for 1 min.
6. Discard the flow-through and collection tube(s).
7. Place each column into a provided 0.5 mL microcentrifuge tube and carefully add 16 μL of DNA Elution Buffer (DE) onto the center of each purification column membrane. Gently

touch the tip of the pipette to the surface of the membrane while dispensing the elution buffer to ensure maximum absorption of DE into the membrane. Gently tap each purification column to distribute the buffer. If necessary incubate for 1 min at room temperature.

8. Place the assembly into the centrifuge and centrifuge at 1,000 × g for 1 min, followed immediately by 16,000 × *g* for 1 min. Discard the column(s) and retain the elution containing cDNA in the micro-centrifuge tube(s) for further processing. The protocol can be stopped at this point. Sample(s) may be stored at –20°C overnight.

3.1.6. In Vitro Transcription (IVT)

1. Thaw all IVT Reaction components on ice (IVT Enzyme Mix does not require thawing and can be put directly on ice (*see* **Note 12**)). To each sample add 8 μL IVT buffer (Blue Vial # 1), 12 μL IVT Master mix (Blue Vial # 2) and 4 μL IVT Enzyme Mix. If multiple samples are processed in parallel, it is advantageous to prepare a complete IVT Reaction Mix using the volumes mentioned above multiplied by the number of samples (*see* **Note 13**). Add 24 μL to each sample being processed.
2. Mix the sample(s) thoroughly, spin down, and incubate at 42°C for 4 h (Optional: 5 h incubation may be used for additional aRNA yield). Cool samples to 4°C.
3. Add 2 μL DNase Mix (Blue-labeled Vial #4). Mix thoroughly and spin down. Incubate at 37°C for 15 min. Cool the sample(s) to 4°C. Continue the purification of aRNA immediately.

3.1.7. aRNA Purification

1. Add 250 μL of RNA Binding Buffer (RB) to a new DNA/RNA purification column and hold for 5 min at room temperature. Centrifuge at 16,000 × *g* for 1 min. Let the flow-through remain in the tube. Prepare one column for each sample being processed.
2. Add 200 μL of RB to each IVT Reaction sample and mix thoroughly. Pipette the entire sample volume into the purification column(s).
3. To bind aRNA to the column(s), centrifuge at 100 × *g* for 2 min (or lowest speed setting available), immediately followed by a centrifugation at 10,000 × *g* for 30 s to remove flow-through.
4. Add 200 μL of RW to each purification column and centrifuge at 10,000 × *g* for 1 min.
5. Add 200 μL of fresh RNA Wash Buffer (RW) to the purification column, and centrifuge at 16,000 × *g* for 2 min. Check the purification column(s) for any residual wash buffer. If any wash buffer remains, re-centrifuge at 16,000 × *g* for 1 min.

6. Discard the flow-through and the used collection tube(s).
7. Place each purification column into a new 0.5 mL microcentrifuge tube provided in the kit and carefully add RNA Elution Buffer (RE) directly onto the center of each purification column membrane. If stopping with one round of amplification add 30 μL or 11 μL if going on to a second round of amplification. Gently touch the tip of the pipette to the surface of each membrane while dispensing the elution buffer to ensure maximum absorption of RE into the membrane. Gently tap the purification column to distribute the buffer, if necessary. Incubate at room temperature for 1 min.
8. Place the assembly into the centrifuge, and centrifuge at 1,000 × *g* for 1 min, immediately followed by 16,000 × *g* for 1 min. Discard the purification column(s) and retain the elution containing the aRNA in the microcentrifuge tube(s) for further processing.

3.1.8. aRNA Analysis

1. Measure the concentration of the aRNA product in a spectrophotometer using a 60 μL quartz micro-cuvette. Wash the cuvette with DEPC-treated water and blow dry with compressed air. Place 59 μL DEPC-treated water or RE in a 60 μL micro-cuvette. Measure OD_{260} and OD_{280} on RE and use this as background. Add 1–2 μL as appropriate of the aRNA product mix well and measure OD_{260} and OD_{280} again. Subtract the background reading from the aRNA measurement (*see* **Note 14**).
2. If sufficient aRNA is obtained, perform electrophoretic analysis to confirm the presence of amplification product. This is most conveniently done on an Agilent 2100 Bioanalyzer™ using the RNA 6000 nano kit. The aRNA will appear as a smear on the gel image from 200 to 2,000 bases in length, with the majority of product around 600 bases in length (*see* **Note 15**).
3. The purified aRNA is ready for use in a reverse transcription reaction or for use as input in a second round of amplification. Purified aRNA may be stored at –80°C until use.

3.1.9. 2nd Round of Amplification (Optional)

1. If a 2nd round of amplification is required, use an appropriate amount of the aRNA product (max 10 μL) as starting material and step 1 in **Section 3.1.3**.

3.2. Reverse Transcription and Aminoallyl-Labeling

3.2.1. Primer Annealing

1. To 2 μg aRNA product (from step 3 above in **Section 3.1.8.**), add 3.35 nmol random pentadecamer primers (2 μL; *see* **Note 1**) and bring the total volume up to 18.5 μL with DEPC-treated water (*see* **Note 16**).
2. Mix well and incubate at 70°C for 10 min.
3. Snap cool on ice for 30 s. Centrifuge briefly at >10,000 × *g* and continue at room temperature.

3.2.2. Reverse Transcription

1. Add the following components to each sample. When processing multiple samples it is advantageous to prepare a master mix consisting of the number of samples multiplied by the volumes listed below. Add 12.6 μL to each sample being processed.

   5X First strand buffer 6 μL
   0.1 M DTT 3 μL
   RNasin™ (RNase inhibitor) 1 μL
   50X labeling mix 0.6 μL (*see* **Note 17**)
   SuperScript™ II RT (200 U/μL) 2 μL

2. Mix and incubate at 42°C for 5 h to overnight. Use overnight incubation to increase cDNA yield.

3.2.3. RNA Hydrolysis

1. To hydrolyze the RNA add 10 μL 1 M NaOH and 10 μL 0.5 M EDTA (*see* **Note 18**) to each sample. Mix and incubate for 65°C for 15 min.
2. To neutralize the pH add 10 μL 1 M HCl to each sample.

3.2.4. cDNA Purification

1. The unincorporated aminoallyl-dUTP and free amines are removed using a Qiagen QIAquick™ PCR purification kit (*see* **Note 3**).
2. Mix each cDNA sample with 300 μL (5X reaction volume) of buffer PB (Qiagen supplied) and transfer for a QIAquick™ column.
3. Centrifuge at 100 × *g* for 1 min followed immediately by 8,000 × *g* for 1 min. Empty collection tube(s).
4. To wash, add 725 μL phosphate wash buffer (*see* **Note 3**) to each column and centrifuge at 8,000 × *g* for 1 min.
5. Empty the collection tube(s) and repeat the wash and centrifugation step (step 4).
6. Empty each collection tube and centrifuge column(s) an additional minute at maximum speed (around 18,000 × *g*).
7. Transfer each column to a new 1.5 mL microcentrifuge tube and carefully add 30 μL phosphate elution buffer (*see* **Note 3**) to the center of each column membrane.
8. Incubate for 1 min at room temperature.
9. Elute by centrifugation at 100 × *g* for 1 min followed immediately by 18,000 × *g* for 1 min.
10. Elute a second time into the same tube by carefully adding another 30 μL phosphate elution buffer and then repeating steps 8–9. The final elution volume should be ~60 μL.

3.2.5. cDNA Drying

1. Dry sample(s) in a speedvac or alternatively in an incubation oven at 60°C. If using the oven, open the lids and cover the front of the oven with aluminum foil to prevent exposure of the sample(s) to light.

3.2.6. cDNA Labeling

1. Re-suspend the dried aminoallyl-labeled cDNA in 4.5 μL sodium carbonate buffer (*see* **Note 4**). Add 4.5 μL of the appropriate Cy-dye dissolved in DMSO (*see* **Note 6**) to each sample. Mix by gentle pipetting. Briefly spin down in a bench-top centrifuge before continuing.
2. Incubate the reaction(s) for 1 h in the dark at room temperature.
3. To stop the reaction(s) add 35 μL 0.1 M sodium acetate solution pH 5.2.

3.2.7. cDNA Purification

1. Removal of uncoupled dye is performed with the Qiagen QIAquick™ PCR Purification Kit.
2. To each reaction add 250 μL (5X reaction volume) Buffer PB (Qiagen supplied).
3. Place a QIAquick™ spin column in a 2 mL collection tube (Qiagen supplied) for each sample being processed, apply each sample to a column, and centrifuge at 8,000 × *g* for 1 min. Empty the collection tube(s).
4. To wash, add 725 μL Buffer PE (Qiagen supplied) to each column and centrifuge at 8,000 × *g* for 1 min.
5. Empty each collection tube and centrifuge column(s) for an additional minute at maximum speed (~18,000 × *g*).
6. Place each column in a clean 1.5 mL microcentrifuge tube and carefully add 30 μL Buffer EB (Qiagen supplied) to the center of each column membrane.
7. Incubate for 1 min at room temperature.
8. Elute by centrifugation at 18,000 × *g* for 2 min.
9. Elute a second time into the same tube by carefully adding another 30 μL Buffer EB and repeating steps 7–8 in **Section 3.2.7**. The final elution volume should be ~60 μL.

3.3. Characterization

In the following section the exposure of the cyanine labeled cDNA targets to light should be minimized or at best eliminated to prevent photo-decomposition of especially the Cy-5 dye (*see* **Note 6**). This will introduce an irreproducible dye bias in the resulting data. Wrap tube(s) in aluminum foil when possible and work fast.

3.3.1. cDNA Characterization

1. Use a 60 μL quartz micro-cuvette to analyze the entire undiluted sample in a spectrophotometer.

2. Wash the cuvette with MilliQ and blow dry with compressed air and pipette sample into cuvette and place cuvette in spectrophotometer.
3. For each sample measure absorbance at 260, 310 nm (background) and either 550 nm for Cy-3 or 650 nm for Cy-5, as appropriate. For each sample calculate the total picomoles of dye incorporation (Cy-3 or Cy-5 as appropriate) using the formulas given in **Note 19** (*see* **Note 19**).
4. Pipette sample from cuvette back into the original sample tube. Store at –20°C until ready to use.
5. Wash the cuvette with MilliQ water and blow dry with compressed air.

4. Notes

1. Order random pentadecamer primers (R-15) in large batches (e.g., 1,000 nmol) as dried desalted oligos from your conventional oligo vendor. They may vary from batch to batch. HPLC purification is not required. Re-dissolve in DEPC-treated water (not just MilliQ water) to a concentration of 1.674 nmol/μL. This result in an equimolar amount of R-15 primers in the reaction compared to 6 μg random hexamers, which is default in many protocols, could also be used in this protocol.
2. Calculation example: If $OD_{289,\ \text{aa-dUTP}}$ is 0.120 and $OD_{289,\ \text{background}}$ is 0.012 the corrected $OD_{289,\text{aa-dUTP}}$ is $0.120–0.012 = 0.108$. The concentration of aa-dUTP is then $0.108 \times 704 - 76.03$ mM.
3. The phosphate buffers are substituted for two of the three supplied Qiagen buffers (Buffer PE and Buffer EB) as they contain free amines which will compete with the Cy-dyes in the coupling reaction and reduce the amount of dye that will bind to the cDNA. The phosphate wash buffer replaces Qiagen buffer PE and phosphate elution buffer replaces the Qiagen Buffer EB. The Qiagen supplied Buffer PB does not contain free amines and is not replaced. This substitution is only relevant prior to Cy-dye labeling. Hence for purification of the labeled cDNA, the Qiagen supplied may be used.
4. Carbonate buffers are stable for less than 1 month. Make sure the buffer is fresh.
5. The dye from one tube is resuspended in 73 μL of DMSO before use. The dye must either be used immediately or aliquotted and stored at –80°C. Aliquots are best prepared in 4.7 μL volumes. This corresponds to one reaction

requirement. Any introduced water to the dye will result in a lower coupling efficiency due to the hydrolysis of the dye. As DMSO is hygroscopic (absorbs water from the atmosphere) buy small bottles and store in desiccant. If the freshness of the DMSO is questionable, buy a fresh bottle rather than compromising the results of a potential large series of experiments simply due to old DMSO.

6. Cy-5 will photo-decompose faster than Cy-3 introducing a bias in downstream microarray experiments. Hence exposure of especially Cy-5 to light should be minimized or even better avoided. Tubes containing the dye should be kept dark at all times by wrapping in aluminum foil.
7. Working with RNA requires clean RNase free surfaces and equipment. Use only dedicated work area, preferably dedicated laboratories, equipment and RNase free labware. Prior to working with RNA the surfaces should be cleaned with 0.5 M NaOH or alternative RNase decontamination product, such as RNase ZAP(R) (Sigma) to minimize RNase contamination.
8. Use the elution buffer rather than standard DEPC-treated water or MilliQ water for a more accurate measurement. Use a measurement of the elution buffer without RNA as background.
9. The quality of the isolated RNA should be assessed prior to amplification, labeling, and subsequent hybridization to avoid poor results due to low RNA quality. Only RNA with RNA Integrity Number (RIN) greater than 9.7 should be used. RNA preparations with lower RIN numbers may result in poor irreproducible results.
10. The RNA amplification process requires multiple heating cooling steps. These can be pre-programmed on a PCR cycler, which is placed on hold (4°C) when the sample(s) are processed in the bench. The program is:

°C	Time	
65	5 min	*1st Strand Synthesis*
4	Hold	
42	45 min	
4	Hold	
37	20 min	
95	5 min	
4	Hold	

(continued)

(continued)

95	2 min	*2nd Strand Synthesis*
4	Hold	
25	5 min	
37	10 min	
70	5 min	
4	Hold	
42	4 h	*IVT*
4	Hold (overnight)	
37	15 min	
4	Hold	

11. Thaw frozen components and mix by gentle vortexing or by inverting the tubes several times, and place on ice. When enzyme mixtures must be removed from –20°C storage for use, always keep them in a cold block or in an ice bucket at the lab bench.
12. Allow IVT reagents (Blue-labeled Vials) to attain room temperature (22–25°C). Dissolve all visible solids and mix prior to use. Place components back onto ice or refreeze immediately after dispensing the reagent. Do not leave reagents at room temperature for any extended period of time. Reagents may be added directly to the sample(s).
13. Components must be added and mixed in the order they are listed. Mix the Complete IVT Reaction Mix gently by brief vortexing at the lowest setting, or by flicking the tube after addition of each component.
14. Measuring OD_{260} and OD_{280} and calculating the OD_{260}/OD_{280} ratio will indicate the purity of the aRNA. An OD_{260}/OD_{280} ratio that approaches 2.0 indicates very pure aRNA. For single stranded aRNA, a measurement of $OD_{260} = 1.0$ corresponds to an aRNA concentration of 40 μg/mL. The yield may be calculated by:

$$(OD_{260}) \times (\text{dilution factor}) \times (40) = \mu g/mL \text{ aRNA}$$

15. The RIN function is only designed for total RNA and the result does not apply for aRNA.

16. The RNA labeling is based upon the Standard Operating Procedure provided by The Institute for Genomic Research (TIGR) and can be obtained at pga.tigr.org/sop/M004_1a.pdf.
17. The 50X labeling mix can be stored at –20°C and thawed just prior to use. Avoid extensive freeze–thaw cycles. If only a few samples are processed at one time, aliquot the 50X labeling mix in smaller volumes.
18. The solution can be prepared in advance by combining 500 μL 2 M NaOH with 500 μL 1 M EDTA in an eppendorf tube. Add 20 μL of this solution to each sample being processed.
19. 1 OD_{260} = 37 ng/μL for cDNA. The average molecular weight of a dNTP is 324.5 pg/pmol. To calculate the amount of Cy-3 and Cy-5 in pmol in the prepared target cDNA use the following formulas:

$$\text{pmol nucleotides} = \frac{OD_{260} \bullet \text{Volume}(\mu L) \bullet 37\text{ng}/\mu L \bullet 1000\ \text{pg/ng}}{324.5\ \text{pg/pmol}}$$

$$\text{pmol Cy} - 3 = \frac{OD_{550} \bullet \text{Volume}\ (\mu L)}{0.15}$$

$$\text{pmol Cy} - 5 = \frac{OD_{650} \bullet \text{Volume}\ (\mu L)}{0.25}$$

$$\text{nucleotides/dye ratio} = \frac{\text{pmol cDNA}}{\text{pmol Cy dye}}$$

More than 200 pmol of dye incorporation per sample and a ratio of less than 50 nucleotides per dye molecule are optimal for efficient hybridization reactions.

5. Safety

Observe standard laboratory safety procedures when working in the laboratory. The edges of the slides may be sharp. Please handle carefully. Pipette tips may be sharp. Please handle carefully and dispose of appropriately. Wear appropriate protective equipment when working in a laboratory.

Acknowledgments

Above all I would like to express my sincere gratitude to Dr. Martin Dufva for giving me the opportunity to share my hard earned experiences with the community. I wished for a collection of relevant protocols including tips and tricks when I started out my experiments, but I found none. I hope some may benefit from reading this.

A grant by the Danish Biotechnology Instrument Center (DABIC) project no. 2014-00-0003 financially supported the project. This is gratefully acknowledged.

Last but not least – my family, especially my wife, Lene. Without her ever present support and tolerance for my physical and mental absence, my completion of this project would not have been possible.

References

1. Schena, M., Shalon, D., Davis, R. W., and Brown, P. O. (1995) Quantitative monitoring of gene expression patterns with a complementary DNA microarray. *Science* **270,** 467–70.
2. Van Gelder, R. N., von Zastrow, M. E., Yool, A., Dement, W. C., Barchas, J. D., and Eberwine, J. H. (1990) Amplified RNA synthesized from limited quantities of heterogeneous cDNA. *Proc Natl Acad Sci U S A* **87,** 1663–7.
3. Phillips, J., and Eberwine, J. H. (1996) Antisense RNA amplification: a linear amplification method for analyzing the mRNA population from single living cells. *Methods* **10,** 283–8.
4. Eberwine, J., Yeh, H., Miyashiro, K., Cao, Y., Nair, S., Finnell, R., Zettel, M., and Coleman, P. (1992) Analysis of gene expression in single live neurons. *PNAS* **89,** 3010–14.
5. Cheetham, J. E., Coleman, P. D., and Chow, N. (1997) Isolation of single immunohistochemically identified whole neuronal cell bodies from post-mortem human brain for simultaneous analysis of multiple gene expression. *J Neurosci Methods* **77,** 43–8.
6. Kamme, F., Salunga, R., Yu, J., Tran, D. T., Zhu, J., Luo, L., Bittner, A., Guo, H. Q., Miller, N., Wan, J., and Erlander, M. (2003) Single-cell microarray analysis in hippocampus CA1: demonstration and validation of cellular heterogeneity. *J Neurosci* **23,** 3607–15.
7. Hughes, T. R., Mao, M., Jones, A. R., Burchard, J., Marton, M. J., Shannon, K. W., Lefkowitz, S. M., Ziman, M., Schelter, J. M., Meyer, M. R., Kobayashi, S., Davis, C., Dai, H., He, Y. D., Stephaniants, S. B., Cavet, G., Walker, W. L., West, A., Coffey, E., Shoemaker, D. D., Stoughton, R., Blanchard, A. P., Friend, S. H., and Linsley, P. S. (2001) Expression profiling using microarrays fabricated by an ink-jet oligonucleotide synthesizer. *Nat Biotechnol* **19,** 342–47.
8. Li, Y., Li, T., Liu, S., Qiu, M., Han, Z., Jiang, Z., Li, R., Ying, K., Xie, Y., and Mao, Y. (2004) Systematic comparison of the fidelity of aRNA, mRNA and T-RNA on gene expression profiling using cDNA microarray. *J Biotechnol* **107,** 19–28.
9. Ma, X. J., Salunga, R., Tuggle, J. T., Gaudet, J., Enright, E., McQuary, P., Payette, T., Pistone, M., Stecker, K., Zhang, B. M., Zhou, Y. X., Varnholt, H., Smith, B., Gadd, M., Chatfield, E., Kessler, J., Baer, T. M., Erlander, M. G., and Sgroi, D. C. (2003) Gene expression profiles of human breast cancer progression. *Proc Natl Acad Sci U S A* **100,** 5974–9.
10. Nygaard, V., and Hovig, E. (2006) Options available for profiling small samples: a review of sample amplification technology when combined with microarray profiling. *Nucl Acids Res* **34,** 996–1014.
11. Ståhlberg, A., Hakansson, J., Xian, X., Semb, H., and Kubista, M. (2004) Properties of the reverse transcription reaction in mRNA quantification. *Clin Chem* **50,** 509–15.
12. Ståhlberg, A., Kubista, M., and Pfaffl, M. (2004) Comparison of reverse transcriptases in gene expression analysis. *Clin Chem* **50,** 1678–80.
13. Stangegaard, M., Dufva, I. H., and Dufva, M. (2006) Reverse transcription using random pentadecamer primers increases yield and quality of resulting cDNA. *Biotechniques* **40,** 649–57.

14. Brooks, E. M., Sheflin, L. G., and Spaulding, S. W. (1995) Secondary structure in the 3' UTR of EGF and the choice of reverse transcriptases affect the detection of message diversity by RT-PCR. *Biotechniques* **19,** 806–12, 14–5.
15. Iturriza-Gomara, M., Green, J., Brown, D. W., Desselberger, U., and Gray, J. J. (1999) Comparison of specific and random priming in the reverse transcriptase polymerase chain reaction for genotyping group A rotaviruses. *J Virol Methods* **78,** 93–103.
16. Brink, A. A., Oudejans, J. J., Jiwa, M., Walboomers, J. M., Meijer, C. J., and van den Brule, A. J. (1997) Multiprimed cDNA synthesis followed by PCR is the most suitable method for Epstein-Barr virus transcript analysis in small lymphoma biopsies. *Mol Cell Probes* **11,** 39–47.
17. Feinberg, A. P., and Vogelstein, B. (1983) A technique for radiolabeling DNA restriction endonuclease fragments to high specific activity. *Anal Biochem* **132,** 6–13.
18. Richter, A., Schwager, C., Hentze, S., Ansorge, W., Hentze, M. W., and Muckenthaler, M. (2002) Comparison of fluorescent tag DNA labeling methods used for expression analysis by DNA microarrays. *Biotechniques* **33,** 620–8, 30.

Chapter 9

Gene Expression Analysis Using Agilent DNA Microarrays

Michael Stangegaard

Abstract

Hybridization of labeled cDNA to microarrays is an intuitively simple and a vastly underestimated process. If it is not performed, optimized, and standardized with the same attention to detail as e.g., RNA amplification, information may be overlooked or even lost. Careful balancing of the amount of labeled cDNA added to each slide reduces dye-bias and slide to slide variation. Efficient mixing of the hybridization solution throughout the hybridization reaction increases signals several fold. The amount of near perfect target–probe hybrids may be reduced by efficient stringency washes of the hybridized microarray slides.

Key words: Gene expression, hybridization, co-hybridization, cDNA, oligo microarray, DNA microarray, mixing, fluorescence, fluorescent labeling.

1. Introduction

An invaluable tool for acquiring an overview of which genes are expressed in a cell culture or even in a single cell relative to a reference is the DNA microarray *(1–3)*. Utilizing a high-density array of nucleic acid probes *(4)* of variable length and sequence located in discrete spots usually on a high optic-grade glass slide cut to the dimensions of the standard microscope slide (3 by 1 inch), a single DNA microarray enables the interrogation of a complex sample for complementary target sequences to all the arrayed probes in relative abundance, all in a single experiment.

DNA microarrays are usually divided in two groups, "cDNA microarrays" and "oligonucleotide microarrays," respectively. In cDNA microarrays the individual probes are derived from a physical source e.g., PCR amplicons from cDNA clones or expressed sequence tags (ESTs) *(1, 5)*. Probes on oligonucleotide

Martin Dufva (ed.), *DNA Microarrays for Biomedical Research: Methods and Protocols, vol. 529*

DOI 10.1007/978-1-59745-538-1_9 Springerprotocols.com

microarrays are systematically prepared based on sequence information and, either synthesized directly on the surface *(6)* or arrayed as individual oligonucleotides using a spotting robot *(7)*. Agilent microarrays contain 60 nucleotide (nt.) long oligo probes synthesized by standard phosphoramidite chemistry. The probes are synthesized base-by-base from digital sequence files using a non-contact industrial inkjet printing process capable of delivering picoliters of each nucleotide. Probes 60 nt. long have previously been found to result in better specificity and sensitivity when compared to shorter oligos *(8)*.

1.1. Hybridization

In the hybridization reaction, two differently labeled target preparations are combined and simultaneously hybridized (co-hybridized) to the arrayed probes. The targets will bind to the arrayed probes in amounts reflecting their concentrations and hence abundance in each of the two analyzed samples. The binding is dependent upon the salt concentration in the hybridization solution and the hybridization temperature. The optimal probe–target hybridization temperature is sequence-dependant and will vary between the arrayed probes.

If the hybridization reaction was to rely exclusively on passive diffusion to allow for each target to come into contact with each probe at the correct orientation, the hybridization time would be measured in years rather than hours *(9)*. Alternatively the concentration of target solution should be increased several fold. To overcome these limitations several mixing strategies has been developed. These strategies include cavitation micro streaming *(10)*, magnetic bar stirring *(11)*, air-driven bladders *(12)*, centrifugal mixing *(13)*, and shear driven mixing *(14)*. The Agilent hybridization chamber employs centrifugal mixing. The Agilent oligo microarray slide is combined with a gasket slide partially filled with hybridization solution. This enables the formation of a large bubble which will mix the hybridization solution when the hybridization chamber is mounted on a rotational device throughout the hybridization reaction.

1.2. Slide Processing

Targets that do not completely match the probe sequence may bind to the probes along with perfect match targets during the hybridization reaction. This will give rise to undesired unspecific hybridization signals. To reduce this effect, the microarrays are subjected to a stringency wash *(8, 15, 16)* that strips off near perfect targets and leaves behind only the perfect probe–target hybrids.

Signals from fluorescently labeled microarrays are detected by exciting the fluorescent molecules conjugated to the hybridized targets and collecting the emitted light for each fluorophore *(1)*. Following excitation, the amount of emitted light is recorded and quantified using a microarray scanner relying on either a Charge

Coupled Device (CCD) or a Photo Multiplier Tube (PMT) for detection. For both types of scanners, the result is stored in a file (usually a 16 bit grey-scale TIFF file). In this format each pixel in the file is assigned a "whiteness" value between 0 and 65,535 ($2^{16} - 1$) reflecting the amount of light collected. If too much of the emitted light is collected, the signal will be oversaturated. Identification of the optimal exposure time for each slide is an empirical process that may require multiple scans. It should be noted that the fluorescent molecules will photo-bleach in a dye-dependent rate.

Subsequently, the ratio between signal intensities of the two fluorescent signals are computed for each probe *(7)*. This ratio can be used as an indication of which genes are regulated up or down relative to a control preparation. This allows for the comparison of two different cell cultures or culturing conditions on a transcription wide scale in a single experiment *(17–19)* or e.g., enables the study of cancer progression in patient material *(20, 21)*. However, critical to correct interpretation of a gene expression experiment is that the starting material represents the sample from which it was originally taken and the transcript abundance is not influenced during the experimental procedure and signal collection.

It should be noted that, much cellular regulation is performed not on the transcriptional level but on the translational level. Furthermore, modifications to the synthesized proteins i.e., glycosylation or (de)phosphorylation are performed in response to certain stimuli. These types of regulation are not reflected in the DNA microarray data. Much of this information could be captured using protein arrays, but this technology is still in its infancy and large-scale protein arrays may not be commercially available for several years *(22)*.

2. Materials

2.1. Hybridization

1. 60 μL quartz micro-cuvette (Hellma, Müllheim, Germany).
2. Agilent oligo microarrays, such as product number G4110A or alternative.
3. Agilent Hybridization Chamber, product number G2534A.
4. Agilent In Situ Hybridization kit-plus, product number 5184-3568.
5. 10X Control target preparation. Included in the In Situ Hybridization kit-plus. Store at −20°C in dark for up to 2 months following re-solubilization.
6. Agilent Gasket Slides, such as product number G2534-60003 or alternative.

7. 20X Saline Sodium Citrate (SSC; AppliChem, Darmstadt, Germany). Store at room temperature (RT)
8. Diethylpyrocarbonate (DEPC)-treated water (RNase-free water, Sigma).
9. cDNA controls and samples both labeled with Cy-3 or Cy-5 (*see* **Chapter 8**).

2.2. Slide Processing

1. Three staining dishes.
2. Slide rack fitting into the staining dishes.
3. DEPC-treated water.
4. Wash solution 1 (6X SSC, 0.005% Triton X-102). Store at RT.
5. Wash solution 2 (0.1X SSC, 0.005% Triton X-102). Store at 4°C.
6. Styrofoam container.

3. Methods

Prior to hybridization, the amount of labeled cDNA product should be normalized. This reduces the amount of over saturation of the microarray in the highly expressed genes and prevents masking of signals from weakly expressed genes. Over saturation and signal masking will lead to inconclusive results for the affected genes. The amount of cDNA required for optimal discrimination may vary between laboratories due to difference in equipment and individual equipment adjustment. Initial empirical investigations may be required. Rather than normalizing based on the amount of cDNA it is more advantageous to normalize based on the amount of each dye added to the hybridization reaction. This is suggested only when the ratio between dye and cDNA (calculated in **Section 3.1** step 1 of this chapter) is comparable for both dyes. As a rule of thumb 200 pmol of each dye and a ratio of less than 50 nucleotides per dye molecule are sufficient for most systems.

3.1. Hybridization

In the following section the exposure of the cyanine-labeled cDNA targets to light should be minimized or at best eliminated to prevent photo-decomposition of especially the Cy-5 dye. This will introduce an irreproducible dye bias in the resulting data. Wrap tube(s) in aluminum foil when possible and work fast.

The Agilent oligo microarray slides are fitted with a unique numerical barcode. Note this barcode in the laboratory notebook and use this number to identify the slides, rather than writing on the slides. Some ink markers may influence the scanning process resulting in elevated background signals. If marking is necessary, use a conventional pencil.

The method with volumes provided below is suitable for Agilent microarrays covering the majority of the slide surface, such as the 1 × 44 k whole human genome array. Other (Agilent) microarrays may require additional steps not mentioned in the current protocol. Refer to the manual accompanying the purchased microarrays for specific details. The protocol provided below may also be useful for other microarrays including custom made microarrays.

Much work may be easily lost if mistakes are made. Read through the entire section and practice with water on dummy-slides before the actual experiment. Make sure to familiarize yourself with the individual steps in the protocol.

1. Prepare 200 pmol of each of the Cy-3 and Cy-5 labeled target cDNA in advance. The concentrations can be estimated by measuring in a spectrophotometer using a 60 μL micro-cuvette.
2. Use a 60 μL quartz micro-cuvette. Wash the cuvette with MilliQ water and blow dry with compressed air.
3. Pipette the entire undiluted sample into cuvette and place the cuvette in spectrophotometer.
4. For each sample measure absorbance at 260, 310 nm (background) and either 550 nm for Cy-3 or 650 nm for Cy-5, as appropriate. For each sample calculate the total amount in picomoles of dye incorporation (Cy-3 or Cy-5 as appropriate) using the formulas given in **Note 1** (*See* **Note 1**).
5. Pipette sample from cuvette back into the original sample tube.
6. Wash the cuvette with MilliQ water and blow dry with compressed air. Repeat step 4–6 for each sample.
7. Re-suspend the combined Cy-3- and Cy-5-labeled cDNA targets to a total volume of 200 μL with DEPC-treated water (*see* **Note 2 and 3**). This is most conveniently performed in 250 μL microcentrifuge tubes.
8. Heat-denature the re-suspended cDNA for 3 min at 98°C and cool to room temperature (RT). This can conveniently be performed in a thermal cycler. When programming, set the cycler to hold at RT (22°C) following the denaturation step.
9. Prepare the 10X control target (*see* **Note 4**) by adding 500 μL DEPC water to the lyophilized control target pellet. Dissolve by gentle vortexing and spin down in a microcentrifuge. Wrap tube in aluminum foil. Store at

−20°C for up to 2 months. After thawing, repeat the vortexing and centrifugation procedures before use.

10. Prepare a cDNA target hybridization solution by adding the following in the order listed to a 1.5 mL eppendorf tube:

    Combined cDNA targets 200 μL
    10X control targets (*see* **Note 4 and 5**) 50 μL
    2X hybridization buffer (*see* **Note 5**) 250 μL

11. Mix well by careful pipetting. Do NOT vortex. Take care to avoid introducing bubbles. The hybridization buffer contains detergent. If bubbles accidentally are introduced, remove by aspiration of air from the individual bubble with a pipette containing a fresh pipette tip. This may be a tedious and laborious process. So take care to avoid bubbles.
12. Spin briefly in a bench-top microcentrifuge to collect the sample at the bottom of the tube. Proceed immediately. Do not store.
13. Remove the gasket slide (NOT an oligo microarray slide) from its protective packaging using either tweezers supplied with the hybridization chamber or cut open the packaging with a scalpel.
14. Place the gasket slide in the hybridization chamber base with the "Agilent" label facing UP and aligned in the rectangular end-section of the base. Ensure that the gasket slide is flush with the chamber base and is not ajar. Confer with the manual accompanying the slides in question for correct orientation.
15. Slowly aspirate 490 μL of the hybridization solution for each microarray without introducing bubbles. SLOWLY dispense the solution onto the gasket slide starting from the top part of the slide and continuously dispensing the solution while slowly moving the pipette towards the bottom part of the gasket slide. This "drag and dispense" method ensures the hybridization solution will not spill when the oligo microarray is placed on top. Do not press the pipette piston completely down. This will introduce bubbles.
16. If bubbles accidentally are introduced, they can be removed with a pipette containing a fresh pipette tip.
17. Do NOT move the chamber base or the gasket slide once the hybridization solution has been dispensed. If the solution is spilled over the gasket, a tight seal might not be possible to obtain and the slide may dry during hybridization.

18. Remove the appropriate Agilent oligo microarray slide from its packaging. Handle the slide only by the ends of the slide to avoid damaging the microarray surface.
19. Turn the oligo microarray slide so that the numeric barcode slide is facing UP as it is lowered onto the gasket slide. Lower carefully and align with the gasket slide. Refer to the manual accompanying the slides in question for correct orientation. The arrayed probes must face down towards the gasket slide.
20. Gently place the oligo microarray slide against the gasket slide to complete the sandwiched slide pair. Quickly assess that the slides are completely aligned and that the oligo microarray is not ajar (ends/sides can get caught on the upper part of the chamber base). Re-align quickly if necessary.
21. Do NOT move the chamber base or the sandwiched slide as this can cause leakage of the hybridization mixture.
22. Correctly place the chamber cover onto the sandwiched slides and then slide on the clamp assembly until it comes to a stopping point in the middle of the chamber base.
23. Tighten the screw by turning it clockwise until it is fully hand tight. The slides will not be harmed by hand-tightening. Do NOT use tools to tighten the assembly.
24. Hold the assembled chamber vertically and slowly rotate it 2–3 times to allow the hybridization solution to wet the entire interior surface. Inspect for leakage. If leakage is observed the slides has not been ajar and the hybridization experiment will most likely fail.
25. Note the bubble formation. A large mixing bubble should have formed. Small stray bubbles may stick to the gasket wall. They may not dislodge during rotation of the assembly. They may be dislodged by GENTLY tapping one corner of the assembled chamber on a hard surface and rotating it vertically again. Continue this gentle tapping followed by rotation until all stationary stray bubbles are moving. It is critical that the stray or stationary bubbles be dislodged before loading the chamber into the hybridization rotator as they will infer with hybridization to the probes below the bubble.
26. Mount the assembled hybridization chambers firmly onto a rotation device and place it in a hybridization oven pre-heated to 60°C. The rotator should rotate at around four rotations per minute.
27. Ensure that the hybridization oven interior is not exposed to light. If required, cover the transparent oven door with aluminum foil using adhesive tape.
28. Hybridize at 60°C for 17 h with rotation.

29. During hybridization prepare two wash solutions in 1 L nuclease-free blue-cap bottles or similar.
30. Wash solution 1 (total volume 1,000 mL; *see* **Note 6**):

    DEPC-treated water 700 mL
    20X SSC 300 mL
    10% Triton X-102 0.5 mL (*see* **Note 5**)
31. Wash solution 2 (total volume 1000 mL; *see* **Note 6**):

    DEPC-treated water 995 mL
    20X SSC 5.0 mL
    10% Triton X-102 0.5 mL (*see* **Note 5**)
32. Pass the solutions through a 0.2 μm sterile filtration unit. Seal the bottle and mix the solution thoroughly by shaking. Store at wash solution 1 at RT and wash solution 2 at 4°C.

3.2. Slide Processing

1. Prior to removal of the slides from the hybridization oven, prepare three staining dishes. Be sure that all three dishes are prepared PRIOR to removal of the hybridization chamber from the oven. The washing steps should be performed as efficiently as possible to maximize reproducibility.
2. The first staining dish is used to facilitate disassembly of the oligo microarray and gasket slide sandwich. Add 250 mL room temperature wash solution 1 (Step 30 in **Section 3.1**)
3. To the second staining dish, add a slide rack and a magnetic stir bar. Cover the rack with room temperature wash solution 1. Place the dish on a magnetic stir plate. Form a cover in aluminum foil that will enclose the container preventing exposure to light.
4. Place a third staining dish into a container filled with ice. A styrofoam container is well-suited to this purpose. Add a magnetic stir bar. Add 4°C wash solution 2 (Step 31 in **Section 3.1**) to a depth sufficient to cover a slide rack. Be sure to replenish the ice in the styrofoam container if required. This will keep the solution as cold as possible. Retain the lid of the styrofoam container and use it to cover the slide(s) during washing.
5. Place the plastic tweezers accompanying the hybridization chamber right beside the beaker containing wash solution 1.
6. Steps 6–13 should be performed as quickly as possible to minimize microarray cool-down. Remove ONLY a single hybridization chamber from the hybridization oven at a

time to avoid chamber cool-down before disassembly. Cool-down will increase unspecific hybridization and generate irreproducible results and elevated background signals.

7. Quickly inspect the chamber and determine if leakage has occurred. If the chamber has leaked the results may be irreproducible and the experiment may be worthless.
8. Inspect the chamber for stationary bubbles by vertical rotation. If stationary bubbles are present, note the approximate location on the array. During data analysis any spots that have been in contact with a stationary bubble should be flagged and not used to base any scientific conclusion(s) upon.
9. Place the hybridization chamber assembly on the bench top and loosen the thumbscrew by turning it counter-clockwise.
10. Slide off the clamp assembly and remove the chamber cover.
11. Manually remove the sandwiched slides from the chamber base by grabbing the slides from their ends. Keep the oligo microarray slide (numeric barcode facing up) as the sandwiched slides and quickly transfer to the first beaker containing wash solution 1.
12. Without letting go of the sandwiched slides, completely submerge the slides in the staining dish containing wash solution 1.
13. With the sandwiched slides completely submerged in the wash solution, pry the slides apart from the **barcode end only**. Do this by inserting one of the blunt ends of the tweezers between the slides and then SLOWLY rotating the tweezers to separate the slides.
14. Let the gasket slide drop to the bottom. Used gasket slides can be used in combination with used oligo microarray to practice the assembly of the hybridization chamber or for demonstration purposes. They should not be re-used for actual experiments.
15. Remove the oligo microarray slide and quickly place into the slide rack contained in the second staining dish containing wash solution 1 at RT. Minimize exposure to air i.e., do not carry the slide across the lab, but work in close vicinity of the staining dishes. Touch ONLY the barcoded portion of the slide or the edges.
16. Once the first chamber has been disassembled and the oligo microarray slide is placed in the second staining dish in wash solution 1 at room temperature, retrieve the next hybridization chamber (if relevant) from the hybridization oven and repeat steps 6–15. Continue this process for disassembling the remainder of the hybridization chamber assemblies.

17. After all oligo microarray slides have been collected in the slide rack contained in the second staining dish filled with wash solution 1, set the magnetic stir plate to medium speed. The speed should be sufficient to allow for thorough washing of all the slides and gentle enough not to generate a huge vortex in the center of the staining dish.
18. Cover the staining dish with aluminum foil and wash the slides for 10 min at this speed.
19. Stop the magnetic stir plate and remove the slide rack. Allow it to drip for a few seconds and transfer it to the third staining dish in the styrofoam container containing wash solution 2 at 4°C. Set the magnetic stir plate to the same speed as in steps 17–18 and place a lid on top of the container.
20. Wash the slides for 5 min.
21. Remove one slide at a time and dry the slides with a gentle flow of N_2 (NOT O_2).
22. Place the dried slides in a light proof container.
23. The slides are ready to be scanned in a microarray scanner or can be stored in the dark under nitrogen at RT until ready to scan. To maximize reproducibility, scan the slides as soon as possible. Do not store for extended period of time prior to scanning.
24. After scanning, store slides in polypropylene slide boxes without cork or foam inserts, in a vacuum desiccator or a nitrogen purge box, in the dark. It is recommended to store the slides under nitrogen if possible, as the dyes quench during vacuum storage due to desiccation. For short periods of time, slides may be stored in the supplied slide-box wrapped in aluminum foil.

4. Notes

1. 1 OD_{260} = 37 ng/μL for cDNA. The average molecular weight of a dNTP is 324.5 pg/pmol. To calculate the amount of Cy-3 and Cy-5 in pmol in the prepared target cDNA use the following formulas:

$$\text{pmol Cy}-3=\frac{OD_{550}\bullet \text{Volume }(\mu L)}{0.15}$$

$$\text{pmol Cy}-5=\frac{OD_{650}\bullet \text{Volume }(\mu L)}{0.25}$$

$$\text{nucleotides/dye ratio}=\frac{\text{pmol cDNA}}{\text{pmol Cy dye}}$$

More than 200 pmol of dye incorporation per sample and a ratio of less than 50 nucleotides per dye molecule are optimal for efficient hybridization reactions.

2. The volume of DEPC water varies for different Agilent microarrays. For 11 K microarrays the DEPC water volume is 80 μL. Confer with the manual accompanying the Agilent microarray for correct volume prior to performing the experiment. If the volume of the combined Cy-3 and Cy-5 target cDNA preparations exceeds the total volume, dry in a speedvac before proceeding.
3. If the concentration of Cy dye is too low. Discard the cDNA and repeat the reverse transcription and labeling procedure (**Chapter 8** in this book).
4. The 10X control targets can be substituted with DEPC water. This is however not recommendable as the control spots on the slide allow for correct orientation of the slide and hence correct assignment of significant genes.
5. Provided with the In situ Hybridization Kit product number 5184-3568.
6. Prepare the solution by adding the components in the order listed to avoid precipitation.

5. Safety

Observe standard laboratory safety procedures when working in the laboratory. The edges of the slides may be sharp. Please handle carefully. Pipette tips may be sharp. Please handle carefully and dispose of appropriately. Wear appropriate protective equipment when working in a laboratory.

The recommended Agilent hybridization buffer contains lithium chloride (LiCl) and lithium lauryl sulfate (LLS).

- **LiCl is toxic.**
- Its target organ is the central nervous system.
- It is a potential teratogen, and may cause harm to breast-fed babies.
- It may also impose a risk of impaired fertility.
- LiCl is harmful by inhalation, by contact with the skin, and if swallowed. Wear suitable protective equipment.
- **LLS** is harmful by inhalation and is irritating to the eyes, the respiratory system, and the skin. Again, please wear suitable protective equipment.

The recommended buffer in the Agilent In situ Hybridization kit-plus contains Triton X-102:

- **Triton X-102** is harmful if swallowed, and may be harmful if inhaled or absorbed through the skin.
- There is a risk of serious damage to your eyes in the event of contact. If either compound contacts your eyes, rinse with copious amounts of water for at least 15 min. Seek medical advice.

For the Material Safety Data Sheets (MSDS) for these compounds, please visit www.agilent.com/chem/msds.

Acknowledgments

Above all I would like to express my sincere gratitude to Dr. Martin Dufva for giving me the opportunity to share my hard earned experiences with the community. I wished for a collection of relevant protocols including tips and tricks when I started out my experiments, but I found none. I hope some may benefit from reading this.

A grant by the Danish Biotechnology Instrument Center (DABIC) project no. 2014-00-0003 financially supported the project. This is gratefully acknowledged.

Last but not least – my family, especially my wife, Lene. Without her ever present support and tolerance for my physical and mental absence, my completion of this project would not have been possible.

References

1. Schena, M., Shalon, D., Davis, R. W., and Brown, P. O. (1995) Quantitative monitoring of gene expression patterns with a complementary DNA microarray. *Science* **270,** 467–70.
2. Dixon, A. K., Richardson, P. J., Lee, K., Carter, N. P., and Freeman, T. C. (1998) Expression profiling of single cells using 3 prime end amplification (TPEA) PCR. *Nucl Acids Res* **26,** 4426–31.
3. Eberwine, J., Yeh, H., Miyashiro, K., Cao, Y., Nair, S., Finnell, R., Zettel, M., and Coleman, P. (1992) Analysis of gene expression in single live neurons. *PNAS* **89,** 3010–14.
4. Southern, E., Mir, K., and Shchepinov, M. (1999) Molecular interactions on microarrays. *Nat Genet* **21,** 5–9.
5. Duggan, D. J., Bittner, M., Chen, Y., Meltzer, P., and Trent, J. M. (1999) Expression profiling using cDNA microarrays. *Nat Genet* **21,** 10–4.
6. Lipshutz, R. J., Fodor, S. P., Gingeras, T. R., and Lockhart, D. J. (1999) High density synthetic oligonucleotide arrays. *Nat Genet* **21,** 20–4.
7. Cheung, V. G., Morley, M., Aguilar, F., Massimi, A., Kucherlapati, R., and Childs, G. (1999) Making and reading microarrays. *Nat Genet* **21,** 15–9.
8. Hughes, T. R., Mao, M., Jones, A. R., Burchard, J., Marton, M. J., Shannon, K. W., Lefkowitz, S. M., Ziman, M., Schelter, J. M., Meyer, M. R., Kobayashi, S., Davis, C., Dai, H., He, Y. D., Stephaniants, S. B., Cavet, G., Walker, W. L., West, A., Coffey, E., Shoemaker, D. D., Stoughton, R., Blanchard, A. P., Friend, S. H., and Linsley, P. S. (2001) Expression profiling using microarrays fabricated

by an ink-jet oligonucleotide synthesizer. *Nat Biotechnol* **19,** 342–47.
9. Vainrub, A., and Pettitt, B. M. (2003) Surface electrostatic effects in oligonucleotide microarrays: Control and optimization of binding thermodynamics. *Biopolymers* **68,** 265–70.
10. Liu, R. H., Lenigk, R., Druyor-Sanchez, R. L., Yang, J., and Grodzinski, P. (2003) Hybridization enhancement using cavitation microstreaming. *Analytical Chemistry* **75,** 1911–17.
11. Yuen, P. K., Li, G. S., Bao, Y. J., and Muller, U. R. (2003) Microfluidic devices for fluidic circulation and mixing improve hybridization signal intensity on DNA arrays. *Lab on A Chip* **3,** 46–50.
12. Adey, N. B., Lei, M., Howard, M. T., Jensen, J. D., Mayo, D. A., Butel, D. L., Coffin, S. C., Moyer, T. C., Slade, D. E., Spute, M. K., Hancock, A. M., Eisenhoffer, G. T., Dalley, B. K., and McNeely, M. R. (2002) Gains in sensitivity with a device that mixes microarray hybridization solution in a 25-microm-thick chamber. *Anal Chem* **74,** 6413–7.
13. Bynum, M. A., and Gordon, G. B. (2004) Hybridization enhancement using microfluidic planetary centrifugal mixing. *Anal Chem* **76,** 7039–44.
14. Pappaert, K., Vanderhoeven, J., Van Hummelen, P., Dutta, B., Clicq, D., Baron, G. V., and Desmet, G. (2003) Enhancement of DNA micro-array analysis using a shear-driven micro-channel flow system. *J Chromatogr A* **1014,** 1–9.
15. Schena, M., Shalon, D., Heller, R., Chai, A., Brown, P. O., and Davis, R. W. (1996) Parallel human genome analysis: microarray-based expression monitoring of 1000 genes. *Proc Natl Acad Sci U S A* **93,** 10614–9.
16. Dufva, M., Petronis, S., Jensen, L. B., Krag, C., and Christensen, C. B. (2004) Characterization of an inexpensive, nontoxic, and highly sensitive microarray substrate. *Biotechniques* **37,** 286–92, 94, 96.
17. Zhang, L., Zhou, W., Velculescu, V. E., Kern, S. E., Hruban, R. H., Hamilton, S. R., Vogelstein, B., and Kinzler, K. W. (1997) Gene expression profiles in normal and cancer cells. *Science* **276,** 1268–72.
18. Stangegaard, M., Petronis, S., Jorgensen, A. M., Christensen, C. B., and Dufva, M. (2006) A biocompatible micro cell culture chamber (μCCC) for the culturing and online monitoring of eukaryote cells. *Lab Chip* **6,** 1045–51.
19. Stangegaard, M., Wang, Z., Kutter, J. P., Dufva, M., and Wolff, A. (2006) Whole genome expression profiling using DNA microarray for determining biocompatibility of polymeric surfaces. *Mol Biosyst* **2,** 421–8.
20. Alizadeh, A. A., Eisen, M. B., Davis, R. E., Ma, C., Lossos, I. S., Rosenwald, A., Boldrick, J. C., Sabet, H., Tran, T., Yu, X., Powell, J. I., Yang, L., Marti, G. E., Moore, T., Hudson, J., Jr., Lu, L., Lewis, D. B., Tibshirani, R., Sherlock, G., Chan, W. C., Greiner, T. C., Weisenburger, D. D., Armitage, J. O., Warnke, R., Levy, R., Wilson, W., Grever, M. R., Byrd, J. C., Botstein, D., Brown, P. O., and Staudt, L. M. (2000) Distinct types of diffuse large B-cell lymphoma identified by gene expression profiling. *Nature* **403,** 503–11.
21. Ma, X. J., Salunga, R., Tuggle, J. T., Gaudet, J., Enright, E., McQuary, P., Payette, T., Pistone, M., Stecker, K., Zhang, B. M., Zhou, Y. X., Varnholt, H., Smith, B., Gadd, M., Chatfield, E., Kessler, J., Baer, T. M., Erlander, M. G., and Sgroi, D. C. (2003) Gene expression profiles of human breast cancer progression. *Proc Natl Acad Sci U S A* **100,** 5974–9.
22. Dufva, M., and Christensen, C. B. (2005) Diagnostic and analytical applications of protein microarrays. *Expert Rev Proteomics* **2,** 41–8.

Chapter 10

Target Preparation for Genotyping Specific Genes or Gene Segments

Jesper Petersen, Lena Poulsen and Martin Dufva

Abstract

Generation of single stranded target is of high importance for hybridization reactions on oligonucleotide microarrays. Several methods have been established for production of single stranded DNA and in vitro transcribed RNA. Here we describe three robust methods for target amplification from purified genomic DNA or pre-amplified DNA. The protocols include incorporation of biotin labels in the target molecules and allow for biotin/streptavidin chemistry to be utilized for flexibility in choice of visualization strategy.

Key words: PCR, microarray, DNA, RNA, polymerase, transcription, target preparation, fragmentation, labelling.

1. Introduction

Preparation of target for hybridization on DNA microarrays can be performed in numerous ways, many of which rely on exponential amplification of minute samples by PCR. The main disadvantage of PCR is that the resulting DNA is double stranded (dsDNA) and will display a high degree of self-competitiveness during hybridization. Therefore, this chapter will focus on target preparation methods resulting in single stranded target. dsDNA can function as target in DNA microarray analysis *(1)*, however it usually yields weak signals in hybridization reactions *(2)*. Several techniques can turn dsDNA into single stranded DNA (ssDNA). One of the simplest methods is to perform asymmetric PCR where the exponential amplification is stopped after a few thermal cycles. This is done by having a limited quantity of one of the amplification primers (sense or anti-sense) in each primer pair or entirely

Martin Dufva (ed.), *DNA Microarrays for Biomedical Research: Methods and Protocols, vol. 529*

DOI 10.1007/978-1-59745-538-1_10 Springerprotocols.com

leaving out one primer. Genomic DNA as a template can result in relatively low yields of target for some microarray applications. However, asymmetric PCR tends to function reasonably well when performed on pre-amplified samples where one of the primers is omitted in the second round of amplification. The major limitation of an elongation reaction is that exponential amplification is not possible and the number of thermal cycles, which in turn is limited by exhaustion of the enzyme, might become a restrictive factor. Another drawback is the limited output when using multiple primers in a single reaction.

A potentially better alternative to asymmetric PCR is strand specific degradation; the use of phosphorothioate (PTO)-modified PCR products *(2–4)*. In this assay one of the primers is protected against enzymatic degradation by four PTO modifications at the 5′-end. These are readily available in primers from most oligonucleotide vendors. After standard PCR has been performed, the non-modified strand can by degraded by incubating with an enzyme with 5′–3′ exonuclease activity (e.g. T7 Gene 6 exonuclease) *(4)*.

An alternative way of generating large amounts of single stranded target is by performing linear RNA transcription. This chapter describes the use of a major bacteriophage promoter sequence specifically recognized by the T7 RNA Polymerase *(5)*, but other polymerases and promoters are readily available (e.g. T3 and SP6). The protocol has two main steps. In the first step, the bacteriophage promoter sequence is incorporated into PCR products by PCR amplification of the fragment(s) of interest using either the sense or anti-sense primer with the promoter sequence at the 5′-end. Depending on the type of probe selected for the array (i.e. sense or anti-sense) the promoter sequence should simply be included in the opposite primer, i.e. with sense probes the anti-sense primer is modified. The second step is T7 in vitro transcription (T7-IVT), where the obtained PCR products serve as a template for the polymerase. The polymerase specifically recognizes its promoter sequence in double stranded DNA and makes thousands of single stranded RNA copies *(6)*.

When target is prepared by PCR there are generally three ways to perform labelling: (i) using a pre-modified primer (e.g. a 5′-end attached fluorochrome or biotin); (ii) random incorporation of labelled nucleotides during amplification, e.g. by substituting a fraction of dCTP with biotin-dCTP; and (iii) using random incorporation of a desired label post-PCR (e.g. by performing RNA transcription or using a commercial random labelling kit). When choosing a labelling strategy it is important to consider the potential impact on the final assay. A pre-modified primer with only a single label might be sufficient if short targets in high concentrations are used, while for longer or lower concentrations of target the assay can be improved by incorporating several labels per target (personal observations). It has been reported that too many labels can cause an assay to malfunction. Whether this is caused by steric hindrance during amplification or hybridization, or any other mechanism is not known *(7,8)*.

2. Materials

The protocols described are based on a simple setup. A single 300 base pair DNA fragment is amplified by the forward primer (BCF) and the reverse primer (BCR) or either a T7 promoter-modified or 5′-end PTO modified reverse primer (T7-BCR or PTO-BCR, respectively; *see* **Note 1**). Furthermore, the protocols use TEMPase Hot Start DNA polymerase (Bie & Berntsen, Rødovre, Denmark) and T7 RNA Polymerase-Plus™ (Ambion, Austin, TX, USA; *see* **Note 2**) for RNA transcription. It should be noted that most of these methods can easily be adapted to comprise several DNA fragments.

1. For standard and asymmetric PCR a 100 μM stock solution of each primer BCF and BCR (**Table 10.1**) is diluted with milliQ water to 10 μM. The 12.5 mM stock of dNTPs (Bie & Berntsen) is diluted to 1 mM. The TEMPase Hot Start DNA polymerase (5 U/μL; Bie & Berntsen) is used with the supplied 10 × PCR buffer (TEMPase Buffer II). DNA template for the PCR can be either PCR product or genomic DNA.

2. For verification of PCR products by agarose gel electrophoresis, prepare a 2% (w/v) agarose gel in 0.5 × Tris/Borate/EDTA (TBE) buffer containing 0.5 μg ethidium bromide/mL agarose solution. Remember to wear nitril gloves as ethidium bromide is a potent mutagen. Alternatively, separation can be performed on an Agilent Bioanalyzer 2100 (Agilent

Table 10.1
List of primers

Name	Sequence, 5′- to 3′-end
BCF	AGCAGGGAGGGCAGGAGCCA
BCR	AGAGTCAGTGCCTATCAGAAAC
T7-BCR[1]	**GAAATTAATACGACTCACTATAGGGAGA**AGAGTCAGTGCCTATCAGAAAC
PTO-BCR[2]	$A^SG^SA^SG^S$TCAGTGCCTATCAGAAAC

[1] The T7 promoter sequence is indicated in bold. [2] The superscript ‘S’ indicates a phosphorothioate bond.

Technologies) using a DNA 1000 LabChip kit. The Bioanalyzer is also capable of estimating RNA concentrations by an RNA 6000 Nano Kit (Agilent Technologies).

3. For PCR meant for strand degradation, a 100 μM stock solution of each primer BCF and PTO-BCR (**Table 10.1**) is diluted with milliQ water to 10 μM. The 12.5 mM stock of dNTPs (Bie & Berntsen) is diluted to 1 mM. The TEMPase Hot Start DNA polymerase (5 U/μL; Bie & Berntsen) is used with the supplied 10 × PCR buffer (TEMPase Buffer II). DNA template for the PCR can be either PCR product or genomic DNA.
4. Digestion of PCR products where one primer is PTO-modified requires the T7 Gene 6 exonuclease (50 U/μL; GE Healthcare) and the standard 5 × reaction buffer provided with the enzyme.
5. For reactions where random labelling is desired, a 1.0 mM Biotin-14-dCTP should be prepared from a 10 mM stock (Invitrogen).
6. For the T7 IVT reaction, it is possible to directly use PCR products as DNA template, without purification. The reagents include 2.5 mM of each NTP (GE Healthcare, Chalfont St Giles, UK) diluted from a 100 mM stock, 1.0 mM Biotin-14-CTP (Invitrogen) diluted from a 10 mM stock, inorganic pyrophosphatase (0.05 U/μL) prepared according to manufacturers instructions, and the enzyme T7 RNA Polymerase-Plus^{TM} (20 U/μL; Ambion) together with the supplied 10 × T7 transcription buffer.

3. Methods

The PCR protocols presented below are based on random labelling of the single stranded target DNA by adding Biotin-14-dCTP to the amplification mixture. For assays relying on non-labelled target DNA such as minisequencing or allele specific primer extension *(9,10)*, the Biotin-14-dCTP can be omitted and a corresponding volume of MilliQ water added. This is also the case for T7 RNA transcription, where Biotin-14-CTP can be substituted by MilliQ water. Refer to **Note 3**.

3.1. PCR Amplification

1. For a 20 μL PCR, mix in a PCR tube 2 μL 10 × TEMPase Buffer II, 2 μL each of primer BCF (10 μM) and BCR (10 μM), 4 μL dNTPs (each 1 mM), 1 μL TEMPase Hot Start DNA polymerase (5 U/μL), and 0.5 μL template DNA (genomic DNA or PCR product). Finally add 11.5 μL MilliQ water and

mix gently by pipetting. The PCR cycling conditions are an initial hot-start of 15 min at 95°C followed by 35 amplification cycles at 95°C for 30 s, 60°C for 45 s, 72°C for 1 min and a final extension at 72°C for 10 min and stored at 4°C until use.

2. Double stranded amplification products can be confirmed by electrophoresis on a 2–3% agarose gel visualized by ethidium bromide staining. Alternatively PCR products can be separated and quantified on an Agilent Bioanalyzer 2100 using a DNA 1000 LabChip kit.

3.2. Asymmetric PCR

1. For a 20 μL asymmetric PCR; mix in a PCR tube 2 μL 10 × TEMPase Buffer II, 0.1 μL BCF (10 μM) and 2.0 μL BCR (10 μM), 4 μL dNTPs (each 1 mM), 0.25 μL Biotin-14-dCTP (1 mM), 1 μL TEMPase Hot Start DNA polymerase (5 U/μL), and 1.0 μL template DNA (genomic DNA or PCR product). Finally, add 12.65 μL MilliQ water and mix gently by pipetting. The PCR cycling conditions are an initial hot-start of 15 min at 95°C followed by 40 amplification cycles at 95°C for 30 s, 60°C for 45 s, 72°C for 1 min and a final extension at 72°C for 10 min and stored at 4°C until use. Suggestions for improvements can be found in **Note 4**.

3.3. Single Stranded DNA Preparation from PTO Protected DNA Fragments

1. For a 20 μL PCR; mix in a PCR tube 2 μL 10 × TEMPase Buffer II, 2 μL each of primer BCF and PTO-BCR (10 μM), 4 μL dNTPs (each 1 mM), 0.25 μL Biotin-14-dCTP (1 mM), 1 μL TEMPase Hot Start DNA polymerase (5 U/μL), and 0.5 μL template DNA (genomic DNA or PCR product). Finally, add 11.5 μL of MilliQ water and mix gently by pipetting. The PCR cycling conditions are an initial hot-start of 15 min at 95°C followed by 35 amplification cycles at 95°C for 30 s, 60°C for 15 s, 72°C for 1 min and a final extension at 72°C for 10 min and stored at 4°C until use.

2. When strand degradation is used to create ssDNA the amplified PCR product should be prepared by adding 5 μL of 5 × T7 Gene 6 exonuclease buffer (for 20 μL of PTO modified PCR mixture). Afterwards, the dsDNA is processed by adding 0.5 μL of the T7 Gene 6 exonuclease (50 U/μL) and the mixture is incubated at 37°C for 60 min followed by an inactivation step at 85°C for 15 min. The mixture should be stored at 4°C until use. The resulting ssDNA can be directly used in a standard hybridization protocol.

3. Strand degradation can be verified by analyzing a sample taken before the strand degradation step and one taken after. Run them side by side on a 2% agarose gel electrophoresis visualized by ethidium bromide. The visible band from the non-degraded sample should not be present if one strand is completely degraded.

3.4. Target Amplification by RNA Transcription

1. For making a 20 μL PCR ready for T7 RNA transcription; mix in a PCR tube 2 μL 10 × TEMPase Buffer II, 2 μL each of primer BCF (10 μM) and T7-BCR (10 μM), 4 μL dNTPs (each 1 mM), 1 μL TEMPase Hot Start DNA polymerase (5 U/μL), and 0.5 μL template DNA (genomic DNA or PCR product). Finally, add 11.5 μL MilliQ water and mix gently by pipetting. The PCR cycling conditions are an initial hot-start of 15 min at 95°C followed by 35 amplification cycles at 95°C for 30 s, 60°C for 45 s, 72°C for 1 min and a final extension at 72°C for 10 min and stored at 4°C until use.
2. Double stranded amplification products can be confirmed by electrophoresis on a 2–3% agarose gel visualized by ethidium bromide staining. Alternatively PCR products can be separated and quantified on an Agilent Bioanalyzer 2100 using a DNA 1000 LabChip kit. Note that the T7 promoter sequence adds another 28 basepairs to the fragment.
3. The PCR products obtained when using the BCF and T7-modified reverse primer are directly used as template DNA for the T7 IVT without purification. For a final volume of 20 μL the following reagents are mixed in a PCR tube: 2 μL 10 × T7 transcription buffer, 4 μL NTPs (2.5 mM each), 0.25 μL biotin-14-CTP (1.0 mM; *see* **Notes 5 and 6**), 1 μL inorganic pyrophosphatase (0.05 U/μL), 1 μL T7 RNA Polymerase-Plus^{TM} (20 U/μL), 2 μL template DNA. Lastly, 9.75 μL MilliQ water are added and the reaction is mixed by

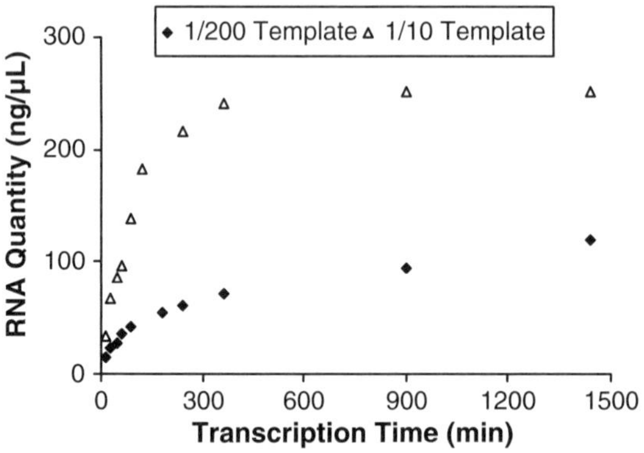

Fig. 10.1. Effect of template concentration and transcription time on aRNA generation. Two different T7 transcription mixtures were set up on ice and incubated in a heating block at 37°C. Samples were removed and immediately frozen. The resulting aRNA was heat denatured and quantified on a 2100 Bioanalyzer using aRNA 6000 Nano Kit (Agilent Technologies). The resulting quantities were adjusted according to a t=0 sample to avoid influence from the DNA template. Each data point represents the average of two individual quantifications. *Diamonds* (♦) represent a 200-fold dilution of the DNA template (0.28 ng/μL) in the final transcription solution and *triangles* (Δ) represent a 10-fold dilution (5.5 ng/μL).

gently pipetting. The T7 reaction mixture is incubated in a heat incubator or thermal cycler at 37°C for 2–16 h (*see* **Notes 7–9** and **Fig. 10.1**). If not used immediately, the amplified RNA target can be stored at –80°C.

4. Notes

1. Oligonucleotide primers are easily ordered from several vendors of life sciences products. In order to facilitate the ordering of oligonucleotides, many vendors have web-interfaces which allow online ordering (e.g. Sigma-Aldrich, Invitrogen, TAG Copenhagen, etc.). In these web interfaces it is possible to enter the sequence of interest, synthesis scale, modifications (e.g. 5'-end Cy3, Cy5, Biotin, etc.) and to select a name to be printed on the tube. After entering these values it is often possible to get some data about the oligonucleotide, including the T_m (which should be matched for all primer pairs used in the same PCR) and indication of the tendency to form secondary structures and primer dimers. Most of these websites also provide online help about how to order at their site. Special guides for ordering multiple oligonucleotides by uploading a spreadsheet can be beneficial.
2. The suggested enzymes can be exchanged by most of the available enzymes of the same kind, e.g. TEMPase by Taq DNA polymerase and T7 RNA Polymerase by T3 RNA Polymerase. Furthermore, the T7 RNA Polymerase-Plus™ that is suggested comes with RNase inhibitors and reduces considerations concerning working RNase free.
3. A few alternatives to random labelling using biotin labelled nucleotides are to use 5′-end labelled primers (in this case, e.g. Biotin-BCR, Cy3-BCR, etc.), post-PCR labelling or no labelling. The latter is generally being used for enzyme-assisted assays, e.g. in mini sequencing. If 5′-end labelled primers are used, these should simply substitute the regular primers in the target preparation protocol. Post-PCR/transcription labelling can most easily be performed by commercially available kits (e.g. for RNA the miRCURY™ LNA Array Power Labelling Kit (Exiqon, Denmark) can be used). *See also* **Note 6**.
4. The amount and complexity of the chosen target DNA can influence the amount of ssDNA resulting from an asymmetric PCR. When performing asymmetric PCR on a genomic DNA sample it is advisable to investigate the effect of raising the concentration of the forward primer in order to gain more resulting ssDNA, e.g. change from 0.1 μL to 0.2 μL BCF (10 μM).

5. The additional step required to visualize target because of the biotin/avidin colouring can be circumvented if interest is purely in a fluorescent readout. In this case the biotin-14-dCTP or biotin-14-CTP can be substituted by a fluorescently marked nucleotide in either DNA or RNA, e.g. Cy3-dCTP or Cy3-CTP, respectively.
6. The fraction of labelled CTP vs. non-labelled CTP should be considered according to the length and cytosine content in the amplified DNA fragment. We have obtained best results by using a theoretical labelling degree corresponding to ~1 label per 50 nucleotides in DNA fragments while 1 in 100 seems enough to produce good results when using RNA. The most investigated fragment was 300 nucleotides long however the above guidelines held true for transcription of 12 fragments ranging from ~170 to 350 nt in a single reaction. Hacia et al. described a similar assay where they simultaneously transcribed 23 fragments *(6)*.
7. If the amount of DNA template is low, the transcription time can be raised, e.g. to an overnight incubation, to increase the resulting RNA concentration. However we have found that the amount of template limits the yield of the IVT reaction. Therefore, loss of signal in hybridization reactions can be caused by too little template from the PCR (*see* **Fig.** 10.1). It should be noted that we seldom observe failures in IVT target preparation/labelling.
8. The inorganic pyrophosphatase can be omitted from the T7 IVT reaction but it has been shown to increase the yield *(11)*. We have likewise observed increases in the RNA yield of roughly 50% and we thus recommend the use of inorganic pyrophosphatase during IVT.
9. RNA forms strong secondary structures which can negatively influence its ability to hybridize to immobilized targets. When long RNA is used as target for hybridization assays it can, therefore, be considered to perform a fragmentation step. To do so, prepare a batch (e.g. 50 mL) of 5 × fragmentation solution (200 mM Tris–acetate, 500 mM potassium acetate, 150 mM magnesium acetate, adjusted to pH 8.4 by acetic acid), and add the proper volume to the transcribed RNA (e.g. 5 μL for 20 μL RNA mixture) and incubate at 75°C for ~10–30 min followed by snap cooling on ice *(12)*. Fragmented samples can be directly used in hybridization assays. It can be advantageous to optimize the fragmentation time by comparing the hybridization signal and/or the fragment size distribution, e.g. on a Bioanalyzer 2100 using an RNA 6000 Nano kit or equivalent. In our experience, fragments ranging from ~50 to 200 nucleotides perform optimally in hybridization assays.

References

1. Keramas, G., Bang, D. D., Lund, M., Madsen, M., Rasmussen, S. E., Bunkenborg, H., Telleman, P., and Christensen, C. B. (2003) Development of a sensitive DNA microarray suitable for rapid detection of Campylobacter spp. Molecular and Cellular Probes 17, 187–196.
2. Erdogan, F., Kirchner, R., Mann, W., Ropers, H. H., and Nuber, U. A. (2001) Detection of mitochondrial single nucleotide polymorphisms using a primer elongation reaction on oligonucleotide microarrays. Nucleic Acids Research 29, E36.
3. Nikiforov, T. T., Rendle, R. B., Goelet, P., Rogers, Y. H., Kotewicz, M. L., Anderson, S., Trainor, G. L., and Knapp, M. R. (1994) Genetic bit analysis – a solid-phase method for typing single nucleotide polymorphisms. Nucleic Acids Research 22, 4167–4175.
4. Nikiforov, T. T., Rendle, R. B., Kotewicz, M. L., and Rogers, Y. H. (1994) The use of phosphorothioate primers and exonuclease hydrolysis for the preparation of single-stranded PCR products and their detection by solid-phase hybridization. PCR Methods and Applications 3, 285–291.
5. Rosa, M. D. (1979) 4 T7 RNA-polymerase promoters contain an identical 23 bp sequence. Cell 16, 815–825.
6. Hacia, J. G., Woski, S. A., Fidanza, J., Edgemon, K., Hunt, N., McGall, G., Fodor, S. P. A., and Collins, F. S. (1998) Enhanced high density oligonucleotide array-based sequence analysis using modified nucleoside triphosphates. Nucleic Acids Research 26, 4975–4982.
7. Shendure, J., Mitra, R. D., Varma, C., and Church, G. M. (2004) Advanced sequencing technologies: methods and goals. Nature Reviews. Genetics 5, 335–344.
8. Tebbutt, S. J., Mercer, G. D., Do, R., Tripp, B. W., Wong, A. W. M., and Ruan, J. (2006) Deoxynucleotides can replace dideoxynucleotides in minisequencing by arrayed primer extension. Biotechniques 40, 331–338.
9. Pastinen, T., Kurg, A., Metspalu, A., Peltonen, L., and Syvanen, A. C. (1997) Minisequencing: a specific tool for DNA analysis and diagnostics on oligonucleotide arrays. Genome Research 7, 606–614.
10. Pastinen, T., Raitio, M., Lindroos, K., Tainola, P., Peltonen, L., and Syvanen, A. C. (2000) A system for specific, high-throughput genotyping by allele-specific primer extension on microarrays. Genome Research 10, 1031–1042.
11. Cunningham, P. R. and Ofengand, J. (1990) Use of inorganic pyrophosphatase to improve the yield of invitro transcription reactions catalyzed by T7 RNA-polymerase. Biotechniques 9, 713–714.
12. Mehlmann, M., Townsend, M. B., Stears, R. L., Kuchta, R. D., and Rowlen, K. L. (2005) Optimization of fragmentation conditions for microarray analysis of viral RNA. Analytical Biochemistry 347, 316–323.

Chapter 11

Genotyping of Mutations in the Beta-Globin Gene Using Allele Specific Hybridization

Lena Poulsen, Jesper Petersen, and Martin Dufva

Abstract

The use of DNA microarrays for genotyping is economically favorable when compared to real-time PCR and DNA sequencing. Here, we demonstrate a DNA microarray-based assay using allele-specific oligonucleotide (ASO) probes for genotyping mutations in the beta-globin gene. The assay makes use of agarose film coated glass slides as substrate, unmodified melting temperature matched ASO probes, target amplification and labeling using T7 in vitro transcription, mixing during hybridization and finally visualization using a fluorescent scanner. In this chapter we will emphasize on probe design, optimization, and validation of an allele-specific hybridization assay.

Key words: DNA microarray, genotyping, mutation detection, beta-globin, probe design, allele-specific hybridization, mixing, fluorescent detection.

1. Introduction

DNA microarray is a powerful platform for high-throughput genotyping *(1,2)*. Commercial arrays from Affymetrix and Illumina allow parallel genotyping of about one million different single nucleotide polymorphisms (SNPs). The high parallelism of DNA microarray-based assays also offer rapid and cost efficient mutation analysis in genes where ten to several hundred disease causing mutations can be assayed at the same time. The most commonly used methods for genotyping using the DNA microarray platform include allele-specific hybridization (ASH) and enzyme assisted reactions such as mini-sequencing and allele-specific primer extension *(3)*. By use of these technologies, DNA microarrays have been developed to detect mutations in

Martin Dufva (ed.), *DNA Microarrays for Biomedical Research: Methods and Protocols, vol. 529*

DOI 10.1007/978-1-59745-538-1_11 Springerprotocols.com

the beta-globin gene *(4–7)*. Mutations in the beta-globin gene lead to impaired production of the beta-globin chain in hemoglobin, and are the cause of the genetic disease beta-thalassemia. Beta-thalassemia is part of the disorder thalassemia, which is hereditary anemia caused by defect in hemoglobin production, and is the most common genetic disorder world-wide. There are a few hundred well described mutations in the beta-globin genes that give rise to thalassemia of various severities. Most mutations are point mutations and small (few bases) insertions or deletions *(8)*.

For the purpose of developing a "beta-thalassemia" array and for making microarrays generally more assessable our research group has made efforts to provide inexpensive solutions for microarray genotyping assays. These include fabrication of agarose coated slides *(9)* and TC tagged ASH probes for genotyping *(10)*. We use ASH for genotyping monogenetic diseases, including beta-thalassemia, because the method is simple to perform, inexpensive, and scalable. Genotype using ASH relies on the differences in thermodynamic stability of a perfect match duplex and a mismatch duplex. Hybridized target will melt off (dissociate) at a lower temperature in a mismatch duplex than in a perfect match duplex.

It is well established, that a mismatch in the center of a ASH probe has more destabilizing effect on probe–target duplex than mismatched in the extremities *(11)*. Hence, to obtain discriminating probes the variant sequence should be placed as central in the probe sequence as possible. When designing probes for an ASH assay, a commonly used approach is to use melting temperature (Tm) matched probes. For this purpose thermodynamic calculations based on hybridization in solution are used. However, attachment of probes to surfaces dramatically changes the melting temperature *(12)*. We have seen weak correlation between calculated Tm and the temperature that was optimal for genotyping (unpublished results), which is in agreement with previous observations *(13)*. Optimization of an ASH assay for genotyping is therefore mainly a trial and error process. Here, we will present allele specific oligonucleotide hybridization as a method to detect mutations in the beta-globin gene. We describe the steps used in development, optimization and validation of a genotyping assay based on allele specific hybridization to microarrays.

2. Materials

2.1. DNA Probes

1. DNA oligonucleotide probes are available from different vendors (**Chapter 10**).

2.2. Fabrication of Slides

1. Mix 150 mL milliQ water with 1.5 g agarose (Invitrogen, Taastrup, Denmark), Boil in a microwave oven until the agarose is completely dissolved. Finally add 0.32 g sodium (meta) periodate $INaO_4$(Sigma-Aldrich, Brøndby, Denmark) and mix. The glass slides used are 26 × 76 mm cut edges SuperFrost® slides or cut edges microscope slides without frosted area (VWR International, Rødovre, Denmark).

2.3. Spotting the Slides

1. DNA Probes are listed in **Table 11.1**. Pipette 15 μL of 100 μM probe solution in the respective wells in a 384-well microtiter plate. Spotting is performed using a Genetix Q-array spotting

Table 11.1
DNA probes sequences and calculated Tm

Probe name	Sequence	(T_m)
CD5 mt	TTTTTTTTTTCCCCCCCCCCGCACCTGACTCGAGGAGAAGT	54.9
CD5 wt	TTTTTTTTTTCCCCCCCCCCGCACCTGACTCCTGAGGAGAA	54.7
CD8 mt	TTTTTTTTTTCCCCCCCCCCCCTGAGGAGGTCTGCCG	54.0
CD8 wt	TTTTTTTTTTCCCCCCCCCCCCTGAGGAGAAGTCTGCCG	53.7
CD8–9 mt	TTTTTTTTTTCCCCCCCCCCGAGGAGAAGGTCTGCCGTTAC	53.5
CD8–9 wt	TTTTTTTTTTCCCCCCCCCCGAGGAGAAGTCTGCCGTTACTG	53.4
CD15 mt	TTTTTTTTTTCCCCCCCCCCACTGCCCTGTAGGGCAAGGT	57.4
CD15 wt	TTTTTTTTTTCCCCCCCCCCTGCCCTGTGGGGCAAGG	57.6
CD17 mt	TTTTTTTTTTCCCCCCCCCCCCCTGTGGGGCTAGGTGA	55.4
CD17 wt	TTTTTTTTTTCCCCCCCCCCCCCTGTGGGGCAAGGTG	55.4
CD24 mt	TTTTTTTTTTCCCCCCCCCCAAGTTGGAGGTGAGGCCCT	54.8
CD24 wt	TTTTTTTTTTCCCCCCCCCCGAAGTTGGTGGTGAGGCCC	54.9
CD27–28 mt	TTTTTTTTTTCCCCCCCCCCGTGAGGCCCCTGGGC	55.7
CD27–28 wt	TTTTTTTTTTCCCCCCCCCCGTGAGGCCCTGGGCAG	55.3
IVS I +5 mt	TTTTTTTTTTCCCCCCCCCCGGCAGGTTGCTATCAAGGTTACA	53.5
IVS I +5 wt	TTTTTTTTTTCCCCCCCCCCGGCAGGTTGGTATCAAGGTTACA	53.3
IVS I +6 mt	TTTTTTTTTTCCCCCCCCCCGGCAGGTTGGCATCAAGG	53.1
IVS I +6 wt	TTTTTTTTTTCCCCCCCCCCGGCAGGTTGGTATCAAGGTTACA	53.3

robot (Genetix, New Milton Hampshire, UK) with Chipmaker CMP3B pins (Telechem, Sunnyvale, CA, USA).

2. 0.1 × sodium saline citrate (SSC) is prepared from stock solution of 20 × SSC (Promega, Madison, WI, USA) diluted in MilliQ water. Sodium dodecyl sulfate (SDS; Sigma-Aldrich) is added to a final concentration of 0.5% (w/v). A 5 L 0.1 × SSC+0.5%SDS solution is made by adding 25 g SDS and 25 mL 20 × SSC. Fill up with milliQ water and mix thoroughly. A 0.1 × SSC solution is prepared in the same manner, however omitting SDS.

2.4. Target Preparation

1. See **Chapter 10** for reagents for PCR, T7 in vitro transcription (T7-IVT), and gel electrophoresis.

2.5. Hybridization Station

1. The hybridization station (**Fig. 11.1**) we use is a hybrid between a commercial gasket slide (Agilent Technologies, Palo Alto, CA, USA) and in-house produced support for gasket slide, and lid to assemble the structure. The support and lid are produced by micromachining polymethylmethacrylate (PMMA) plates, 3 mm thickness, using a laser ablation system (Synrad Inc., Mukilteo, WA, USA) controlled by CAD software (winMark Pro, Synrad). The support has the dimensions 50 × 90 mm and in the center has a 1 mm deep alignment void of 26 × 76 mm to fit the 25 × 75 mm gasket slide. The six screws and nuts, which are placed in holes in the support, are used for clamping of the lid and support. The lid consists of two parts. (1) An upper part, which is made to fit the screws in the support and (2) a lower part, which has the dimensions 24 × 74 mm and functions to press down on a mounted microarray slide. The two parts of the lid are heat bonded together between two glass bonding blocks (Gustav Sørensen

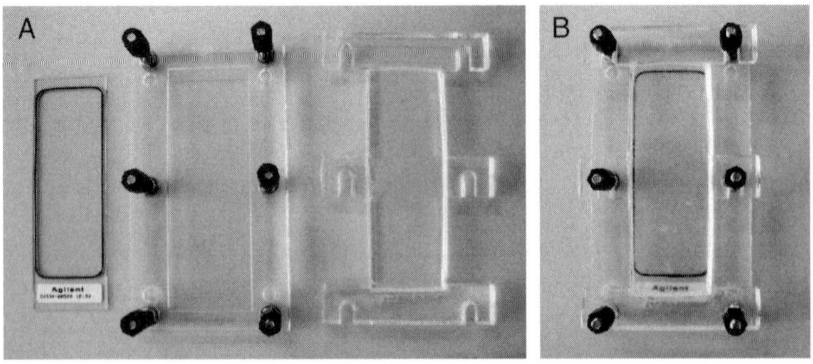

Fig. 11.1. Photographs of the hybridization station. (**A**) The individual parts of the hybridization station; (from left to right) gasket slide, support with alignment void for gasket slide and crews and nuts to connect to the lid. (**B**) An assembled hybridization station with a microarray slide mounted.

& Søn, Rødovre, Denmark) at 100°C for about 90 min with a pressure of 100 cNm obtained with a torque screwdriver (Lindstrom, Orange, CA, USA).

2.6. Hybridization and Stringency Wash Buffers

1. Hybridization buffer is easiest prepared by making a 2 × hybridization buffer by mixing 1% w/v SDS and 10 × SSC. Note, that the salt will precipitate the SDS immediately and stay insoluble at room temperature. Prior to use, put the flask with the 2 × hybridization buffer in the microwave oven and heat *gently* until a clear solution is obtained. To avoid boil over heat and swirl the bottle. If heating a bottle, the lid should not boil over, heat and be tightened firmly, because steam can thereby escape from the bottle.
2. Five liters of the stringency wash buffer 0.1 × SSC+0.1%SDS is prepared by adding 5 g of SDS and 25 mL of 20 × SSC, filling up with milliQ water and mixing thoroughly. Alternative stringency wash buffers are prepared with same concentration of SDS, but varying the added volume of 20 × SSC. For 0.2 × SSC add 50 mL, 0.35 × SSC (87.5 mL), 0.7 × SSC (175 mL), 1 × SSC (250 mL) and 3 × SSC (750 mL). To avoid precipitation of SDS, first mix 20 × SSC and almost all volume of milliQ water. Subsequently add SDS and add water up to 5 L and mix the buffer.
3. Five liters of 2xPBS buffer is prepared by mixing 1 L of 10 × Phosphate Buffered Saline (PBS; Bie & Berntsen A/S, Rødovre, Denmark) with 4 L of milliQ water.

2.7. Scanners and Softwares Used for Hybridization and Stringency Wash Buffers

1. Scanners used are arrayWoRx (Applied Precision, WA, USA) or Packard ScanArray Lite (PerkinElmer Life Sciences, Boston, MA, USA).
2. Microarray image analysis software used are the free program ScanAlyze 2.5 (http://rana.lbl.gov/EisenSoftware.htm) or commercial GenePix Pro 6.0 (Molecular Devices, Union City, CA, USA).

3. Methods

3.1. Probe Design

1. Using the genomic DNA sequence of the beta-globin gene, a probe-pair consisting of a wild-type (Wt) and mutant (Mt) probe is designed for each mutation to be genotyped. We have designed probes on the sense strand, which are complementary to antisense target. For maximum discrimination the variant base/bases is positioned as close to the center of the probe as possible. For calculating the melting temperature (Tm) we use the freely available oligonucleotide analyzer software provided by Integrated DNA technologies at http://www.idtdna.com/analyzer/Applications/OligoAnalyzer/ Settings are chosen to

be an oligonucleotide concentration of 0.1 μM and with a monovalent salt concentration of 36.7 mM. All probes have a TC-tag (5′-TTTTTTTTTTCCCCCCCCCC-3') in the 5′ end *(10)* (**Note 1**). A T_m-matched probe-set, for genotyping nine different beta-globin mutations, is then designed based on a the following criteria; the probes should be at least 15 nucleotides (nt) long *(14)* in order to yield sufficiently high hybridization signal and the probes should have as equal a T_m over the entire set as possible (**Table 11.1**; **Note 2**).

3.2. Fabrication of Agarose Coated Slides

1. Place 100–150 Superfrost glass slides on a clean planar surface. Pour gel onto the slides using a 1 mL pipette. It is simplest if the tip of the pipette is held approximately where the frost starts and a rapid ejection will push the melted hot agarose towards the area of the slides that is not frosted. Normally if done correctly, the solution will form a layer of agarose over the whole slide surface. If frosted slides are not desired, an entire glass slide (76 × 26 mm) can be coated with 1.5 mL gel (**Note 3**). Let the agarose polymerize for 30 min and transfer the slides to a water bath of deionized water or milliQ water. When submerging the slides, tilt the slides a little and let the slide slip very gently into the water. The slides should not be placed on top of each other. Incubate for 3 h.
2. Dry the slides air overnight or for a couple of hours at 37°C. The slides should be covered with a thin film (about 8 μm) of agarose.

3.3. Fabrication of Microarrays by Contact Printing

1. Place the agarose-coated slides in the spotter. Prepare a microtiter plate with probes where the probes are diluted in water to 100 μM. Position the plate in the spotter and spot the slides using 70% humidity. Alternatively, spot at 10°C without humidifier (**Note 4**).
2. After spotting, UV crosslink the agarose slides for 4 min in a Stratalinker 2400 equipped with 254 nm bulbs *(15)* or similar.
3. Remove unbound probes by washing the slides in 0.1 × SSC+0.5% SDS. Place the slides in a Microarray Wash Station (Arrayit, Sunnyvale, CA) and put the washing station in a 400 mL beaker. Add a stirring magnet and 400 mL 0.1 × SSC+0.5% SDS. Stir vigorously using a magnetic stirrer for 10 min. Subsequently wash slides in 0.1 × SSC for 10 min. Finally dry the slides by centrifuge or by gently blowing air or nitrogen over the slide.

3.4. Target Preparation

1. Target preparation comprising PCR and T7-IVT is performed as described in **Chapter 10**, however the reaction volume for T7-IVT is 80 μL, and target is randomly labeled using Cy3-CTP. PCR products are confirmed by gel electrophoresis as in **Chapter 10**.

3.5. Hybridization and Mixing

1. Hybridization is performed by bubble mixing in a hybridization station (**Section 2.5**, **Fig. 11.1** and **Note 5**). Before assembly, the nuts should be positioned at the end of the screws (away from the support). Mix amplified RNA target (80 μL) with 170 μL of MilliQ water and 250 μl of preheated hybridization buffer (**Section 2.6**). The final hybridization solution contains 5 × SSC and 0.5% SDS. Place a gasket slide in the alignment void of the support. Pipette 490 μL of the hybridization mixture along the center of the gasket slide. To avoid leakage, the liquid should not touch the ring of the gasket. Place a slide with microarray facing the gasket gently down on the gasket slide. Next, slide the lid on the screws, just beneath the nuts. The lower part of the lid (**Section 2.5**) should be facing the microarray slide and support. Gently lower slide onto the microarray slide and tighten the nuts as equally as possible to avoid leakage. Vertically turn around the assembled hybridization station to wet the gasket and mount it on a rotator in heat incubator, using rubber bands and/or tape. Set rotation at the minimum setting. After about 10–30 min inspect the hybridization station, and confirm that the liquid is moving freely around, with no stationary bubbles. If stationary bubbles are observed, these can be removed by gently tapping the support of the hybridization station. The hybridization conditions are 37°C, for 2 h (**Note 6**).
2. Following hybridization, the hybridization station is disassembled by unscrewing the nuts, lifting of the lid and gently taking the sandwiched microarray slide and gasket slide out of the support. Having the gasket slide at the bottom, lower the sandwiched slides horizontally in beaker with 2xPBS. Holding the "sandwich" under buffer separate the slides using for instance a plastic tweezer. Gently insert the tweezer in between the two slides at the barcoded area of the gasket slide, and move it until liquid separates the slides, and allow the gasket slide to fall to the bottom of the beaker, while holding the edges of the microarray slide. Dry the microarray slide as described in **Section 3.3**, Point 3.

3.6. Stringency Wash

1. Hybridized slides are washed for 30 min in beakers using magnetic stirring as described in **Section 3.3**, Point 3. A beaker for each stringency condition is filled with stringency buffers (**Section 2.6**, Point 2). The buffer is heated to 37°C in a microwave. To obtain and sustain a constant temperature of 37°C, add buffer from two additional small beakers with hot (60–90°C) and cold buffer (cooled with ice and water). Temperature of the buffer is continuously monitored using a thermometer.

2. Post-stringency wash of the slides is performed in a beaker containing 2xPBS using magnetic stirring for 5 min at room temperature, subsequently slides are dried.

3.7. Scanning, Quantification and Data Analysis

1. Scanning of slides is performed using either the CCD-based scanner, arrayWoRx or the laser scanner Packard ScanArray Lite. Slides are scanned at Cy3 excitation wavelengths. The scanners generate 16-bit greyscaled, tagged image files (TIFF).
2. Quantification of TIFF files is performed according to software manual using ScanAlyze 2.5 (http://rana.lbl.gov/EisenSoftware.htm) or GenePix Pro 6.0.
3. The quantified data is copied into excel (GenePix Pro), or imported as a tab delimited file into excel (ScanAlyze) for further analysis.
4. For genotyping classification, calculate a ratio as signal from a wild-type probe divided by the sum of the signal from the

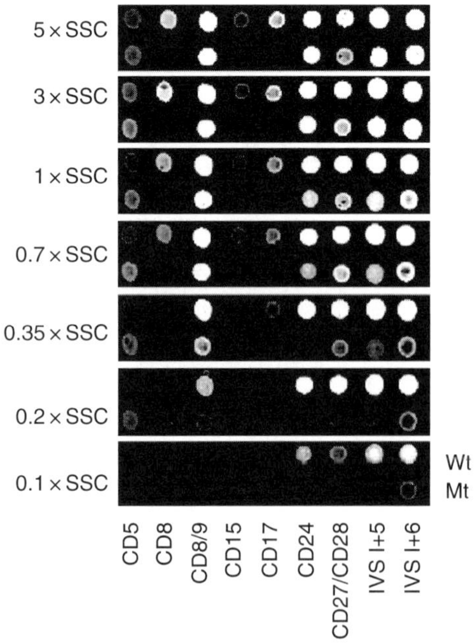

Fig. 11.2. Scanning images of microarrays of T_m-matched probes (**Table 11.1**), hybridized with amplified target from an individual heterozygous for the CD5 mutation and washed at different conditions. All slides were washed at 37°C but using various stringency buffers. The SSC content of the buffers is denoted to the left and the identity of the probes is denoted beneath the panels. All buffers were supplemented with 0.1% SDS, except the 5 x SSC buffer, which was supplemented with 0.5% SDS. Wild-type probes for the respective mutation are at the top of each panel and the corresponding mutant probes are at the bottom of each panel.

wild-type probe and the corresponding signal from the mutant probe (**Note 7**).

3.8. Assay Optimization and Validation

1. The first step in evaluating a new set of allele specific probes is washing hybridized slides at different conditions (**Fig. 11.2** and **Note 8**). For initial testing, amplified target from a control subject with no mutations can be used. The first criteria for inclusion of a probe in an assay, is that it functions adequately as capture probe, hence provides sufficient signal (**Note 9**).
2. The next inclusion criterion for a probe-pair is discrimination between signals from wild-type and mutant probe. This should be achieved after testing a wide range of stringencies (**Note 10**).
3. After probes with sufficient signal and discriminatory power are selected, the assay is validated (**Note 11**) based on ability to correctly classify subjects as homozygote wild-types, heterozygotes or homozygote mutant (**Note 12**). This is performed at optimized stringency and using a patient material (**Fig. 11.3**). The steps involved in troubleshooting a new genotyping assay are summarized in **Table 11.2**.

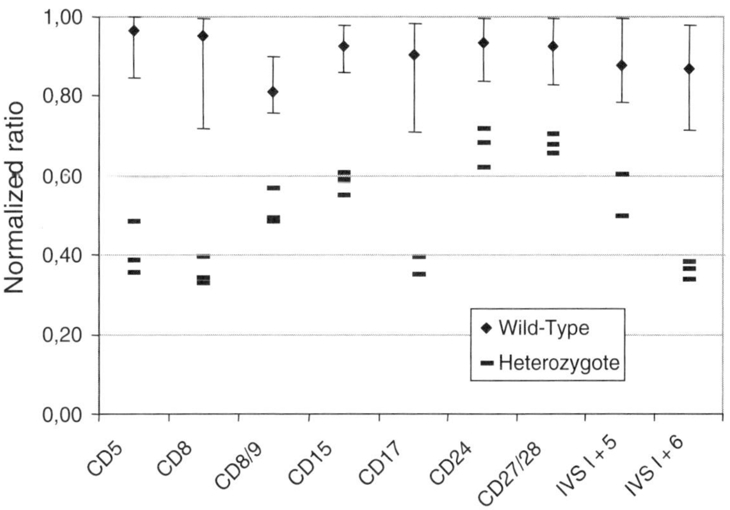

Fig. 11.3. Genotyping nine different mutations in the beta-globin gene using a patient material. Hybridized slides were washed at an optimized stringency condition (0.2xSSC+0.1%SDS at 37°C). The normalized ratio (y-axis) is seen for each mutation (x-axis). *Diamonds* represent the average value of all samples carrying the wild-type DNA sequence on both alleles (homozygote wild-types). *Error bars* represent the minimum and maximum observed ratios. *Dashes* represent the normalized ratio of heterozygotes.

Table 11.2
Troubleshooting probe design

Issue	Potential cause	Solution
Low signals	Too short or AT-rich probe, intra-molecular hybridization	Longer probe, fragment target
Poor discrimination	Too long or CG-rich probe, cross-hybridization	Shorter probe, use different tag
Classification	Wt and Mt probe show different hybridization behaviour	Make one probe shorter or longer

4. Notes

1. The TC tag can also be employed for quality control of the fabricated DNA microarray, because all the probes contain the sequence in the 5′ end. Although the added tag sequence makes the probes relatively long, the price is only between a forth to a fifth of amino modified probes. The TC tag was selected from a subset of tags for its property to link DNA covalently to an agarose film *(16)*. However, care should be taken to check for cross hybridization of the TC tag and the targets used prior to fabrication of large arrays. Another loci we have genotyped, the phenylalanine hydroxylase (*PAH*) gene, shows extensive cross-hybridization towards the TC tag. For *PAH* we screened different 20–25 nt long sequences to serve as 5′ tag. A tag with the sequence 5′-AAG-TATTCGTTCACTTCCGATATGC-3′ was selected, showing low hybridization to *PAH* fragments and good immobilization to agarose.
2. The first probe-pair to be designed in a Tm matched probe-set can be for a mutation in a CG rich sequence, and about 15 bp. First the Wt and Mt probe is Tm matched, by varying length of sequence, while keeping the variant base in the center. Next, probe-pairs for genotyping the remaining mutation are designed to have Tm-matched Tm Wt and Mt probes, and have a Tm as closely matched as possible for the entire probe-set.
3. Larger microarrays can be accommodated on slides without frosted area, although there are some considerations when choosing this solution. Coating an entire microscope glass slide with agarose is somewhat more challenging, than to coat a Superfrost glass slide, and therefore might require some practice. Furthermore handling off "un-frosted" slides is more difficult, as they can only be held at the edges. Another

consideration is labeling of slides without frosted area. Each slide can be labeled by etching a number in the corner of the slide with a diamond pen. An alternative is to use laminated labels that can sustain all aspects of slide processing.

4. Perfect spots are homogenous and often round if the arrays are fabricated using a spotter. Most spotting procedures require humidity to reduce the evaporation rate. Additions, such as polymers and salts (phosphate buffers), can be mixed with the DNA to reduce evaporation. We have found that spotting at low temperatures results in low evaporation rate and acceptable spot morphology even when spotting DNA without such buffers. It should be noted that other spotting schemes might be necessary if other substrates are used instead of the agarose slides.
5. The hybridization station we use is developed to accommodate the 76 × 26 mm format of the agarose coated slides. Agilent's hybridization station, which is designed for 75 × 25 mm, does not fit the slightly larger slides. We reuse Agilent gasket slides several times for hybridization. After each run the gasket slides are cleaned with MilliQ water, dried by gently blowing air over the slide and stored.
6. Agarose coated slides should not be subjected to long hybridizations (more than 4 h), as target can diffuse into the agarose layer and therefore be impossible to wash off during a stringent wash.
7. The simplest ratio to construct is to divide the signal from the wild-type spot with the signal from the mutant spot. However, if the mutant signal is low it will have a large influence on the obtained ratio. In the special case where the signal of the mutant probe is zero, a ratio cannot be constructed. To circumvent this problem, the signal ratio of Wt/(Wt+Mt) is used. Using this ratio, ideally homozygote wild-types are close to 1, heterozygotes close to 0.5 and homozygote mutants have a ratio close to 0.
8. Washing the slides at stringent condition is needed to dehybridize mismatch hybrids but not the perfect match hybrids. It is presently difficult to predict the optimal stringency and it must therefore be determined experimentally. The optimal stringency for the array is easiest determined by washing at different conditions. Stringency wash can be performed keeping temperature or concentration of monovalent ions (often Na^+) constant, and varying the other parameter. Here we use a fixed temperature and different concentrations of salt, by varying the SSC content. A low concentration of SDS (0.1%) is included as it results in lower background signals.

9. Lack of signals for a probe/probe-pair can be due the length or sequence (AT-rich) of the probe or intra-molecular folding of the probe or target. A too short probe (or AT-rich probe) can be redesigned to be longer, however long probes are less discriminating. Hair-pin formation of the probe is difficult to circumvent in site-specific assays, such as mutation analysis. For optimal hybridization to probes, the target fragments should to be too long (preferably <500 nt). Long targets or targets with extensive intra-molecular folding can be fragmented (chemically or enzymatically) to enhance hybridization.
10. Lack of discrimination is observed as similar signals from Mt and Wt probes at all conditions, although the subject is homozygote wild-type. Lack of discrimination can due to probe length or sequence or cross-hybridization. In the first case the probe can be shortened, however a too short probe can result low signals, because of its poor ability as capture molecule. If cross-hybridization is due to hybridization to the tag sequence this can be replaced (**Note 1**).
11. There are many ways to evaluate a genotyping microarray. One way is to simply look at the spots (**Fig. 11.2**). At all stringency washes it is seen that the subject is homozygote wild-type at position CD8, because the signal from the wild-type probe is strong while the signal from the mutant probe is weak/absent. Other probe pairs (e.g. IVSI+5 and IVSI+6) require more stringent conditions ($\leq$0.35xSSC) for accurate genotyping (homozygote wild-type). In contrast, signals from both the wild-type probe and the mutant probe for CD5 are equally strong at all stringencies, indicating that the subject is heterozygous for this mutation. Statistical methods are often required to evaluate if an assay is robust or not and then the quantified spots must be converted into a ratio (**Note 7**). For validation of a genotyping array all mutations must be tested with a patient material consisting of homozygote wild-types, heterozygotes, and if available homozygote mutated. In case of beta-globin we run one or more heterozygotes for each mutation and the available homozygotes and some controls with no mutation. As heterozygotes only have one mutation, and are homozygote wild-types for the remaining mutations tested in the beta-globin gene, these are used as homozygote wild-types. However, for overlapping probes (e.g. CD8 and CD8/9), both probe-pairs are affected by a mismatch, and therefore not included as homozygote wild-types when one of the sites has mutation.
12. One approach to assign genotype, is to have a classification key for each mutation. This would be the case for the

genotyping assay shown in **Fig. 11.3**. Heterozygotes are clearly separated from homozygotes at all positions, however heterozygote ratios are widely distributed from about 0.3 (CD8) to 0.7 (CD24 and CD27/28). Hence a classification key for the above mutation would include that the heterozygotes are expected to be low or high, respectively. A more strict classification criterion is to require that the normalized ratios fall into these classes; Average homozygote wild-type ratio more than 0.70, heterozygote ratio between 0.35 and 0.65 and homozygote mutant ratio less than 0.30, and of course no overlap between the three possible genotypes at each site. A more strict classification simplifies assignment of genotypes, however it requires that the wild-type and mutant probes have similar thermodynamic properties. For some mutations it might require testing of several probe lengths and combinations of wild-type and mutant probes.

References

1. Matsuzaki, H., Dong, S., Loi, H., Di, X., Liu, G., Hubbell, E., Law, J., Berntsen, T., Chadha, M. and Hui, H. et al. (2004) Genotyping over 100,000 SNPs on a pair of oligonucleotide arrays. *Nat Methods*, **1**, 109–111.
2. Syvanen, A.C. (2005) Toward genome-wide SNP genotyping. *Nat Genet*, **37 Suppl**, S5–10.
3. Syvanen, A.C. (2001) Accessing genetic variation: genotyping single nucleotide polymorphisms. *Nat Rev Genet*, **2**, 930–942.
4. Bang-Ce, Y., Hongqıong, L., Zhuanfong, Z., Zhengsong, L. and Jianling, G. (2004) Simultaneous detection of alpha-thalassemia and beta-thalassemia by oligonucleotide microarray. *Haematologica*, **89**, 1010–1012.
5. Gemignani, F., Perra, C., Landi, S., Canzian, F., Kurg, A., Tonisson, N., Galanello, R., Cao, A., Metspalu, A. and Romeo, G. (2002) Reliable detection of beta-thalassemia and G6PD mutations by a DNA microarray. *Clin Chem*, **48**, 2051–2054.
6. Petersen, J., Stangegaard, M., Birgens, H. and Dufva, M. (2006) Detection of mutations in the beta-globin gene by colorimetric staining of DNA microarrays visualized by a flatbed scanner. *Anal Biochem*, **360,** 169–171.
7. Chan, K.M., Wong, M.S., Chan, T.K. and Chan, V. (2004) A thalassaemia array for Southeast Asia. *Br J Haematol*, **124**, 232–239.
8. Rund, D. and Rachmilewitz, E. (2005) Beta-thalassemia. *N Engl J Med*, **353**, 1135–1146.
9. Dufva, M., Petronis, S., Jensen, L.B., Krag, C. and Christensen, C.B. (2004) Characterization of an inexpensive, nontoxic, and highly sensitive microarray substrate. *Biotechniques*, **37**, 286–296.
10. Dufva, M.P.J., Stoltenborg, M., Birgens, H. and Christensen, C.B. (2006) Detection of mutations using microarrays of poly(C)10-poly(T)10 modified DNA probes immobilized on agarose films. *Anal Biochem*, **352**, 188–197.
11. Wick, L.M., Rouillard, J.M., Whittam, T.S., Gulari, E., Tiedje, J.M. and Hashsham, S.A. (2006) On-chip non-equilibrium dissociation curves and dissociation rate constants as methods to assess specificity of oligonucleotide probes. *Nucleic Acids Res*, **34**, e26.
12. Vainrub, A. and Pettitt, B.M. (2003) Surface electrostatic effects in oligonucleotide microarrays: control and optimization of binding thermodynamics. *Biopolymers*, **68**, 265–270.
13. Kajiyama, T., Miyahara, Y., Kricka, L.J., Wilding, P., Graves, D.J., Surrey, S. and Fortina, P. (2003) Genotyping on a thermal gradient DNA chip. *Genome Res*, **13**, 467–475.

14. Koehler, R.T. and Peyret, N. (2005) Thermodynamic properties of DNA sequences: characteristic values for the human genome. *Bioinformatics*, **21**, 3333–3339.
15. Dufva, M., Petronis, S., Bjerremann Jensen, L., Krag, C. and Christensen, C. (2004) Characterization of an inexpensive, non-toxic and highly sensitive microarray substrate. *Biotechniques*, **37**, 286–296.
16. Dufva, M., Petersen, J., Stoltenborg, M., Birgens, H. and Christensen, C.B. (2006) Detection of mutations using microarrays of poly(C)10-poly(T)10 modified DNA probes immobilized on agarose films. *Anal Biochem*, **352**, 188–197.

Chapter 12

Microarray Temperature Optimization Using Hybridization Kinetics

Steve Blair, Layne Williams, Justin Bishop and Alexander Chagovetz

Abstract

In any microarray hybridization experiment, there are contributions at each probe spot due to the match and numerous mismatch target species (i.e., cross-hybridizations). One goal of temperature optimization is to minimize the contribution of mismatch species; however, achieving this goal may come at the expense of obtaining equilibrium reaction conditions. We employ two-component thermodynamic and kinetic models to study the trade-offs involved in temperature optimization. These models show that the maximum selectivity is achieved at equilibrium, but that the mismatch species controls the time to equilibrium via the competitive displacement mechanism. Also, selectivity is improved at lower temperatures. However, the time to equilibrium is also extended, so that greater selectivity cannot be achieved in practice. We also employ a two-color real-time microarray reader to experimentally demonstrate these effects by independently monitoring the match and mismatch species during multiplex hybridization. The only universal criterion that can be employed is to optimize temperature based upon attaining equilibrium reaction conditions. This temperature varies from one probe to another, but can be determined empirically using standard microarray experimentation methods.

Key words:: Hybridization kinetics, temperature effects, cross-hybridization, SNP detection, real-time detection, two-color, evanescent wave.

1. Introduction

When setting up a new microarray-based assay, experimentalists inevitably go through the stage of empirical optimization of reaction conditions, most notably temperature optimization. Depending on the goals of an experiment, the definition of the optimal temperature may be different: one may target minimizing the time of the experiment to increase throughput, enhance sensitivity (signal to background ratio), enhance specificity of target

Martin Dufva (ed.), *DNA Microarrays for Biomedical Research: Methods and Protocols, vol. 529*

DOI 10.1007/978-1-59745-538-1_12 Springerprotocols.com

recognition, achieve better quantitative reproducibility of data, or any combination of the above. The purpose of this chapter is to rationalize temperature optimization of common DNA microarray-based experiments to minimize cross-hybridizations. This approach is based on analysis of multiplex surface hybridization kinetics using real-time fluorescence detection, and is applicable to both the transient and, in the limit, equilibrium regimes. Using this approach, temperature effects are manifested as corresponding changes in various rate constants (association, dissociation, folding etc.) considered in the models.

There are several commonsensical considerations when choosing a preferred temperature: (1) higher temperatures accelerate both mass transport and surface reactions; (2) higher temperature resolves multiple meta-stable states (intra-molecular secondary structures and heteroduplex formation), which is advantageous for quantitative analysis; (3) equilibrium surface concentration of duplexes is reduced at higher temperatures; and (4) the signal is decreased due to the temperature dependence of the fluorescence proper, while fluorescence background often becomes higher. Since there are considerations justifying both lower and higher experimental temperatures, it is reasonable to assume that there should be a temperature optimum for each experiment, and in particular, a temperature optimum for each probe.

2. Kinetics and Thermodynamics of Hybridization at a Surface

To develop a rational approach to temperature optimization, we first consider a simple model system where a solution-based target interacts with a surface bound probe. For simplicity of practical discussion we omit description of the surface specific effects (geometry of the sensing zones, electrostatic interactions, effects of surface density of the probes, and linkage chemistry), which are discussed elsewhere *(1–5)*. We further assume that the binding of targets does not alter the solution concentration, i.e., the well-mixed assumption *(6)*.

With these assumptions, the kinetic equation of binding is

$$\frac{dB}{dt} = k_a C(R_T - B) - k_d B, \qquad (12.1)$$

where k_a and k_d are the association and dissociation rate constants, C is the concentration of the target, R_T is the initial surface concentration of probes, B is the surface concentration of the hybrid, and R_T-B is the accessible concentration of probe.

The temperature dependence of hybridization is reflected in the rate constants:

$$k_{a,d} = e^{-\Delta G^*_{a,d}/RT} \tag{12.2}$$

where ΔG^* is a reaction activation energy. Here, R is the gas constant (1.987 cal/K-mol) and T is temperature in Kelvin. Association and dissociation activation energies are connected through a simple equation

$$\Delta G^*_d = \Delta G^*_a + \Delta G \tag{12.3}$$

where ΔG is the Gibbs free energy of hybridization. It is clear that the activation energy of dissociation is higher than the activation energy of association, which translates into stronger temperature dependence of dissociation. In practical terms it means that the dissociation rate changes with temperature more than the association rate *(7,8)*. The equilibrium constant is defined here $K_{eq} = k_d/k_a$, or

$$K_{eq} = e^{-\Delta G/RT}, \tag{12.4}$$

which has units of concentration (M). The equilibrium concentration of hybrids on the surface is a function of free energy through the equilibrium constant, and exponentially decreases with temperature. The free energy of hybridization may be written as a linear function of temperature:

$$\Delta G = \Delta H - T\Delta S, \tag{12.5}$$

so that the free energy diminishes linearly with temperature growth.

Throughout this chapter, we will use a few example sequences to illustrate the temperature dependence of hybridization. For these sequences, we employ thermodynamic properties calculated for solution target-probe hybridization for a number of reasons. First, this remains a common practice in the microarray community. Second, there is a well-known correspondence between ΔG in solution and observed signal intensity on a microarray for the same duplex *(9–12)*. Third, because the reaction conditions vary considerably from one microarray platform to another (and from solution), there is no standard reference point for surface reactions. The target and probe sequences, and the respective thermodynamic stabilities of their hybrids, are listed in **Table 12.1**. These parameters were obtained from the DINAMelt server *(13)* at a temperature of 298 K, using 1 μM target and probe concentrations and 600 mM Na^+ salt concentration in solution.

Table 12.1
Example sequences used in calculations and experiments

	Sequence	ΔH (kcal/mol)	ΔS (cal/mol/K)
1t	5′-CGAGGGCAGCAATAGTACAC-3′	171.1	498.3
1p	3′-GCTCCCGTCGTTATCATGTG-5′		
2t	5′-CGAGGGCAGCATTAGTACAC-3′	161.3	477.2
2p	3′-GCTCCCGTCGTTATCATGTG-5′		
3t	5′-CGAGGGCAGCATAAGTACAC-3′	152.8	453.4
3p	3′-GCTCCCGTCGTTATCATGTG-5′		
4t	5′-AAGAAAATATCATCTTTG-3′	136.2	411.2
4p	3′-TTCTTTTATAGTAGAAAC-5′		
5t	5′-AAGAAAATACCATCTTTG-3′	117.9	366.1
5p	3′-TTCTTTTATAGTAGAAAC-5′		
6t	5′-AAGAAAATACCATCTTTG-3′	136.9	410.0
6p	3′-TTCTTTTATGGTAGAAAC-5′		
7t	5′-AAGAAAATATCATCTTTG-3′	122.6	375.4
7p	3′-TTCTTTTATGGTAGAAAC-5′		

3. Real-Time Microarray Instrumentation

In order to obtain kinetic data, we used a custom real-time detection system that records data over the entire duration of a hybridization experiment. There are two common types of real-time detection methods, based upon whether bound mass or fluorescence is transduced. Mass transduction is the preferred method by many researchers because extra target labels are typically not necessary, i.e., it is a label-free method. Typically, this type of sensing relies on a shift in an electrical *(14–16)*, mechanical *(17)*, or optical *(18,19)* resonance in proportion to the concentration of targets bound to the sensor. Traditional techniques include the quartz crystal micro balance (QCM) *(20,21)* and surface plasmon resonance (SPR) *(22,23)*. While these instruments have become dominant in protein binding studies, they are less widely used in the

DNA community, because (1) mass instruments lose sensitivity with small molecules such as PCR products, and (2) until recently, they have been limited in the number of independent detection zones. For these reasons, we focus on fluorescence systems.

3.1. Real-Time Fluorescence Detection

Real-time fluorescence detection systems have the following characteristics in common:

1) The target to be detected must be labeled with either a fluorescent molecule or a fluorescence quenching molecule (note that light scattering labels can also be employed) *(24)*;
2) The method of optical illumination does not excite fluorescence from species in solution; otherwise, the background level may be too high to discriminate the surface-bound species; and/or
3) The method of optical collection only detects fluorescence from surface-bound species, excluding fluorescence from species in solution.

It should be noted that either (2) or (3) is sufficient for real-time systems. The primary method used to selectively excite at the surface is to rely on the phenomenon of total internal reflection (TIR). In particular, an evanescent electromagnetic field will form upon reflection from the interface between a high refractive index medium and a low refractive index medium, as shown in **Fig. 12.1**. The penetration depth of the evanescent field into the low refractive index medium can be controlled by the angle of incidence, but is typically 100–200 nm. The first publication where evanescent excitation was used in a biosensor format was in

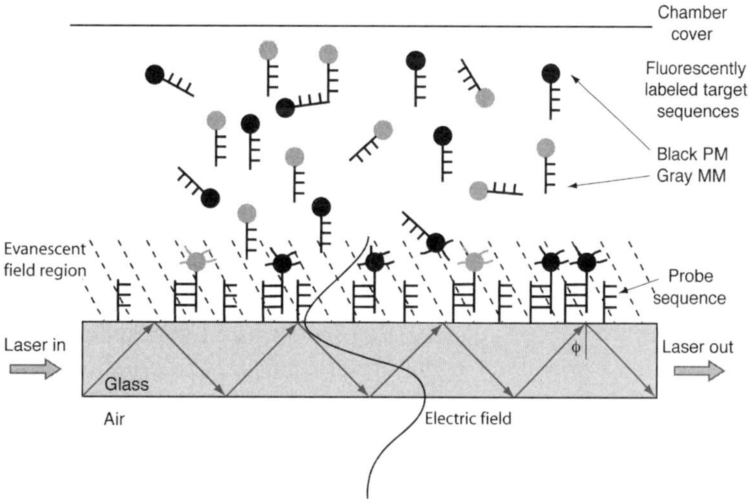

Fig. 12.1. Illustration of evanescent wave fluorescence excitation. The TIR condition is obtained when $sin\phi > n_{cover}/n_{substrate}$, where the refractive indices $n_{cover} \sim 1.33$ and $n_{substrate} \sim 1.43$.

1975 *(25)*. Since then, researchers have explored variations of the technique such as: total internal reflection fluorescence (TIRF) arrangement *(26)* and a dielectric waveguide *(27–29)*. It should be noted that at least two commercial platforms exist – zeptosens *(30)* and TIRF Technologies *(31)* – for real-time microarray experimentation based upon planar waveguides.

Other real-time systems are based upon the use of scanning confocal imaging, which uses an aperture in an intermediate image plane to isolate a small detection volume. This method is not as surface-selective as evanescent-wave excitation, but it offers very high spatial resolution and can be implemented with many fluorescence microscopes. Companies offering platforms include PamGene *(32)* and IMSTAR *(33)*.

3.2. Planar Waveguide Format

There are two important design considerations when setting up a real-time fluorescence system with planar waveguide architectures. The first is how the excitation light is coupled into the waveguide. There are in general three approaches: prism, grating, and end-fire coupling. While prism and grating coupling techniques have higher coupling efficiencies and are easier to use with thin film planar waveguide structures than end-fire coupling, they are more tedious to work with. For simplicity, we will keep the discussion to end-fire coupling methods with thick waveguides which, while not as elegant as prism and grating coupling with thin films waveguides, are more rugged and less sensitive to small variation in alignment and distance changes. The second question is how the fluorescence emission is collected. For medium to high-density microarrays, the most common solution is the use of a scientific charge-coupled device (CCD) array.

End-fire coupling of light into a waveguide is straightforward. The light from the laser source is focused to a line on the edge of the planar waveguide. As long as the angle at which the light strikes the core/cladding interface is larger than the critical angle, total internal reflection will occur and a surface-selective evanescent field will be produced. However, one issue that must be addressed is the edge quality when end-fire coupling is used. While most current microarray technologies use glass microscope slides that make suitable multi-mode planar waveguides, not all slides have edges that are polished. It is possible to couple light with some efficiency into slides that have frosted edges; however, slides that have snapped edges must be ground and polished. In any case, without smoothly polished surfaces, substantial scattering results, which is a source of background. Another significant source of background is the autofluorescence of the substrate itself, which cannot be distinguished from that of the fluorophores. As a result, low fluorescence glasses or quartz should be used. While the use of end-fire coupling directly into a microscope slide format substantially simplifies experimental conditions, many researchers have developed custom substrates which enhance detection sensitivity *(28)*. For a more complete review of planar waveguides the reader is directed to a 2005 review article *(29)*.

A CCD array provides a method to image multiple zones in parallel using off-the-shelf camera lenses or microscope objectives. Considerations for CCD imagers include: resolution, dynamic range, and noise. The resolution (i.e., number of pixels) ultimately determines the number of independent zones that can be imaged. Since real-time detection is used to capture a kinetic trace of hybridization, low fluorescence levels will be expected initially. As hybridization progresses with time, large fluorescence levels are likely. Therefore, the CCD should have large dynamic range, as determined by the well depth and the noise in readout. The signal level from each pixel is also converted into a digital signal, typically with 12 or 16 bits of resolution. Due to the readout noise in most CCDs, the lowest 1–2 bits do not contribute to actual dynamic range. Also, because low light situations are probable, thermal noise can be a major problem. For this reason most cameras used for fluorescence detection have a means of thermally cooling the CCD array, either Peltier and/or circulating coolant.

3.3. Experimental Setup

The experimental results in this chapter were obtained using a thick waveguide setup that utilized end-fire coupling and a CCD to collect fluorescence images. The details of the setup and experimental methods have been described elsewhere *(34,35)* and therefore will only be described briefly here. As shown in **Fig. 12.2**, the setup uses two lasers (532 and 635 nm) that were end-fire coupled into the microarray substrate (quartz) which acts as an optical waveguide. Both lasers pass through bandpass filters to cut out any spontaneous emission that might interfere with fluorescence detection. The 532 and 635 nm laser outputs are brought into a coincident optical path using a dichroic beam splitter. The beams were then expanded, to cover the 1″ width of the slide, and focused to a size less than the 1 mm thickness of the substrate edge used for coupling. The fluorescence intensity from the waveguide surface was detected with a CCD camera (Santa Barbara Instrument Group ST-7XMEI). Attached to the camera was an electronically controlled filter wheel, allowing alternating frames to be acquired with Cy3 and Cy5 emission filters. For a Cy5 frame, the interference filter had a 690 nm center with 40 nm bandpass, and for a Cy3 frame the filter was a 580 nm interference filter with 30 nm bandpass. Besides the Cy5 and Cy3 filters, a neutral density 3.0 filter in the third position was used to produce a dark frame image. A TTL signal from the filter wheel was used to modulate the laser outputs via electronic shutters to synchronize with the respective emission filter. During the acquisition of the dark frame, both laser outputs were blocked. The temperature of the microarray substrate was maintained by a Peltier heated

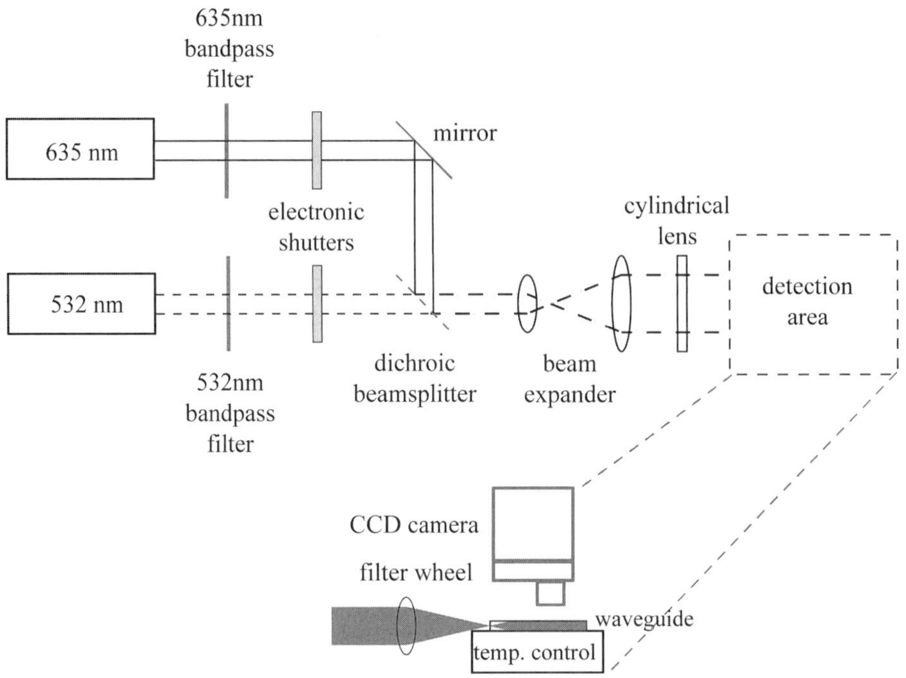

Fig. 12.2. Illustration of the optical components in the real-time two-color microarry reader platform. Note that the fluid handling and computer control systems are not shown.

mount, which was electronically controlled using a feedback circuit and thermocouple. The absolute temperature accuracy was estimated to be 2°C.

Prior to probe immobilization, quartz microscope slides were washed with Alconox, rinsed with DI water, and dried with ultra high purity N_2. The slides were then cleaned in O_2 plasma for 100 minutes at a pressure of 400 mTorr oxygen and power of 100 W. The slides were transferred to a vacuum oven where deposition of glycidoxypropyltrimethoxysilane (GPS, Sigma-Aldrich, product number 440167) was performed in vapor phase for 8 h at 120°C. The processed slides were then spotted with amine-modified oligonucleotide probes using a home-built single pin non-contact spotter. The spotting volume was 100 nL, and spots were allowed to dry down during spotting. Spotted slides were stored overnight in a humid chamber at room temperature. After rinsing with cold DI water and drying with N_2, the slide was ready for assembly.

Sandwiching a piece of adhesive tape between the spotted quartz slide and a top glass slide formed the hybridization chamber. To build the chamber, the inlet and outlet holes were drilled in the top glass slide, then it underwent the same GPS surface treatment as the quartz slide to promote adhesion of the tape layer. Microfluidic access was achieved using Nanoport connections (Upchurch Scientific Inc., Model N-333). The top of the glass cover was treated with AP8000 adhesion promoter (Dow Chemicals) and then the Nanoports were glued in place with DP-190 epoxy (3M). The cover and ports were placed in an oven at 70°C for 2 h to cure the epoxy. The chamber was made from 250 μm thick VHB adhesive tape (3 M, model 9460PC). The adhesive tape chamber was cut with a sharp blade and placed on the bottom of the cover glass slide, and sandwiched with the spotted quartz slide.

The reactive epoxy sites on the quartz and glass slides were blocked with 0.5% BSA. The BSA solution was heated to 50°C for 1 h and then 5 mL was pushed through the microfluidic chamber with a syringe pump (KD Scientific, model KDS120), followed by 5 mL of 90°C DI water. The DI water was removed from the chamber and the sample solution loaded into the system. The system was placed in a vacuum desiccator and degassed for 1 min to eliminate bubbles in the chamber. The assembled slide was then placed on the temperature-controlled mount described previously. The temperature was ramped slowly (10°C steps every 5 min) up to 70°C to melt any hybrids formed during sample loading. Data collection was started at this point and then the temperature was quickly lowered to the desired set point with the Peltier element, enabling capture of the beginning of the hybridization reaction.

Raw images obtained from a typical experiment are shown in **Fig. 12.3**. The left image is through the green filter (green laser on) and the right image is through the red filter (red laser on). Red and green fluorescence control spots were implemented by immobilizing Cy5 and Cy3 labeled probe sequences designed so as not to interact with any target sequences. Two columns of hybridization spots are immobilized on the array, one column being wild type probes (wtp) and the other being SNP probes (snpp). Note from the images below that cross-hybridization is clearly evident. These raw images are used to obtain intensity signals from each hybridization spot that are representative of the bound target

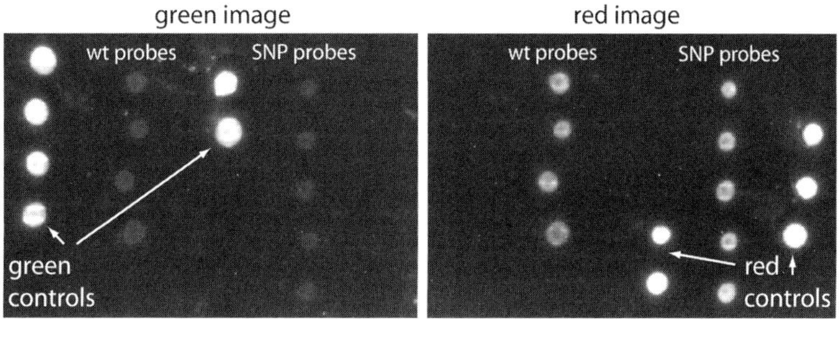

Fig. 12.3. Raw fluorescence images after 1 h of hybridization at 30°C with 3 nM Cy3-wtt and 3 nM Cy5-snpt solution concentrations.

concentrations. For each spot, a neighboring background level is subtracted, and then this signal is divided by a background corrected average of the respective control spots.

4. Multiplex Hybridization

Due to the complexity of sample composition in a microarray experiment, there is a subset of targets with significant affinity to a particular sensing zone that contribute to the observed signal (i.e., cross-hybridization). In other words, the signal is always a composite of several labeled species. We consider the simplest multiplex situation of two components. As we have shown recently *(36)*, this model can be readily expanded to include multiple targets interacting with a sensing zone on the surface. In these models, we disregard self and cross-hybridization of target species in solution *(37)* and secondary structure formation *(38)*. These effects can be incorporated via effective concentrations based upon the fraction of solution concentration available for surface hybridization *(34)*, but this partitioning is temperature dependent. Higher temperatures do tend to reduce these effects. Similarly, self-complementary in probe sequences reduces the accessible concentration for binding; the impact of this effect can be mitigated through probe design and temperature as well. The full description of surface hybridization, and it's temperature dependencies incorporating these effects, is beyond the scope of this chapter, but is nevertheless an important consideration.

The corresponding system of kinetic equations for surface capture is

$$\begin{aligned}\frac{dB_1}{dt} &= k_{a,1} C_1 (R_T - B_1 - B_2) - k_{d,1} B_1 \\ \frac{dB_2}{dt} &= k_{a,2} C_2 (R_T - B_1 - B_2) - k_{d,2} B_2\end{aligned} \tag{12.6}$$

The coupling of the two competing target species is represented by the concentration of accessible probes, which is given by R_T-B_1-B_2. To simplify our analysis, we assume that since the association rate constants have weaker dependence on temperature than dissociation rate constants, and were demonstrated experimentally to be within an order of magnitude for different mismatched duplexes *(7,39)*, they are equal and temperature independent; therefore, we assume that $k_{a,1} = k_{a,2} \equiv k_a$. This shifts the discussion on specificity of target recognition and effects of temperature to the equilibrium constant, which is directly proportional to the dissociation rate constant.

4.1. The Competitive Displacement Mechanism

This system of kinetic equation predicts biphasic behavior in hybridization. In the first phase, binding occurs in a quasi-independent manner, where the total bound concentration can be predicted by the sum of two single-component hybridizations. This phase is governed by association kinetics (which includes mass transport). Transition to the second phase occurs when the total bound concentration approaches the equilibrium bound concentration. In this phase, the total bound concentration undergoes slow growth (i.e., what appears to be equilibrium), but the partitioning of the two bound species still changes with time, with the higher-affinity species displacing the lower affinity one. We have termed this process "competitive displacement." Competitive displacement is in fact the mechanism for specificity in multiplex hybridization, and is largely governed by dissociation kinetics.

Figure 12.4 shows the kinetics of competitive displacement using the following target sequences – (wtt) Cy3-CGAGGG-CAGCATTAGTACAC-3′ and (snpt) Cy5-CGAGGGCAGCAA-TAGTACAC-3′. Complementary probe sequences (wtp) and (snpp), respectively, were spotted onto the microarray surface. Experiments were performed at 30°C for 1 nM wtt concentration

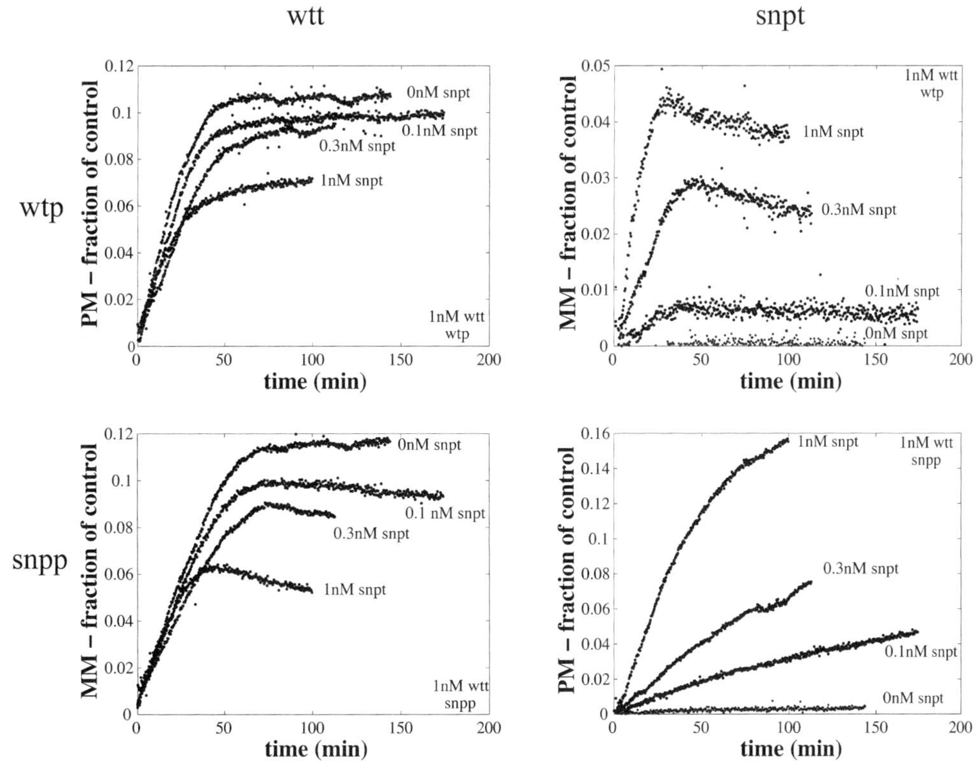

Fig. 12.4. Experimental hybridization curves showing the competitive displacement of the MM hybrid in favor of the PM hybrid. In these experiments, the wtt concentration is fixed at 1 nM, but the snpt concentration is varied from 0 to 1 nM.

and varying snpt concentrations. In the case of the wt probes (wtp), the wtt sequence is the perfect match (PM), while the snpt sequence is the mismatch (MM). The situation is reversed for the snp probes (snpp).

In the case of wtp, it is clear that the MM hybrid is displaced by the PM hybrid. However, the MM concentration has significant effect on the overall kinetics. For example, the rate of growth of the PM hybrid is influenced by the MM concentration. In other words, with high (or comparable) MM concentration, it takes longer to reach an equilibrium condition. Also, the equilibrium bound PM concentration is reduced, such that the overall signal represents a greater proportion of MM hybrids, potentially giving rise to a false positive signal. This latter point is seen more clearly on the snpp figures. Here, even in the absence of the PM sequence (snpt in this case), significant signal growth is obtained due to the MM sequence (wtt). As the snpt concentration increases, the MM hybrid becomes displaced.

4.2. Time to Equilibrium

In microarray experimentation with end-point readout, it is important that reactions reach equilibrium *(40,41)*. By definition, the equilibrium condition ensures temporal stability in data acquisition and analysis. Equilibrium is also important for interpreting post-hybridization array processing steps such as washing or melting (i.e., dissociation reactions). Real-time data acquisition can aid in determination of the equilibrium state. In the non-equilibrium regime, analysis can still be performed *(36)*, but that topic will not be addressed in this chapter. Equilibrium is considered a universal requirement for all microarray experiments, and the ramifications of obtaining equilibrium over a given period of time on other metrics such as hybridization efficiency and dynamic range will be explored here.

For a single species, the bound concentration at equilibrium is given by

$$B_{eq} = \frac{k_a C}{k_a C + k_d} R_T = \frac{C}{C + K_{eq}} R_T \text{ or } \frac{B_{eq}}{R_T} = \frac{r}{r+1}, \qquad (12.7)$$

where r is the ratio of C to K_{eq} and B_{eq}/R_T denotes the hybridization efficiency. Under the assumption of a well-mixed sample (i.e., there is no concentration depletion so that the reaction occurs at the chemical rate limit), the time required to reach a given bound concentration B is given by the kinetic limit

$$t_{kl} = -\frac{1/k_a}{C + K_{eq}} \ln\left(1 - \frac{B}{B_{eq}}\right), \qquad (12.8)$$

where, obviously, the time to equilibrium is infinite. However, a mathematically convenient definition of equilibrium time is the time to reach 95% of B_{eq}, so that

$$t_{eq} \sim \frac{3}{k_a\left(C + K_{eq}\right)} \text{ or } k_a K_{eq} t_{eq} = \frac{3}{r+1} = \frac{3B_{eq}}{rR_T}. \tag{12.9}$$

Keep in mind that this equation represents a lower limit, and any concentration depletion (i.e., mass transport effects) will serve to slow the reaction, which can be described by an apparent association rate constant. This equation also illustrates the fundamental relationship between hybridization efficiency – defined by the equilibrium bound fraction $B_{eq}/R_T = k_a t_{eq} C/3$ – and time to equilibrium and concentration. In other words, higher efficiency fundamentally requires longer times and/or higher solution concentration.

In the multiplex regime (represented here by two components), the bound concentration is given by

$$B_{eq,1} = \frac{\alpha_2 C_1}{\alpha_2 C_1 + K_{eq,1}} R_T \text{ or } \frac{B_{eq,1}}{R_T} = \frac{\alpha_2 r_1}{\alpha_2 r_1 + 1}. \tag{12.10}$$

Due to competitive displacement, the time to equilibrium is stretched. This can be expressed by

$$t_{eq} = -\frac{1/k_a}{\alpha_2 C_1 + K_{eq,1}} \ln\left(1 - \frac{B_1}{B_{eq,1}}\right) \sim \frac{3}{k_a\left(\alpha_2 C_1 + K_{eq,1}\right)} \text{ or}$$
$$k_a K_{eq,1} t_{eq} = \frac{3}{\alpha_2 r_1 + 1} = \frac{3B_{eq,1}}{\alpha_2 r_1 R_T} \tag{12.11}$$

In these equations, α_2 is defined as

$$\alpha_2 = \frac{K_{eq,2}}{C_2 + K_{eq,2}} = \frac{1}{r_2 + 1}. \tag{12.12}$$

The equations for the second species can be obtained by exchanging the 1 and 2 subscripts. It is interesting that in the multiplex case, these equations are identical to the single component case, except that they are based upon an effective concentration that is reduced due to the presence of the other species. The time elongation factor due to competition is given by $1/\alpha_2$. Again, there is a fundamental relationship among efficiency, time, and concentration, given by $B_{eq,1}/R_T = \alpha_2 k_a t_{eq} C_1/3$.

A universal way to plot the time to equilibrium is shown in **Fig. 12.5**, where the scaling is chosen such that the concentrations of the two competing species are normalized to their

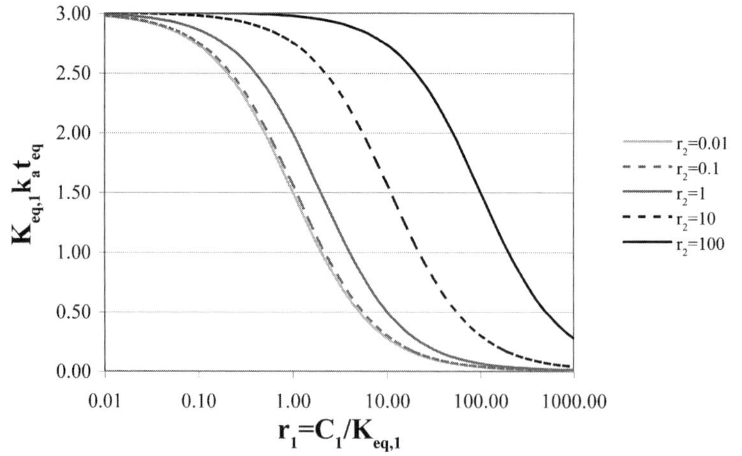

Fig. 12.5. Universal plot of the effect of the solution concentration of two competing target sequences on the time to equilibrium.

respective equilibrium constants. It is clear that the time is reduced when C_1 is greater than $K_{eq,1}$. However, when C_2 is greater than $K_{eq,2}$, then there is a dramatic increase in the time. This is because the time to equilibrium is controlled by the displacement of the lower affinity species, which is a slow process.

The normalization used in **Fig. 12.5** hides the temperature dependence. In order to explicitly demonstrate the effects of temperature, we use the example system of the 1t-1p PM and 2t-1p MM hybrids (hereafter referred to as the 1–2 system). Assuming a 1 nM solution concentration of the PM target, **Fig. 12.6** plots the normalized time to equilibrium versus

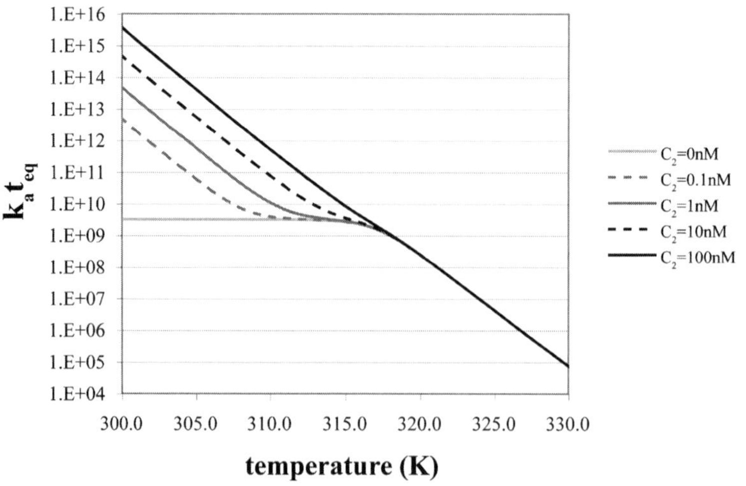

Fig. 12.6. Normalized equilibrium time versus temperature for the 1–2 model system assuming C_1=1 nM.

temperature parameterized by the concentration of the MM target. Time is normalized with the association rate, which is also temperature-dependant, both in terms of the rate of hybridization and the apparent rate due to diffusion. Typical apparent rates will range from 10^3 to 10^6 1/M•s. With no competing MM hybrid, the time saturates at a maximum value for temperatures at which $C_1 > K_{eq,1}$. At higher temperatures, the time to equilibrium drops exponentially. In the presence of the MM hybrid, and at lower temperatures, there is a significant increase in time with C_2. An interesting effect, however, is that at higher temperatures for which $\alpha_2 C_1 < K_{eq,1}$, there is minimal effect of the MM sequence.

The advantage of high temperatures in minimizing the influence of the MM sequence on equilibrium time must be contrasted against the hybridization efficiency, which decreases with temperature. This is shown in **Fig. 12.7** using the same parameters as **Fig. 12.6**. Again, there is a clear transition to low efficiency after 315 K, which is when $C_1 < K_{eq,1}$. Also, it is clear that high MM concentration can suppress formation of PM hybrids. By comparing **Figs. 12.6** and **12.7**, it appears that one might be able to identify an optimal temperature (about 315 K in this example), at which the time to equilibrium is stable against large changes in C_2 and still retain a large fraction of PM hybrids. Before any conclusion can be reached, however, the dynamic range must be considered, as will be done in **Section 5**.

Many of these effects are demonstrated in the experimental results of **Fig. 12.4**. Further experiments were performed using

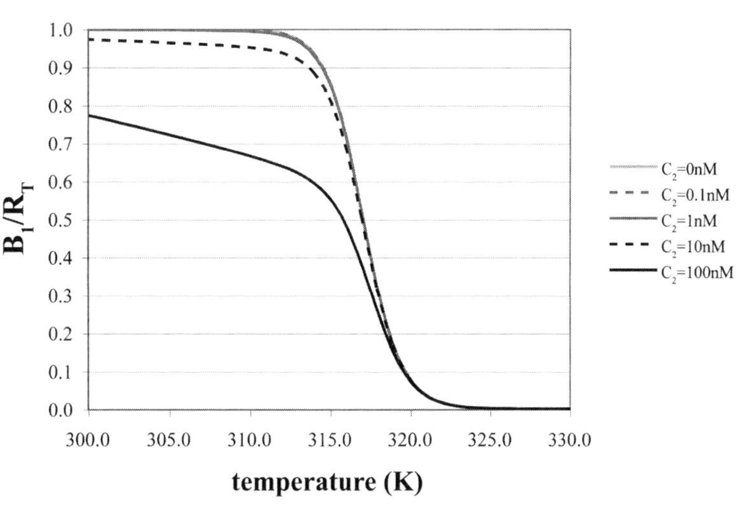

Fig. 12.7. Hybridization efficiency of PM target versus temperature for the 1–2 model system assuming C_1=1 nM.

the sequences of **Table 12.1**, in particular the 4–5 and 6–7 systems. These systems were derived from the CFTR gene, and will be used to demonstrate temperature effects. One caveat of these systems is that each target is predicted to form secondary structures in solution. The calculated melting temperatures for these reactions are about 37°C (the temperatures are nearly the same for each target, so the average is reported). While we do not explicitly incorporate the effects of these solution reactions, we do note that they will reduce the effective target concentrations in solution available for surface binding, in particular, at the lower temperature ranges in the experiments. This reduction results in longer reaction times and reduced bound fraction. Nevertheless, the fits of the kinetic equations to the experimental data are quite good in most cases.

In **Fig. 12.8**, hybridizations at 40°C are shown, with 3 nM:3 nM and 3 nM:15 nM (wtt:snpt) concentrations. Also plotted are the composite curves, which is the sum of the bound PM and MM hybrid concentrations and would be the curves observed if all targets were labeled with the same fluorophore. Note that the composite curves reach a steady state value well in advance of equilibrium conditions. It is also clear that the higher MM concentration stretches the time to equilibrium.

For reference purposes, **Fig. 12.9** shows hybridization with 3 nM:3 nM at 40°C to the SNP probe. It is clear that this system requires longer time to reach equilibrium under the same experimental conditions, owing to the fact that the thermodynamic stabilities of the PM and MM hybrids are more similar (*see* **Table 12.1**).

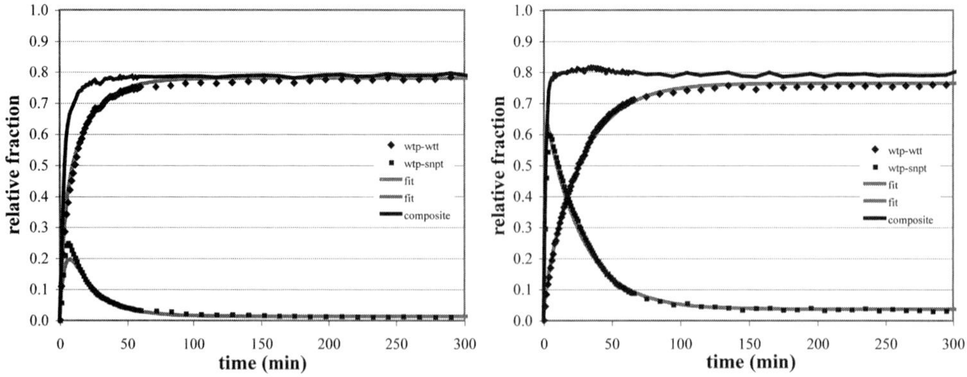

Fig. 12.8. Kinetics of PM and MM hybrids of the 4–5 system at 40°C. The left panel is for 3 nM:3 nM solution concentration, while the right panel is for 3 nM:15 nM concentrations. Fits were based upon direct solutions to the coupled differential equations using six fitting parameters.

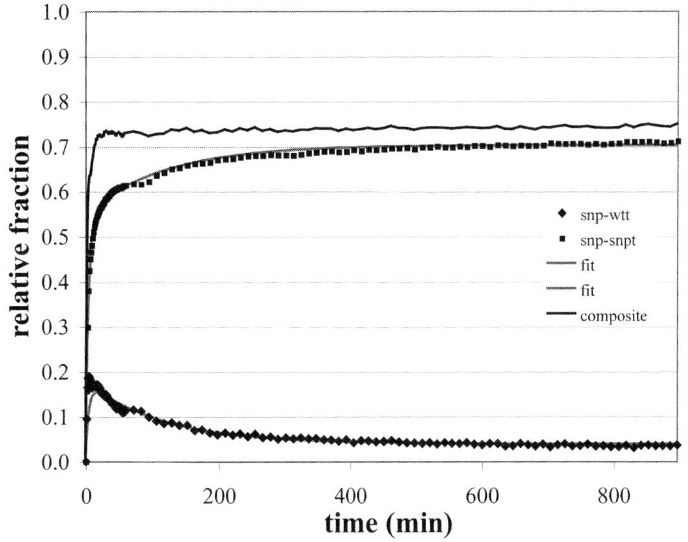

Fig. 12.9. Kinetics of PM and MM hybrids of the 6–7 system at 40°C for 3 nM:3 nM solution concentration.

5. Dynamic Range of Discrimination

There are two metrics to quantify the contribution of cross-hybrids to the observed signal from a microarray spot. The percentage of signal owing to the perfect match sequence can be written

$$\frac{PM_{\%}}{100\%} = \frac{B_1}{B_1 + B_2} = \frac{DR}{1 + DR},$$

where $DR = B_1/B_2$ is defined as the dynamic range of hybridization and $MM_{\%} = 100\% - PM_{\%}$.

5.1. Equilibrium Conditions

Solving the coupled kinetic equations at equilibrium, the ratio of bound concentrations can be expressed by

$$DR = \frac{C_1 K_{eq,2}}{C_2 K_{eq,1}} = \frac{r_1}{r_2} = \frac{C_1}{C_2} e^{\Delta\Delta G/RT}.$$

Using the definitions

$$\Delta\Delta G \equiv \Delta G_1 - \Delta G_2 = (\Delta H_1 - \Delta H_2) - (\Delta S_1 - \Delta S_2)T$$
$$\equiv \Delta\Delta H - \Delta\Delta S \bullet T$$

the dynamic range becomes

$$DR = \frac{C_1}{C_2} e^{-\Delta\Delta S/R} e^{\Delta\Delta H/RT}.$$

It is desirable to maximize the dynamic range. Typically, this is approached via methods of probe design, where the thermodynamic stabilities of the PM and MM hybrids are manipulated via the choice of probe sequence in order to maximize $\Delta\Delta G$ under the constraint of a fixed hybridization temperature. As seen from the equation above, however, the thermodynamic stability is only one contribution to the dynamic range. The second contribution is the relative concentrations of the species in solution. Since this factor can vary over many orders of magnitude in a non-predictable manner, it is not clear that probe design alone, even with temperature optimization, is sufficient to maximize selectivity.

In order to illustrate the effect of temperature on selectivity, **Table 12.2** shows the predicted dynamic range and $MM_{\%}$ at equilibrium for PM and MM hybrid pairs at different temperatures. These calculations assumed equal concentrations and show a dramatic decrease of dynamic range with increasing temperature. If the MM target concentration is greater than the PM target concentration, then the dynamic range decreases proportionally, with corresponding increase in $MM_{\%}$.

Again using the 1–2 system, **Fig. 12.10** plots the percentage of MM hybrids versus temperature for varying ratios of PM to MM target concentrations. Note that these results are independent of absolute concentrations. **Figure 12.10** clearly shows the effects of high concentrations of the MM species in reducing the fraction of PM hybrids. At the apparent optimal temperature based upon hybridization efficiency and time considerations

Table 12.2
Predicted dynamic range and *MM* bound percentage at different temperatures for the sequences of Table 12.1

PM-MM	$\Delta\Delta H$	$\Delta\Delta S$	*DR* 300 K	$MM_{\%}$	*DR* 310 K	$MM_{\%}$	*DR* 320 K	$MM_{\%}$	*DR* 330 K	$MM_{\%}$
1–2	9.8	21.1	340	0.29	200	0.50	121	0.82	76	1.30
1–3	18.3	44.9	3192	0.03	1186	0.08	469	0.21	196	0.51
2–3	8.5	23.8	9.4	9.62	5.9	14.4	3.9	20.6	2.6	28.0
4–5	18.2	45.1	2,754	0.04	1026	0.10	407	0.25	171	0.58
6–7	14.3	34.6	767	0.13	353	0.28	170	0.58	86	1.15

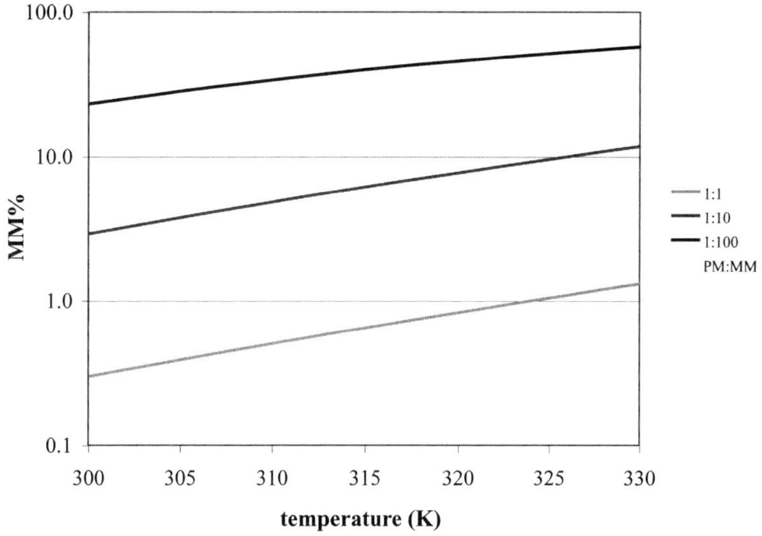

Fig. 12.10. Calculated percentage of MM hybrids as a function of temperature and PM:MM solution concentration ratio for the 1–2 model system.

from **Section 4.2**, it is clear that the dynamic range of selectivity is not high enough to avoid the possibility of quantification errors when the MM target is in significant excess concentration compared to the PM target. The selectivity can only be improved through the use of lower temperatures *(7)*, which also improves efficiency, but results in a dramatic increase in time to equilibrium, as well as sensitivity in that time to the MM target concentration.

The decrease in dynamic range with increasing temperature can be directly measured using the two-color real-time setup. The results of a two-color equilibrium melt are shown in **Fig. 12.11**. In the experiment, the solution concentrations were 3 nM wtt (PM) and 15 nM snpt (MM), denoted 3:15. Higher MM concentration was used in order to increase the signal level from the

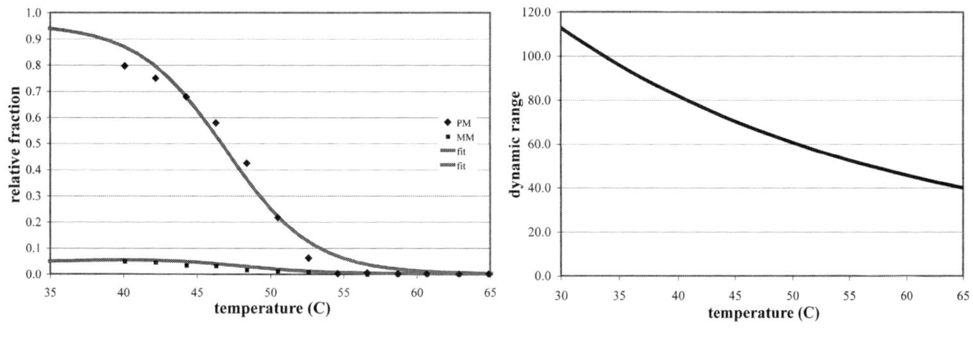

Fig. 12.11. (*Left*) PM and MM hybrid concentrations for the 4–5 system as a function of temperature for 3 nM wtt and 15 nM snpt concentrations. (*Right*) Calculated dynamic range from fitting the experimental data, adjusted to represent 1:1 solution concentrations.

MM hybrids. The theoretical result of equation 10 (and it's complement for the MM hybrid) was fitted to the experimental data by adjusting the thermodynamic parameters for each hybrid. The fitted parameters are (PM) $\Delta G = 70.6 - 0.182T$ and (MM) $\Delta G = 64.5 - 0.171T$. Note that these values are significantly different from the solution predictions, as would be expected. Note also that the fitted curve is extremely sensitive to these values, illustrating both the difficulty in obtaining this information from surface reactions and their susceptibility to experimental conditions and solution reactions (which were not taken into account here). The dynamic range calculated from the fitted curves is also plotted in **Fig. 12.11** demonstrating the predicted decrease. This dynamic range curve was adjusted to represent DR for a 1:1 solution concentration by multiplying the DR curve obtained for 3:15 by five.

In practical terms, these results indicate that the equilibrium dynamic range of discrimination drops as the temperature is elevated, a conclusion that appears contradictory to what is accepted in common microarray practice. However, this contradiction may be resolved by examining dynamic range as a function of temperature in the non-equilibrium regime.

5.2. The Non-Equilibrium Regime

It is clear that increasing temperature has three effects on hybridization: (1) decreased dynamic range, (2) decreased bound concentration, and (3) decreased time to equilibrium. The latter effect is the one that drives experiments to higher temperatures, with the apparent benefit of increased dynamic range. In order to understand this, we need to consider hybridization in the non-equilibrium regime.

Figure 12.12 plots the hybridization and dynamic range curves as a function of time for the 1–2 system using temperatures of 35, 40, and 45°C and equal 10 nM solution concentrations for PM and MM. On the hybridization plot, a log-log scale is used in

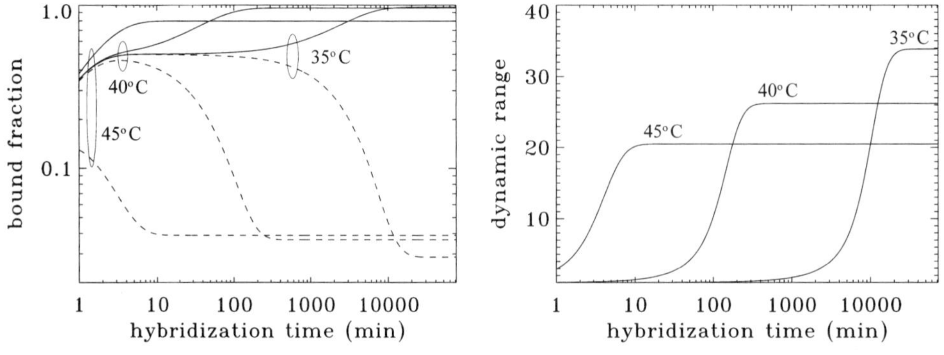

Fig. 12.12. (*Left*) Calculated PM (solid linestyle) and MM (dashed linestyle) hybridization curves for the 1–2 system at temperatures of 35, 40, and 45°C. (*Right*) Calculated dynamic range as a function of time.

order to represent all of the curves on the same graph. The dynamic range graph shows two important results: the dynamic range at equilibrium decreases with increasing temperature, but, at a fixed hybridization time, the dynamic range can be greater at higher temperatures. The latter effect can be explained by the influence of temperature on time to equilibrium, where the dynamic range reaches its maximum for any given reaction. Even though the lower temperatures have greater dynamic range, these reactions take significantly longer to reach equilibrium, and thus is the origin of the apparent discrepancy. In experimentation with temperature, the greater dynamic range regimes would not be observed due to the limited time of the experiment, and therefore, practitioners opt towards the higher temperature conditions.

These effects can be observed experimentally, as shown in **Fig. 12.13**. Here, hybridization for the 4–5 system is shown at 35°C with 3 nM concentrations. It is clear that the time to equilibrium is increased significantly over the 40°C hybridization (**Fig. 12.8**). Because hybridization was only performed up to 16 h, equilibrium was not reached. However, by fitting the hybridization curves (where the exponential decay and end-point values are the most relevant; errors in the fit at early times are due to the fact that diffusion is not modeled), the dynamic range at equilibrium can be extracted, as also plotted in **Fig. 12.13** for 35 and 40°C. The noise in the 40°C data is due to the MM hybrid being present at very low signal level, and the *DR* calculated from the curve fits provides a better indication of the actual *DR*. *DR* for the 6–7 system at 40°C is also plotted. It is clear that the 6–7 system takes longer to reach equilibrium than the 4–5 system, with lower dynamic range. This observation shows that a single temperature cannot always be used to optimize all relevant reactions *(42)*.

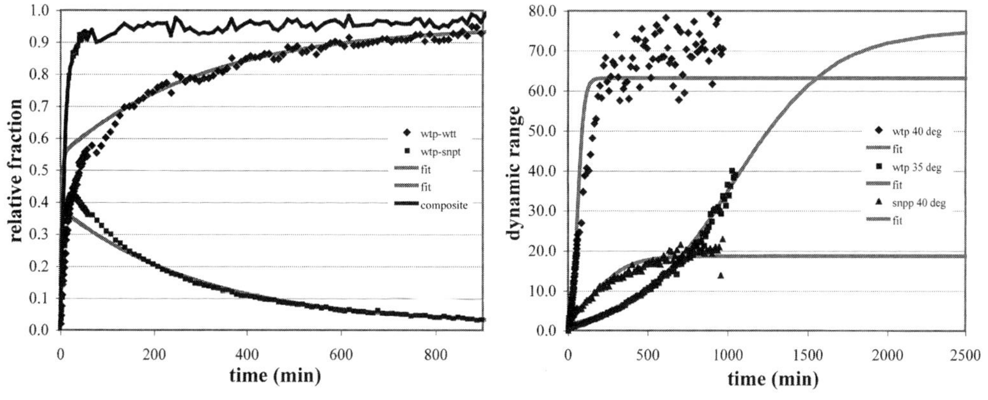

Fig. 12.13. (*Left*) Kinetics of PM and MM hybrids of the 4–5 system at 35°C for 3 nM:3 nM solution concentration. (*Right*) Dynamic range versus time for experimental hybridization curves.

6. Discussion

6.1. Optimal Temperature

The only meaningful universal consideration for temperature optimization is the temperature at which equilibrium can be obtained over a desired period of time (i.e., the duration of the assay); otherwise, any subsequent analysis, melting, or washing steps cannot be interpreted. At temperatures lower than this, equilibrium may not be obtained, making quantitative analysis difficult. At higher temperatures, efficiency and dynamic range suffer. Likewise, across an array, no single temperature optimizes all reactions. Too low, and some reactions will not reach equilibrium, and too high, and some may not be detectable.

Based on the equilibrium criterion, determining the optimal temperature is complicated by the fact that the concentration of the mismatch species influences the time to equilibrium. Since this effect diminishes with increasing temperature, it is advantageous to err on the side of higher temperatures. The important issue now is how this temperature can be empirically determined. Clearly, experiments can be performed using a multi-color real-time platform, as demonstrated here. Equilibrium is most easily determined as the time at which the dynamic range stops changing, as can be directly obtained from **Fig. 12.13**.

However, with more common experimental procedures, real-time data can only be acquired as a series of end-point experiments at different times, making this method of temperature optimization cumbersome, even though multi-color imaging is common (for ratiometric assays) and can be used to determine the bound fractions of multiple species at each probe spot, where each species has a different fluorescent label. Nevertheless, by performing a series of end-point experiments (over a fixed duration) at different temperatures, an optimal temperature can be determined. The temperature at which the dynamic range reaches its maximum value is the lowest temperature at which equilibrium is reached over that period of time. Below that temperature, the reaction is not in equilibrium, and the discussion of **Section 5.2** applies. Above that temperature, the reaction is in equilibrium but the dynamic range decreases, as per **Section 5.1**. **Figure 12.14** plots the dynamic range as a function of temperature for fixed periods of time using the 1–2 system, illustrating this discussion. As can be expected, for shorter periods of time, higher temperatures are needed, and the dynamic range decreases.

6.2. Towards Optimized Detection – A Microheater Array Chip

A microheater array device was designed by our group and fabricated at Sandia National Laboratories using the SwIFT^{TM} process. The device contains 18 individually controllable microheaters, each square, 250 μm on a side, and is shown in **Fig. 12.15**. The heaters are isolated from each other through a deep silicon etch, so as to prevent thermal cross-talk and improve temperature uniformity.

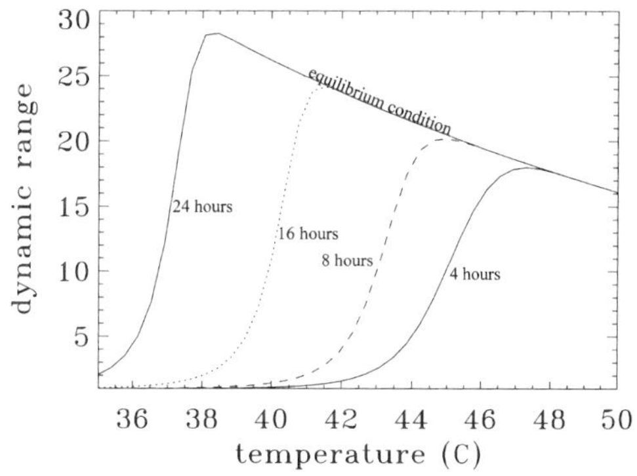

Fig. 12.14. Calculated dynamic range versus temperature for the 1–2 model system over different, fixed, periods of hybridization time. Concentrations for both targets are 10 nM and the association rate was assumed k_a=10^6 1/M•s.

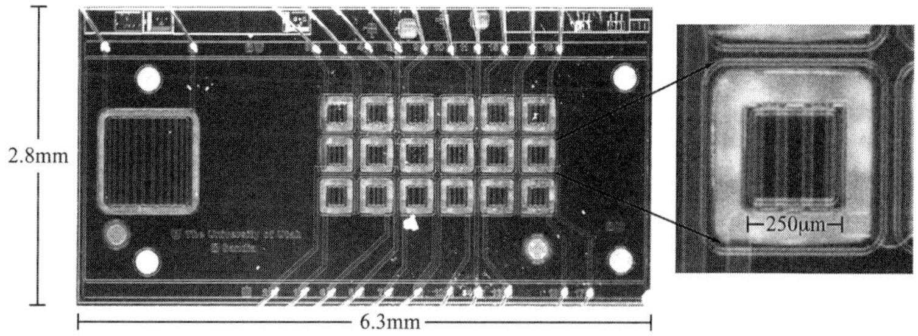

Fig. 12.15. Optical image of a microheater array chip.

A similar microheater array chip has been demonstrated previously *(43)* for the purposes of SNP genotyping. These chips have a number of advantages relative to temperature optimization. One is to determine the optimal temperature for a particular reaction by virtue of being able to monitor hybridization simultaneously at a number of different temperatures. Another is that, when the optimal temperatures are determined for a number of reactions, these reactions can be performed simultaneously under their optimal conditions. Even though the number of reaction spots is somewhat limited due to microfabrication and packaging considerations, these types of chips may become very useful for highly optimized diagnostic assays.

7. Conclusions and Prospects

In this chapter we have demonstrated that there are no simple criteria for unambiguous temperature optimization. Indeed, temperature optimization should be performed on a case by case basis.

Considerations for the overall goals of the experiment, sample preparation procedures, probe design, etc should be evaluated before a specific optimization procedure is determined. Our goal was to provide experimentalists with the tools for a rational approach to temperature optimization of the hybridization step during a microarray experiment, while putting the main emphasis on the importance of reaching equilibrium and expanding the dynamic range of target discrimination. The same approach can (and should) be applied to optimize the wash step of microarray experiments.

Our results were based on the study of binding of two highly homologous target species to a surface-immobilized probe. These interactions cannot be presented as independent events. There are significant consequences of our results for the interpretation of SNP detection (genotyping) assays using microarray approaches. The same considerations may be expanded to include a subset of targets with lower homology (down to 75%), which may have comparable affinity to the probe *(44)*. This latter fact has to be taken into consideration when performing expression profiling experiments. A similar model can be applied to the partitioning of a target among multiple homologous probes, which may become relevant for interpretation at low target concentrations *(45)*.

One of the possible solutions to resolve problems associated with the differences in thermodynamic parameters of probes and targets in the context of the same array is to use a microheater approach, which allows for independent temperature optimization on each sensing spot. While preliminary results of microheater approaches provide an optimistic outlook *(42,43)*, this approach can be applied only to small to medium size arrays. On the other hand, theoretical considerations and experimental results of kinetic studies open opportunities for emerging microarray-based technologies, most notably in real-time melting analysis *(46–48)* and real-time hybridization analysis *(6,36,49,50)*. Both of these approaches have the potential to drastically improve quantitative microarray analysis.

References

1. Oh, S. J., Hong, B. J., Choi, K. Y., and Park, J. W. (2006) Surface modification for DNA and protein microarrays. OMICS: A Journal of Integrative Biology 10, 327–343.
2. Dandy, D. S., Wu, P., and Grainger, D. W. (2007) Array feature size influences nucleic acid surface capture in DNA microarrays. Proceedings of the National Academy of Sciences 104, 8223–8228.
3. Erickson, D., Li, D., and Krull, U. (2003) Modeling of DNA hybridization kinetics for spatially resolved biochips. Analytical Biochemistry 317, 186–200.
4. Peterson, A. W., Wolf, L. K., and Georgiadis, R. M. (2002) Hybridization of mismatched or partially matched DNA at surfaces. Journal of the American Chemical Society 124, 4601–4607.

5. Hagan, M. F., and Chakraborty, A. K. (2004) Hybridization dynamics of surface immobilized DNA. Journal of Chemical Physics 120, 4958–4968.
6. Myszka, D. G., He, X., Dembo, M., Morton, T. A., and Goldstein, B. (1998) Extending the range of rate constants available from BIACORE: Interpreting mass transport-influenced binding data. Biophysical Journal 75, 583–594.
7. Livshits, M. A., and Mirzabekov, A. D. (1996). Theoretical analysis of the kinetics of DNA hybridization with gel-immoblized oligonucleotides. Biophysical Journal 71, 2795–2801.
8. Okahata, Y., Kawase, M., Niikura, K., Ohtake, F., Furusawa, H., and Ebara, Y. (1998). Kinetic measurements of DNA hybridization on an oligonucleotide-immobilized 27-MHz quartz crystal microbalance. Analytical Chemistry 70, 1288–1296.
9. Fish, D. J., Todd H. M., Brewood G. P., Goodarzi J. P., Alemayehu S., Bhandiwad A., Searles R. P., and Benight A. S. (2007) DNA multiplex hybridization on microarrays and thermodynamic stability in solution: a direct comparison. Nucleic Acids Research 35, 7197–7208.
10. Wick, L. M., Rouillard, J. M., Whittam, T. S., Gulari, E., Tiedje, J. M., and Hashsham, S. A. (2006) On-chip non-equilibrium dissociation curves and dissociation rate constants as methods to assess specificity of oligonucleotide probes. Nucleic Acids Research 34, e26.
11. Carlon, E. and Heim, T. (2006) Thermodynamics of RNA/DNA hybridization in high-density oligonucleotide microarrays. Physica A 362, 433–449.
12. Held, G. A., Grinstein, G. and Tu, Y. (2003) Modeling of DNA microarray data by using physical properties of hybridization. Proceedings of the National Academy of Sciences 100, 7575–7580.
13. www.bioinfo.rpi.edu/applications/hybrid/hybrid2.php/
14. Landry, J. P., Zhu, X. D., and Gregg, J. P. (2004) Label-free detection of microarrays of biomolecules by oblique-incidence reflectivity difference microscopy. Optics Letters 29, 581–583.
15. Swami, N. S., Chou, C., and Terberueggen, R. (2005) Two-potential electrochemical probe for study of DNA immobilization. Langmuir 21, 1937–1941.
16. Cheng, Y. T., Tsai, C. Y., and Chen, P. H. (2007) Development of an intergrated CMOS DNA detection biochip. Sensors and Actuators B 120, 758–765.
17. Ilic, B., Yang, Y., Aubin, K., Reichenbach, R., Krylov, S., and Craighead, H. G. (2005) Enumeration of DNA molecules bound to a nanomechanical oscillator. Nano Letters 5, 925–929.
18. Kim, H., Kim, J., Kim, T., Oh, S., and Choi, E. (2005) Optical detection of deoxyri-bonucleic acid hybridization using an anchoring transition of liquid crystal alignment. Applied Physics Letters 87, 143901–143903.
19. McKendry, R., Zhang, J., Arntz, Y., Strunz, T., Hegner, M., Lang, H. P, Baller, M. K., Certa, U., Meyer, E., Guntherodt, H., and Gerber, C. (2002) Multiple label-free biodetection and quantitative DNA-binding assays on a nanomechanical cantilever array. Proceedings of the National Academy of Sciences 99, 9783–9788.
20. Roederer, E., and Bastiaans, G. J. (1983) Microgravimetric immunoassay with piezoelectric crystals. Analytical Chemistry 55, 2333–2336.
21. Ngeh-Ngwainbi, J., Suleiman, A. A., and Guilbault, G. G. (1990) Piezoelectric crystal biosensors. Biosensors and Bioelectronics 5, 13–26.
22. Liedberg, B., Nylander, C., and Lundstrom, I. (1983) Surface plasmon resonance for gas detection and biosensing. Sensors and Actuators 4, 299–304.
23. Bianchi, N., Rustigliano, C., Tomassetti, M., Feriotto, G., Zorzato, F., and Gambari, R. (1997) Biosensor technology and surface plasmon resonance for real-time detection of hiv-1 genomic sequences amplified by polymerase chain reaction. Clinical and Diagnostic Virology 8, 199–208.
24. Stimpson, D. I., Hoijer, J. V., Hsieh, W. T., Jou, C., Gordon, J., Theriault, T., Gamble, R., and Baldeschwieler, J. D. (1995) Real-time detection of DNA hybridization and melting on oligonucleotide arrays by using optical waveguides. Proceedings of the National Academy of Sciences 92, 6379–6383.
25. Kronick, M. N., and Little, W. A. (1975) New immunoassay based on fluorescence excitation by internal-reflection spectroscopy. Journal of Immunological Methods 8, 235–240.
26. Reichert, W. M. (1989) Evanescent detection of adsorbed protein films: assessment of

optical considerations for absorbance and fluorescence spectroscopy at the crystal solution and polymer solutions interfaces. Critical Reviews in Biocompatability 5, 173.

27. Zhou, Y., Laybourn, P. J., Magill, J. V., and Rue, R. M. D. L. (1991) An evanes-cent fluorescence biosensor using ion-exchanged buried waveguides and the enhancement of peak fluorescence. Biosensors and Bioelectronics 6, 595–607.
28. Plowman, T. E., Reichert, W. M., Peters, C. R., Wang, H. K., Christensen, D. A., and Herron, J. N. (1996) Femtomolar sensitivity using a channel-etched thin film waveguide fluoroimmunosensor. Biosensors and Bioelectronics 11, 149–160.
29. Taitt, C. R., Anderson, G. P., and Ligler, F. S. (2005) Evanescent wave fluorescence biosensors. Biosensors and Bioelectronics 20, 2470–2487.
30. http://www.zeptosens.com/
31. http://www.tirftechnologies.com/
32. http://www.pamgene.com/
33. http://www.imstarsa.com/
34. Bishop, J., Wilson, C., Chagovetz, A. M., and Blair, S. (2007) Competitive displacement of DNA during surface hybridization. Biophysical Journal 92, L10-L12.
35. Bishop, J., Wilson, C., Chagovetz, A. M., and Blair, S. (2007) Real-time optical detection of competitive hybridization on microarrays, Proceedings of the SPIE: Advanced Biomedical and Clinical Diagnostic Systems 6430, 643002.
36. Bishop, J., Chagovetz, A. M., and Blair, S. (2008) Kinetics of Multiplex Hybridization: Mechanisms and Implications. Biophysical Journal 94, 1726–1734.
37. Horne, M. T., Fish, D. J., and Benight, A. S. (2006) Statistical thermodynamics and kinetics of DNA multiplex hybridization reactions. Biophysical Journal 91, 4133–4153.
38. Gao, Y., Wolf, L. K., and Georgiadis, R. M. (2006) Secondary structure effects on DNA hybridization kinetics: a solution versus surface comparison. Nucleic Acids Research 34, 3370–3377.
39. Sekar, M. M. A., Bloch, W., and John, P. M. S. (2005) Comparative study of sequence-dependent hybridization kinetics in solution and on microspheres. Nucleic Acids Research 33, 366–375.
40. Bhanot, G., Louzoun, Y., Zhu, J., and Delisi, C. (2003) The importance of thermodynamic equilibrium for high throughput gene expression arrays. Biophysical Journal 84, 124–135.
41. Sator, M., Schwanekamp, J., Halbleib, D., Mohamed, I., Karyala, S., Medvedovic, M., Tomlinson, C. R. (2004). Microarray results improve significantly as hybridization approaches equilibrium. Biotechniques 36, 790–796.
42. Petersen, J., Poulsen, L., Petronis, S., Birgens, H., Dufva, M. (2007). Multi-thermal washer for DNA microarrays simplifies probe design and gives robust genotyping assays. Nucleic Acids Research, Advance Access doi:10.1093/nar/gkm1081
43. Kajiyama, T., Miyahara, Y., Kricka, L., Wildeng, P., Graves, D., Surrey, S., and Fortina, P. (2003) Genotyping on a thermal gradient DNA chip. Genome Research 13, 467–475.
44. Wren, J. D., Kulkarni, A., Joslin, J., Butow, A., and Harold, R. G. (2002). Cross-hybridization on PCR-spotted microarrays. IEEE Engineering in Medicine and Biology 21, 71–75.
45. Fish, D. J., Horne, M. T., Searles, R. P., Brewood G. P., Benight, A. S. (2007). Multiplex SNP discrimination. Biophysical Journal 92, L89–L91.
46. Meuzelaar, L. S., Hopkins, K., Liebana, E., and Brookes, A. J. (2007) DNA diagnostics by surface-bound melt-curve reactions. Journal of Molecular Diagnostics 9, 30–41.
47. Fiche, J. B., Buhot, A., Calemczuk, R., and Livache, T. (2007) Temperature effects on DNA chip experiments from surface plasmon resonance imaging: isotherms and melting curves. Biophysical Journal 92, 935–946.
48. Russom, A., Haasl, S., Brookes, A. J., Andersson, H., and Stemme, G. (2006) Rapid melting curve analysis on monolayered beads for high-throughput genotyping of single nucleotide polymorphisms. Analytical Chemistry 78, 2220–2225.
49. Vijaynedran, R. A., Ligler, F. S., and Leckband, D. E. (1999) A computational reaction-diffusion model for the analysis of transport limited kinetics. Analytical Chemistry 71, 5405–5412.
50. Sadana, A., and Ramakrishnan, A. (2001) A fractal analysis approach for the evaluation of hybridization kinetics on biosensors. Journal of Colloid and Interface Science 234, 9–18.

Chapter 13

Whole-Genome Genotyping on Bead Arrays

Kevin L. Gunderson

Abstract

In this review, we describe the laboratory implementation of Infinium® whole genome genotyping (WGG) technology for whole genome association studies and copy number studies. Briefly, the Infinium WGG assay employs a single tube whole genome amplification reaction to amplify the entire genome; genomic loci of interest are captured on an array by specific hybridization of picomolar concentrations amplified gDNA. After target capture, single nucleotide polymorphisms (SNPs) are genotyped on the array by a primer extension reaction using hapten-labeled nucleotides. The resultant hapten signal is amplified by immunhistochemical sandwich staining and the array is read out on a high resolution confocal scanner. We have combined this Infinium assay with high-density BeadChips to create the first array platform capable of genotyping over 1 million SNPs per slide. Additionally, the complete Infinium assay is automated using Tecan GenePaint™ slide processing system. Hybridization, washing, array-based primer extension and staining are performed directly in the Tecan capillary gap Te-Flow Through chambers. This automation process greatly increases assay robustness and throughput while enabling Laboratory Information Management System (LIMS) control of sample tracking. Finally, we give several examples of how this advance in genotyping technology is being applied in whole genome association and copy number studies.

Key words: Arrays, genotyping, SNP, whole genome association, whole genome amplification (WGA), Infinium®, DNA copy number.

1. Introduction

Whole genome association studies (WGAS) were originally proposed over a decade ago by Risch and Merikangas *(1)*. However, the genomic information and the technology to conduct such studies have only recently come to fruition *(2)*. The strategy underlying WGAS is to employ linkage disequilibrium (LD) within the genome to type genetic markers linked to nearby causal variants. Regions in the genome in which the markers show statistically

Martin Dufva (ed.), *DNA Microarrays for Biomedical Research: Methods and Protocols, vol. 529*

DOI 10.1007/978-1-59745-538-1_13 Springerprotocols.com

significant (χ^2 test) differences in allele frequency are flagged as potential regions harboring a candidate disease locus *(3)*. Two breakthroughs have enabled WGAS; *(1)* Identification of SNPs and characterization of the LD structure of the human genome by the International HapMap Project (www.hapmap. org) *(4)*, and *(2)* development of WGG technology *(5)*. These breakthroughs have enabled analysis of hundreds of thousands to over a million SNPs on a single sample for a price of a few hundred dollars. This represents a cost reduction of greater than 100-fold within the past 5 years. In this chapter, we describe the technological innovations that have enabled WGG along with a detailed protocol for implementation of WGG in the laboratory. Second chapter in this book describes the bioinformatics selection strategy for choosing SNPs for the Infinium tag SNP HapMap product line.

1.1. Development of a WGG Platform

Illumina has developed a whole genome SNP genotyping assay, called the Infinium assay, which enables virtually unlimited multiplexing in SNP genotyping. The design of the assay ensures that its multiplexing ability scales purely with the number of features that can be placed on the array. Besides the array, the overall success of the Infinium assay derives from the quality of probe features found on our bead array format – the BeadChip (**Fig. 13.1**). The BeadChip provides an ideal substrate for array-based enzymatic assays. First of all, DNA probes on beads are all full length by virtue of the immobilization process in which oligonucleotides are attached to beads by a 5′ reactive moiety synthesized in the last step of a 3′->5′ oligonucleotide synthesis reaction. This also ensures that all probes have intact 3′ termini suitable for primer

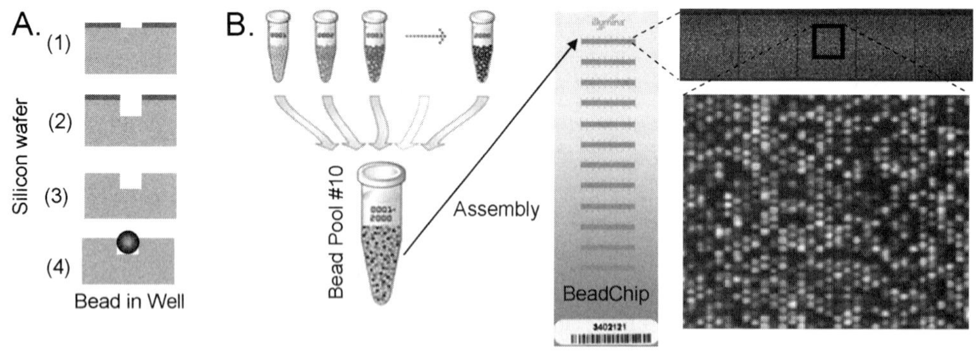

Fig. 13.1. BeadChip design. **(A)** Standard MEMS technology is used to create silicon-patterned substrates used in BeadChip product. Microwells are created by removal of photolithographically patterned regions of a photoresist material, and then plasma etching of the underling silicon substrate. **(B)** Individual bead pools are loaded onto separate regions (patterned areas) of the BeadChip substrate. The 1 M substrate contains 20 regions accommodating 20 independent bead pools each with ~60,000 different bead types. Each region can accommodate ~1 million beads generating greater than 16-fold redundancy per bead type. An image of individual beads developed with the Infinium assay on the Human 1 M BeadChip is shown.

extension. In contrast, arrays created through "in situ" synthesis are usually synthesized 3′–5′ so that there are no free 3′ termini, and if synthesized in the reverse direction the probes contain a significant degree of 3′ truncations, again not suitable for primer extension. BeadChips are inherently more manufacturable than ordered arrays created by spotting, ink jetting, or photolithographic processes since beads of a particular type are created in "one" immobilization event from which tens of thousands of arrays can be made. This leads to improved array-to-array feature consistency. Finally, each BeadChip is decoded (identity of each bead determined) by hybridization in the factory, providing a functional QC of each array *(6)*.

The specificity and scalability of Infinium WGG assay originates from its intrinsic design elements consisting of four simple well-controlled steps: (1) whole-genome amplification (WGA), (2) locus-specific hybridization capture of WGA products, (3) array-based enzymatic SNP scoring, and (4) immunohistochemical-based signal development and amplification (**Fig. 13.2**). The WGA reaction is used to amplify, in a relatively unbiased manner, the input gDNA (10–750 ng) by at least 1000-fold, generating several hundred micrograms of final amplified DNA product. This large amount of product translates into a final target concentration of several pM in the hybridization reaction which effectively drives hybridization capture to the array. The capture probes consist of 5'-end immobilized 80-mer oligonucleotides of which the first 30-bases (5' sequence) are used for decoding, and the remaining 50 bases constitute the query sequence for the locus of interest. After a stringent hybridization of the WGA product to the capture array, an array-based polymerase extension step is used to score the SNP directly on the array. This bipartite combination

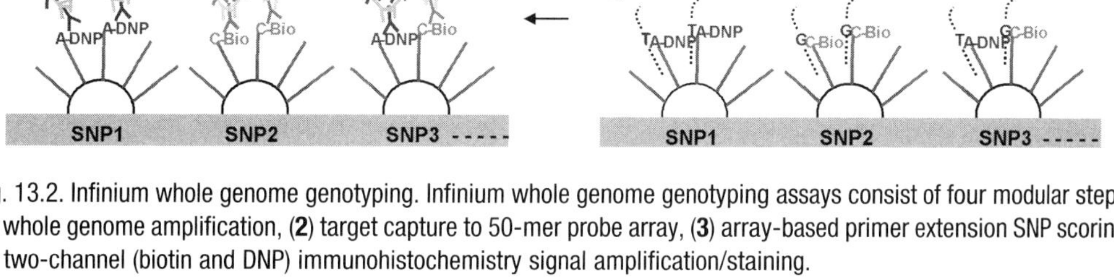

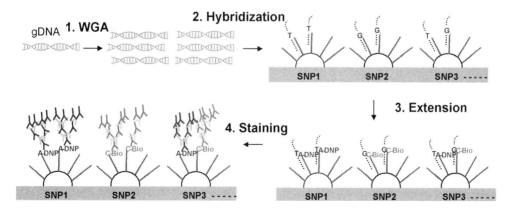

Fig. 13.2. Infinium whole genome genotyping. Infinium whole genome genotyping assays consist of four modular steps: (**1**) whole genome amplification, (**2**) target capture to 50-mer probe array, (**3**) array-based primer extension SNP scoring, (**4**) two-channel (biotin and DNP) immunohistochemistry signal amplification/staining.

of hybridization and primer extension greatly increases the overall specificity of the assay.

The primer extension step, besides increasing locus specificity, also provides single base discrimination of the SNP alleles. Two different modalities of primer extension are employed in the Infinium assay: (1) Infinium I – allele-specific primer extension (ASPE), and (2) Infinium II – single base extension (SBE) *(7–9)*. The Infinium I (ASPE) assay employs two probe pairs (designated bead types A and B corresponding alleles A and B) per locus (**Fig. 13.3A**). The two probes are identical in their 50 bases of sequence except at their 3′ terminal base. Each probe is designed to match and extend on its cognate allelic target sequence. All possible SNP classes can be designed with the Infinium I assay. In contrast, the Infinium II (SBE) assay employs single probe per assay effectively increasing information content on the array and discriminates between alleles by incorporating differentially labeled dideoxynucleotide terminators in the primer extension step (**Fig. 13.3B**) *(10)*. For ease of both assay and instrument design, we employ only two hapten labels – biotin and dinitrophenol (DNP). In our current extension mix, ddATP and ddUTP nucleotides are labeled with DNP, and ddCTP and ddGTP are labeled with biotin. After completion of the primer extension step, a two-channel immunohistochemistry-based sandwich staining and signal amplification step are used to boost the overall signal-to-noise of the assay.

The use of two hapten configuration in the assay restricts the types of SNPs that can be scored to [A/C], [A/G], [C/T], and [T/G]. Nonetheless over 83% of the SNPs in dbSNP can be scored using this design format. Moreover, for selection of tag SNPs, the

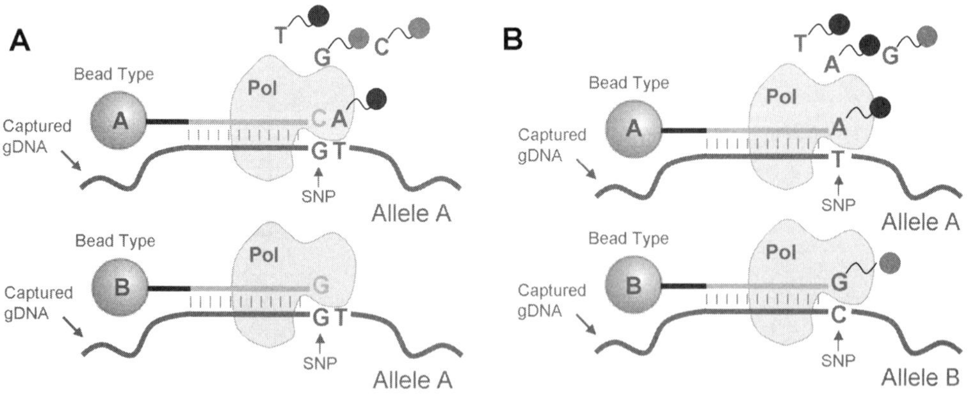

Fig. 13.3. Primer extension with ASPE and SBE probes using Infinium II assay. **(A)** Allele-specific primer extension (ASPE) uses two bead types and one color per SNP assay. The primer extension incorporated label is read out in a single color channel for both bead types, either green or red depending on the base adjacent to the SNP. **(B)** Single base extension (SBE) employs a single bead type and two haptens (colors) per SNP assay. Allelic discrimination is achieved by using two color differentially labeled terminators. In this manner, both ASPE and SBE genotyping probes can be read out on a single array using Infinium II (SBE) biochemistry.

redundancy of tag SNPs allows judicious selection of only tag SNPs compatible with SBE probe design. Finally, this limitation has been completely eliminated in the 1 M and iSelect™ products by combining both ASPE and SBE probe designs on a single array and using the Infinium II (SBE) biochemistry to simultaneously score both assay designs. This is accomplished by reading out the two ASPE probes in the color channel corresponding to the base following the SNP. SNPs with next base A or T are read out in the red (DNP) channel, and those with next base C or G in the green (biotin) channel. The challenge to assaying both Infinium I and II probe designs in a single primer extension reaction is to use SBE biochemistry but at limiting nucleotide and polymerase concentrations to ensure adequate ASPE discrimination.

1.2. The 1 M Infinium WGG BeadChip

SNPs are the markers of choice for high-density array-based genotyping assays, with over 10 million SNPs (>1%) present in the human genome *(4)*. Much of the human genome exhibits blocks of linkage disequilibrium with an average size of 20–50 kb *(11, 12)*. Within a block, the phasing of SNPs are highly correlated, and one can identify a minimal set of spanning SNPs called haplotype "tagging" SNPs (htSNPs) that suffice to characterize the common genetic variation within these haplotype blocks *(13)*. Alternatively, rather than characterizing the LD of the genome by haplotype blocks, the LD can be characterized by selected "tag SNPs" that represent LD bins *(3)*. The major goal of the International HapMap Project was to characterize the linkage disequilibrium across the genome for the three major ethnic groups, and provide information for selecting tag SNPs for LD-based association studies.

One of the strengths of the Infinium WGG assay is its relatively unconstrained ability to select "high value" SNPs. Using this feature, the content of the 1 M BeadChip was developed around tag SNPs generated from the International HapMap Consortium. Also, "high value" SNPs such as CNV SNPs, functional SNPs, genic SNPs, and others were also included.

The high density of the 1 M BeadChip was achieved using a MEMS-patterned slide substrate (82.5 mm length vs. standard 75 mm) supporting over 1 million different bead types (**Fig. 13.3**). The BeadChip design consists of 20 sections into which beads are assembled from a pool containing ~60,000 different bead types. Each section receives a different bead pool, thus a different set of 60,000 bead types for a total of over 1 million bead types. Each stripe also holds ~1 million beads generating an average redundancy of greater than 16 beads per bead type. Using our Infinium II assay, the current design can support over 1 million SNP assays. Further reduction in bead size and spacing can further increase the number of SNP assays per slide.

1.3. Multi-Sample BeadChip Formats

The flexibility of the BeadChip substrate supports a variety of different product offerings including both single sample and multi-sample platforms. Simply by partitioning the sections of the BeadChip substrate into separate compartments by application of simple laminated adhesive seals, various products spanning the density vs. sample number matrix can be created. For instance, the 1 M BeadChip substrate is also used as a two sample BeadChip for the HumanHap550 product, and the original HumanHap550 substrate is also used as a 12 sample product with much less features.

2. Materials

1. BeadStation 500G System (Illumina, SC-16-101/102)
2. Infinium Plus Standard-Throughput Option Package (WG-15-113/114; **Fig. 13.4A**)
   a. Custom Hybridization Oven w/ Rocker Attachment
   b. Heat Block
   c. Heat Sealer
   d. Hybex Incubator

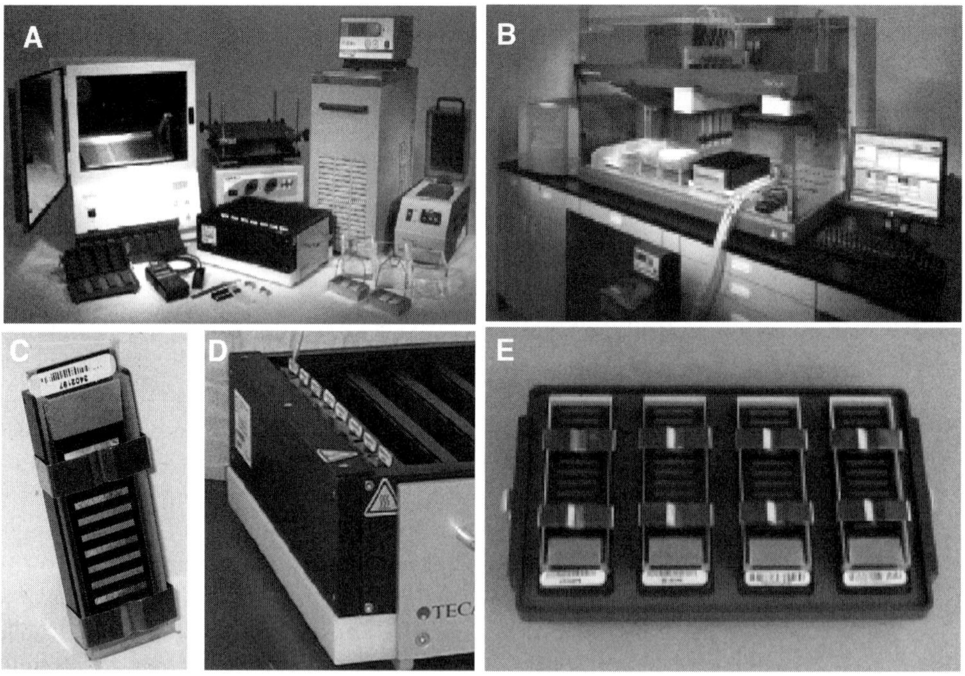

Fig. 13.4. Automated BeadChip processing. **(A)** BeadStation accessories for Infinium WGG assay. **(B)** A complete automated robotics solution for processing the BeadChips on the Tecan Freedom EVO. **(C)** A Tecan capillary gap Te-Flow Through chamber is used for easy reagent exchanges. **(D)** A temperature-controlled Te-Flow Through Chamber Rack and Tecan robot are used for the washing, primer extension, and staining (detection) steps of the Infinium assay. **(E)** Humidified hybridization enclosure housing BeadChips during overnight hybridization.

   e. Microplate Shaker
   f. Re-circulating Water Pump
   g. Vacuum Desiccator with Tubing
   h. BeadChip storage boxes, hybridization chamber gaskets, hybridization chambers, TeFlow Chambers and spacers, TeFlow Chamber Rack, and other accessories.
   i. BeadStudio Data Analysis package
   j. "non-autoclavable" cap mat (ABgene, AB-0566)
3. Infinium Automation Standard package (SC-30-401/402; **Fig. 13.4B**)
   a. Tecan EVO 150 cm with LiHa, Workstation PC, accessories
4. Infinium II WGG Kit (WG-30-502, HumanHap550 Bead-Chips & Reagents)
5. Infinium Assay Reagents (**Table 13.1**)

Table 13.1
Infinium XStain reagents

Name	Description	Part No.
MP1	Neutralization solution	11190751
AMM	Amplification Master Mix	11192044
FRG	Fragmentation solution	11190022
PA1	Precipitation solution	11190031
RA1	Resuspension, hybridization, and wash solution	11191914
PB1	Reagent used to prepare BeadChips for hybridization	11191922
PB2	Humidifying buffer used during hybridization	11191130
XC1	XStain BeadChip solution 1	11208288
XC2	XStain BeadChip solution 2	11208296
TEM	Two-Color Extension Master Mix	11208309
XC3	(80 ml) XStain BeadChip solution 3	11208392
XC3	(240 ml) XStain BeadChip solution 3	11208421
LTM	Labeling Two-Color Master Mix	11208325
ATM	Anti-Stain Two-Color Master Mix	11208317
XC4	XStain BeadChip solution 4	11208430

3. Methods

The Infinium assay is straightforward to implement in most molecular biology labs since most steps are automated and employ pre-formulated reagents which only need to be thawed and mixed before use (**Note 1**). The use of pre-formulated reagents increases overall robustness by minimizing user error and generating more run-to-run consistency. Assay automation is performed on a Tecan Freedom Evo® robot with an integrated GenePaint slide processing system (www.tecan.com). The GenePaint system employs a capillary gap Te-Flow Through chamber coupled to a BeadChip (**Fig. 13.4C**) to enable easy washing and exchange of reagents on the BeadChip. The Te-Flow Through chambers are placed in a temperature-controlled Te-Flow Through chamber rack (**Fig. 13.4D**), allowing precise temperature control of all extension and staining steps. Finally, a Tecan robot performs all reagent transfer steps including pipetting of wash solutions, blocking mixes, extension reagents, and staining reagents. The overall process flow is outline in **Fig. 13.5,** and each process step will be described in detail below. The steps include (1) WGA; (2) BeadChip Hybridization; (3) Extension-Staining (XStain), and (4) Imaging and Data Analysis. Please consult the user's manual for the latest protocol updates before running any experiment.

3.1. Whole Genome Amplification

The Infinium assay employs an overnight (20 h) isothermal WGA protocol to amplify the genome over 1000-fold. The recommended input amount varies from 200 ng for multi-sample assays to 750 ng for single sample BeadChip assays (**Note 2** for lower inputs). DNA should be of high quality (**Note 3**). The Infinium assay uses a pre-formulated WGA kit consisting of proprietary components required for relatively uniform amplification, downstream fragmentation, and processing. Amplification proceeds isothermally at 37°C for ~ 20 h. After amplification, a fragmentation protocol is used to reduce the amplified product size to 200–300 bp. This fragmentation improves both resuspension and hybridization efficiency. After fragmentation, the reaction is stopped using a mix of ammonium acetate and EDTA. The EDTA serves the dual purpose of stopping the fragmentation reaction, and helping to dissolve precipitates formed during amplification, which could interfere in downstream processes. One volume of isopropanol is added to precipitate the final product. This reduces the carry-over of dNTPs, and concentrates the DNA sample. The DNA pellet is incubated in a formamide-containing hybridization

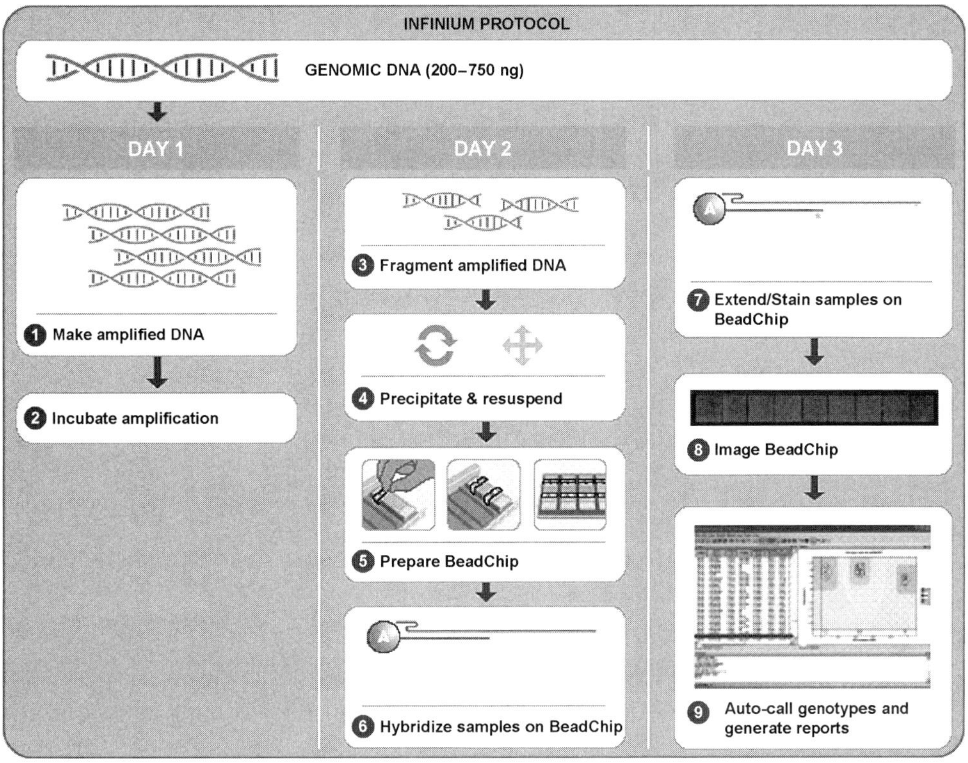

Fig. 13.5. Infinium WGG process flow. The DNA sample is whole-genome amplified in an overnight isothermal reaction (Step 1 and 2). After amplification, the product is fragmented to about 300–500 bases (Step 3). After isopropanol precipitation and resuspension (Step 4), the samples are applied to a BeadChip mated to a capillary gap Te-Flow Through chamber and hybridized overnight for ~ 20 h (Step 5). After the overnight hybridizations, the Te-Flow unit with assembled BeadChip is placed on a GenePaint Te-Flow Through chamber rack equilibrated to 44°C. An automated Extension-Staining (XStain) process is performed directly on the GenePaint system integrated with a Freedom EVO robot. Washing, primer extension, and immunohistochemical staining are completed in about 2.5 h (Step 7). After staining, the BeadChips are read on a high-resolution confocal scanner (0.84 μm resolution). Genotype calls are made automatically by GenCall software, generating reports of the results.

buffer at 48°C for 1 h and then vortexed for 1 min at 1,500 rpm to resuspend the pellet.

The expected yield from the WGA reaction is ~ 1.5 μg/μl, and the final hybridization concentration is 5–6 μg/μl. After resuspension, the sample can be used immediately or frozen down (−20°C or −80°C) and stored at least several months before use. Immediately before hybridization, the sample should be denatured at 95°C for ~20 min and allowed to cool on the bench for 5–10 min before being applied to the BeadChip array according to manufacturer's instructions.

3.1.1. WGA Protocol (For Single Sample/BeadChip)

1. Resuspend DNA sample at 50 ng/μl. Depending on the purity of DNA, DNA can be quantitated by 260 nm absorbance or PicoGreen fluorescence (Invitrogen). PicoGreen

generally gives more consistent result across a broader range of DNA purities.

2. Add 15 μl 0.1 N NaOH to 15 μl DNA sample placed in a MIDI (0.8 ml) 96-well storage plate and incubate for 10 min at room temperature. From one to 24 samples can be processed in a single MIDI plate. Only three columns (1, 5, and 9) are used for samples since the remaining columns will be used to accommodate the extra volume required in the isopropanol precipitation step.
3. Add 270 μl primer/neutralization mix (MP1) to each sample well.
4. Add 300 μl amplification master mix (AMM) to each sample well.
5. Seal plate with "non-autoclavable" cap mat. This type of cap mat seals more effectively.
6. Invert plate 10 times to mix contents and briefly spin down at 280 g.
7. Place MIDI plate in oven at 37°C for ~20 h (*see* **Note 4** for precipitate formation).
8. Briefly spin down plate at 50 *g* for 1 min (light spin avoids compacting precipitate, *see* **Note 4**).
9. Split 600 μl WGA reaction into 4 wells per sample, with each well receiving 150 μl (Col 1 -> 2, 3, and 4; Col 5 -> 6, 7, and 8; Col 9 -> 10, 11, and 12). This creates room in the well to receive isopropanol in the precipitation step.
10. Add 50 μl fragmentation mix (FRG) to each well, seal and vortex at 1,600 rpm in High-Speed Microplate Shaker for 1 min, spin down at 50 *g* for 1 min, and then incubate at 37°C for 1 h. This plate shaker employs Velcro straps for securing the plate.
11. Briefly spin down plate at 50 *g* for 1 min.
12. Add 100 μl precipitation mix (PA1) to each well, seal and vortex at 1,600 rpm for 1 min, spin down at 50 *g* for 1 min, and incubate 5 min at 37°C.
13. Add 300 μl isopropanol to each well.
14. Seal plate with a fresh "non-autoclavable" cap mat (make sure cap mat and plate are dry around rims to ensure good seal; otherwise isopropanol may leak during inversion.). Mix plate by inversion 10 times.
15. Incubate the amplification plate for 30 min at 4°C.
16. Spin down plate at 3,000 *g* for 20 min at 4°C.
17. Immediately after centrifugation, remove supernatant by slowly inverting plate over sink and pouring out the

isopropanol. Blot inverted plate on a paper towel. Blue pellets (due to a blue additive in the precipitation step) should be visible in the bottom of each well at this point.

18. Dry inverted plate on a plastic test tube rack for 1 h at room temperature.
19. Add 42 μl hybridization buffer (RA1) to each well. Heat seal with foil sheet, and incubate for 1 h at 48°C.
20. Resuspend sample by vortexing at 1,800 rpm for 1 min, then centrifuge at 280 g for 1 min.
21. Retrieve the samples from the four dispersed wells and collect into one original well for a total volume of 160 μl per original sample well (*see* **Note 5** for long-term storage).
22. Denature samples at 95°C for 20 min in a custom aluminum heat block on Hybex heating unit.
23. Remove samples from the heat block, pulse centrifuge at 280 *g*, and place on bench for ~ 5–10 min before loading samples onto BeadChips.

3.2. BeadChip Hybridization (Single Sample)

Hybridization of DNA to the BeadChip occurs directly in the Te-Flow Through chamber. To prevent evaporation, these chambers are placed into a humidified enclosure.

1. Assemble BeadChip into the Te-Flow Through chamber as recommended by Infinium user manual.
2. Prepare Illumina® hybridization enclosures (chambers) by pipetting 200 μl humidifying buffer (PB2) into troughs as described in Infinium user manual (This buffer has the same water vapor pressure as the sample to prevent evaporation).
3. To prepare BeadChips-Te-Flow Through chambers for hybridization, slowly dispense 150 μl of 100% formamide into assembled Te-Flow Through chamber placed upright in Te-Flow rack (equilibrated to room temperature). The formamide is used to pre-wet the BeadChip surface and prevent bubbles from forming during DNA sample loading. If bubbles form during formamide loading, disassemble, wash slide in low salt buffer (PA1), spin dry slide (280 *g*) and repeat assembly and formamide loading.
4. Dispense 150 μl of hybridization buffer (RA1) into reservoir of Te-Flow Through chamber and allow to flow through. Repeat with another 150 μl of hybridization buffer (RA1).
5. Dispense 150 μl denatured DNA sample into BeadChip as above.
6. Remove Te-Flow Through chamber from chamber rack, briefly blot residual hybridization buffer from bottom of Te-Flow Through chamber, and place horizontally in

Illumina hybridization enclosures, seal, and place on rocker in Hybex oven equilibrated to 48°C.

7. Hybridize samples to BeadChips overnight at 48°C for 16–24 h.

3.3. BeadChip Hybridization (Multi-Sample)

1. Prepare Illumina hybridization chambers by pipetting 200 μl humidifying buffer (PB2) into troughs as described in Infinium user manual (This buffer has the same water vapor pressure as the sample to prevent evaporation).
2. Place multi-sample BeadChip inserts into hybridization chambers (not needed for single sample Beadchip assembled in Te-Flow Through cells).
3. Pipette samples, manually or robotically, into appropriate ports on multi-sample BeadChips. Evacuation ports or slits are employed on the seal to enable smooth uptake of the sample.
4. Place BeadChips into inserts in hybridization chamber, seal, and hybridize samples at 48°C for 16–24 h.

3.4. Extension/ Staining (XStain)

This step is performed on BeadChips assembled in Te-Flow Through chambers placed on a GenePaint Te-Flow Through Chamber Rack. RA1 reagent is used to wash away unhybridized and non-specifically hybridized DNA targets. TEM is the "single base extension" reagent, and incorporates differentially labeled nucleotides into the extended primers. After extension, 95% formamide/1 mM EDTA is added to remove all hybridized DNA (the probe on the array is now labeled). After neutralization using the XC3 reagent, the labeled extended primers undergo a multi-layer staining process on the Chamber Rack. Next, Te-Flow units are disassembled, washed in low salt PB1 buffer, coated with XC4 protectant, and finally dried. For multi-sample BeadChips, the BeadChips are assembled into the Te-Flow units immediately after hybridization as per user's guide instructions.

3.4.1. Array-Based Primer Extension

Preheat GenePaint rack to 44°C before addition of BeadChips assembled in Te-Flow Through Chamber to the rack.

1. Add 450 μl hybridization buffer (RA1) Te-Flow Through chamber reservoir and allow to drain. Repeat 5 times.
2. Add 450 μl blocking buffer (XB1) and incubate 10 min.
3. Add 450 μl equilibration buffer (XB2) and incubate 5 min.
4. Add 200 μl terminator extension mix (TEM) and incubate 15 min.
5. Add 450 μl 95% formamide/10 mM EDTA and incubate for 1 min. Repeat once.

3.4.2. Multi-Layer IHC Sandwich Staining

Adjust GenePaint rack to 37°C and allow to equilibrate before preceding to next step (this is automated on robotic system).

1. Add 450 μl wash buffer (XB3) and drain. Repeat twice.
2. Add 250 μl staining solution (LMM) and incubate 10 min.
3. Repeat wash in step (1).
4. Add 250 μl anti-stain solution (ASM) and incubate 10 min.
5. Repeat wash in step (1).
6. Repeat steps 2–5 to add a second immunohistochemical sandwich layer.
7. Add final 250 μl staining solution (LMM) and incubate 10 min.
8. Repeat wash in step (1).
9. Remove Te-Flow units from GenePaint Rack and lay horizontally until ready for coating.

3.4.3. Coating of BeadChips

The coating process deposits a protective film over the BeadChip to protect the fluorescent dyes from oxidative damage. Coated BeadChips are stable for weeks in this format.

1. Fill vertical wash tray with 310 ml of PB1.
2. Place staining rack into PB1 wash tray.
3. Disassemble BeadChips from Te-Flow Through chambers and submerge into PB1 by placing in staining rack (do not let BeadChips dry).
4. Wash BeadChips in PB1 by moving staining rack up and down ten times (break surface each time).
5. Soak BeadChips 5 min in PB1.
6. Fill vertical wash tray with 310 ml of XC4 (should be used within 10 min).
7. Remove BeadChips in staining rack from PB1 wash tray and place in XC4 wash tray.
8. Wash in XC4 with physical motion described in step 4 followed by a soak for 5 min.
9. Remove staining rack in one smooth motion going from a vertical orientation to a horizontal orientation with the barcodes on the BeadChips facing upward. Place on tube rack.
10. Individually remove BeadChips from staining rack and place barcode side up on tube rack making sure BeadChips are not touching each other during the drying step.
11. Place tube rack with horizontal BeadChips into vacuum desiccator, turn on vacuum to ~508 mm Hg, and dry for 50 min at room temperature.

3.5. BeadChip Scanning and Data Analysis

BeadChips are scanned at 0.84 μm resolution (one or two colors) on a BeadArray^{TM} Reader that accommodates up to three BeadChips. The intensity values on the BeadChip are extracted after automated registration and grid *(14, 15)*. Genotyping analysis is performed using proprietary BeadStudio software that generates archetypal clusters from a set of training data based upon a population of gDNA samples (**Fig. 13.7**). The A and B intensity information is converted to normalized polar coordinates. Genotype calls are made using a probabilistic model in which probabilities are assigned for membership into the three clusters (AA, AB, or BB). Data points falling between clusters have a low probability of belonging to any one cluster and are scored as a no call (GenCall score threshold of 0.15–0.25). The genotyping data is output in a tabular comma delimited format for further downstream analysis.

Overall genotyping quality of the 1 M BeadChip, as assessed by measuring several genotyping quality parameters across 120 different DNA samples, were all remarkably high. The call rate was > 99.4%, the reproducibility was > 99.99%, and the concordance with HapMap data was > 99.65% (manuscript in preparation). A representative genotyping plot for a single locus across all 120 samples is shown (**Fig. 13.6A–C**). For easier clustering and visualization, the rectangular coordinates (A bead type vs. B bead type signal) are transformed into polar coordinates (log-normalized R intensity, and a theta measurement). In the single locus genotyping plot shown, the three SNP clusters are clearly distinguishable allowing accurate genotype calls.

Detection of chromosomal aberrations (DNA copy aberrations, loss-of-heterozygosity, etc.) is accomplished by comparing the normalized intensity of a subject sample to a reference sample(s). Two analysis modes are available, a single sample mode in which reference values are derived from canonical clusters created from clustering on ~120 normal reference samples (**Note 6**). Alternatively, a paired sample mode allows direct intensity

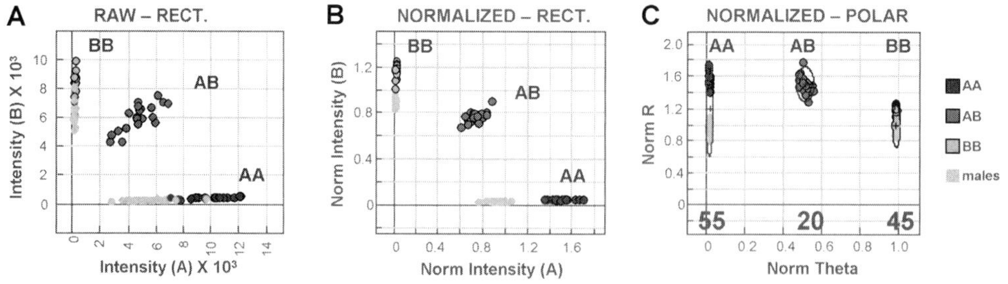

Fig. 13.6. Genotype clustering on high-density Infinium WGG BeadChips. **(A)** An example Genoplot of raw genotyping data from 138 different DNA samples from a representative locus from the 1 M BeadChip. Notice the three distinct clusters corresponding to the three genotype categories (A/A, A/B, and B/B). **(B)** and **(C)** Genotyping data is normalized and converted to polar coordinate genotyping data for a single locus across 120 different DNA samples. The encircled data points indicate the locations of the archetypal clusters generated by Illumina BeadStudio software. R is the normalized intensity and theta (θ) is $(2/\pi)$*arctan(B/A).

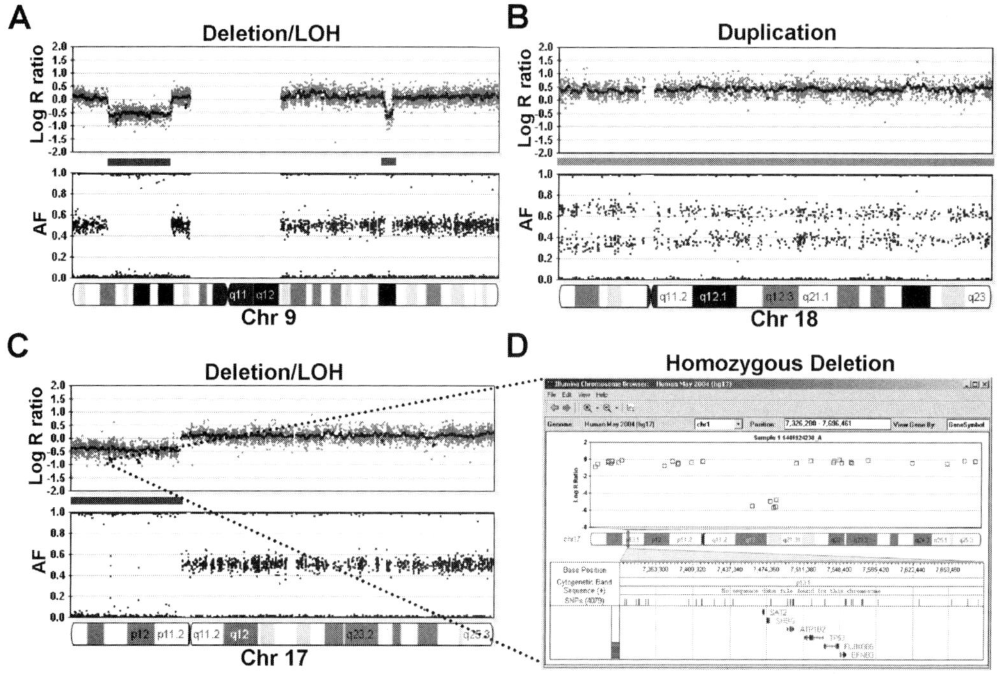

Fig. 13.7. Examples of aberrations detected in HL60. **(A)** Two regions (marked in *grey*) exhibiting LOH are characterized by a downward deflection in the log R ratio. **(B)** Chromosome 18 exhibits trisomy (marked in *light grey*) and is characterized by four clusters in the B allele frequency (AF) plot. **(C)** A ∼20 Mb heterozygous deletion is present on the p arm of Chromosome 17. A small homozygous deletion is just discernable within this region. **(D)** High resolution analysis of the homozygous deletion shown in (C) spanning the ATP1B2 gene. For all plots, the dark smoothing line represents a moving median of 500 kB.

comparisons between a subject sample and its corresponding matched pair. Visualization of copy number and LOH data is performed by plotting the $\log_2 (R_{subject}/R_{reference})$ also known as log R ratio and the allele frequency (AF) across the genome (**Fig. 13.7**). It is these plots that form the foundation of detecting chromosomal aberrations. Using both log ratios of normalized intensity and allelic ratios increases the sensitivity of detecting chromosomal aberrations as well as providing genetic information (haplotypes) about the aberrant locus. Given that SNP arrays provide both copy number and genetic information, "SNP-CGH" may replace array CGH as a standard for measuring genome-wide chromosomal aberrations *(16)*.

4. Notes

1. All reagents should be thoroughly thawed, mixed by inversion ten times, and spun down before use.
2. The recommended amount of input gDNA into the Infinium WGA reaction is 750 ng for large-scale (single and two sample

BeadChips) and 200 ng for small-scale amplifications (12 sample BeadChips). One can use less input of gDNA (as low a 1–10 ng) in the reaction, but to obtain optimal call rates and low noise in the log R ratios, new genotype clusters should be created by clustering (GenTraining) on a similar set of "low input" samples.

3. The use of high quality DNA in the assay is important, since DNA less than 1 kb in size amplifies much less efficiently. DNA quality can be assessed by gel analysis using ~100 ng of gDNA run out on a 6% PAGE gel and stained with SYBR Gold nucleic acid stain (Invitrogen). The major smear of DNA should be above the 1 kb size marker, although we have obtained good performance on samples with an average size as small as 500 bp.
4. After 37°C overnight WGA amplification, a white flocculent material (magnesium pyrophosphate byproduct, $Mg_2P_2O_7 = Mg_2PPi$) should be visible in the wells, indicating that amplification has occurred.
5. After resuspension of samples in RA1 hybridization buffer, the plate can be heat sealed and samples stored for several months at −20°C or for up to a year or longer at −80°C.
6. Paired sample analysis on multi-sample BeadChips is the method of choice for detection of DNA copy alterations or loss-of-heterozygosity (LOH). Paired analysis generates much lower noise than comparing a sample to the centroids of the GenTrain clusters.

Acknowledgments

The technology development summarized here would not have been possible without the efforts of many dedicated individuals. We would like to thank the people at Illumina for their valuable contributions in molecular biology, automation, oligonucleotide synthesis, chemistry, engineering, bioinformatics, software, manufacturing, and process development. Special thanks go to John Stuelpnagel, Cynthia Allred, and Rose Espejo for careful reading of the manuscript. The WGG research was funded, in part, by grants from the NIH/NCI.

References

1. Risch, N., and Merikangas, K. (1996) The future of genetic studies of complex human diseases. *Science* **273,** 1516–7.
2. Fan, J. B., Chee, M. S., and Gunderson, K. L. (2006) Highly parallel genomic assays. *Nat Rev Genet* **7,** 632–44.
3. Carlson, C. S., Eberle, M. A., Rieder, M. J., Yi, Q., Kruglyak, L., and Nickerson, D. A. (2004) Selecting a maximally informative set of single-nucleotide polymorphisms for association analyses using linkage disequilibrium. *Am J Hum Genet* **74,** 106–20.
4. Consortium, T. I. H. (2003) The International HapMap Project. *Nature* **426,** 789–96.
5. Gunderson, K. L., Steemers, F. J., Lee, G., Mendoza, L. G., and Chee, M. S. (2005) A genome-wide scalable SNP genotyping assay using microarray technology. *Nat Genet* **37,** 549–54.
6. Gunderson, K. L., Kruglyak, S., Graige, M. S., Garcia, F., Kermani, B. G., Zhao, C., Che, D., Dickinson, T., Wickham, E., Bierle, J., Doucet, D., Milewski, M., Yang, R., Siegmund, C., Haas, J., Zhou, L., Oliphant, A., Fan, J. B., Barnard, S., and Chee, M. S. (2004) Decoding randomly ordered DNA arrays. *Genome Res* **14,** 870–7.
7. Shumaker, J. M., Metspalu, A., and Caskey, C. T. (1996) Mutation detection by solid phase primer extension. *Hum Mutat* **7,** 346–54.
8. Pastinen, T., Kurg, A., Metspalu, A., Peltonen, L., and Syvanen, A. C. (1997) Minisequencing: a specific tool for DNA analysis and diagnostics on oligonucleotide arrays. *Genome Res* **7,** 606–14.
9. Pastinen, T., Raitio, M., Lindroos, K., Tainola, P., Peltonen, L., and Syvanen, A. C. (2000) A system for specific, high-throughput genotyping by allele-specific primer extension on microarrays. *Genome Res* **10,** 1031–42.
10. Steemers, F., Chang, W., Lee, G., Shen, R., Barker, D. L., and Gunderson, K. L. (2006) Whole genome genotyping (WGG) using single base extension (SBE). *Nat Methods* **3**(1), 31–3.
11. Patil, N., Berno, A. J., Hinds, D. A., Barrett, W. A., Doshi, J. M., Hacker, C. R., Kautzer, C. R., Lee, D. H., Marjoribanks, C., McDonough, D. P., Nguyen, B. T., Norris, M. C., Sheehan, J. B., Shen, N., Stern, D., Stokowski, R. P., Thomas, D. J., Trulson, M. O., Vyas, K. R., Frazer, K. A., Fodor, S. P., and Cox, D. R. (2001) Blocks of limited haplotype diversity revealed by high-resolution scanning of human chromosome 21. *Science* **294,** 1719–23.
12. Gabriel, S. B., Schaffner, S. F., Nguyen, H., Moore, J. M., Roy, J., Blumenstiel, B., Higgins, J., DeFelice, M., Lochner, A., Faggart, M., Liu-Cordero, S. N., Rotimi, C., Adeyemo, A., Cooper, R., Ward, R., Lander, E. S., Daly, M. J., and Altshuler, D. (2002) The structure of haplotype blocks in the human genome. *Science* **296,** 2225–9.
13. Johnson, G. C., Esposito, L., Barratt, B. J., Smith, A. N., Heward, J., Di Genova, G., Ueda, H., Cordell, H. J., Eaves, I. A., Dudbridge, F., Twells, R. C., Payne, F., Hughes, W., Nutland, S., Stevens, H., Carr, P., Tuomilehto-Wolf, E., Tuomilehto, J., Gough, S. C., Clayton, D. G., and Todd, J. A. (2001) Haplotype tagging for the identification of common disease genes. *Nat Genet* **29,** 233–7.
14. Galinsky, V. L. (2003) Automatic registration of microarray images. II. Hexagonal grid. *Bioinformatics* **19,** 1832–6.
15. Galinsky, V. L. (2003) Automatic registration of microarray images. I. Rectangular grid. *Bioinformatics* **19,** 1824–31.
16. Peiffer, D. A., Le, J. M., Steemers, F. J., Chang, W., Jenniges, T., Garcia, F., Haden, K., Li, J., Shaw, C. A., Belmont, J., Cheung, S. W., Shen, R. M., Barker, D. L., and Gunderson, K. L. (2006) High-resolution genomic profiling of chromosomal aberrations using Infinium whole-genome genotyping. *Genome Res* **16,** 1136–48.

Chapter 14

Genotyping Single Nucleotide Polymorphisms by Multiplex Minisequencing Using Tag-Arrays

Lili Milani and Ann-Christine Syvänen

Abstract

The need for multiplexed methods for SNP genotyping has rapidly increased during the last decade. We present here a flexible system that combines highly specific genotyping by minisequencing single-base extension with the advantages of a microarray format that allows highly multiplexed and parallel analysis of any custom selected SNPs.

Cyclic minisequencing reactions with fluorescently labeled dideoxynucleotides (ddNTPs) are performed in solution using multiplex PCR product as template and detection primers, designed to anneal immediately adjacent and upstream of the SNP site. The detection primers carry unique Tag-sequences at their 5′ ends and oligonucleotides complementary to the Tag-sequence, cTags, are immobilized on a microarray. After extension, the tagged detection primers are allowed to hybridize to the cTags, and the fluorescent signals from the array are measured and the genotypes are deduced by cluster analysis of the incorporated labels. The "array of arrays" format of the system, accomplished by a silicon rubber grid to form separate reaction chambers, allows either 80 or 16 samples to be analyzed for up to 200 or 600 SNPs, respectively on a single microscope slide.

Key words: Single nucleotide polymorphism, SNP, genotyping, fluorescent minisequencing, microarray, multiplex PCR, "array of arrays."

1. Introduction

As a result of the successful Human Genome Project we have access to a complete nucleotide sequence of the 3 billion nucleotides that constitute our genome *(1–3)*. Another large effort, the Human Haplotype Mapping (HapMap) project has generated detailed information on over 5 million single nucleotide polymorphisms (SNPs) and the linkage disequilibrium patterns between the SNP alleles in the human genome *(4, 5)*. The Human Genome Project,

Martin Dufva (ed.), *DNA Microarrays for Biomedical Research: Methods and Protocols, vol. 529*

DOI 10.1007/978-1-59745-538-1_14 Springerprotocols.com

and particularly the HapMap project, have been accompanied by a rapid development in genotyping technology, and today there are commercially available systems for SNP genotyping on a very large, and even genome-wide scale. These genotyping systems are enabled by recent developments in technology for production of microarrays and for fluorescence scanning of the arrays in clever combination with known reaction principles for genotyping *(6)*. One of the most frequent reaction principles in multiplex genotyping systems today is minisequencing or single-base primer extension, in which a DNA polymerase is allowed to extend a detection primer by a single nucleotide at the position of the SNP *(7)*. The most spectacular example of single-base extension is the Infinium II assay from Illumina *(8)*. The system, presented in this chapter, combines the highly specific genotyping principle of minisequencing with the advantages of a microarray format that allows medium to highly multiplex and parallel analysis. The system is flexible and can be designed for any panel of SNPs. It is particularly useful for establishing genotyping panels for other organisms than humans, where common predefined assays are not available *(9, 10)*. The specificity of the single-base extension reaction allows quantitative analysis of SNP genotypes in genomic DNA for analysis of copy number variation or genotyping pooled DNA samples *(11)*, and in RNA for allele-specific gene expression analysis *(12, 13)* or relative quantification of alternatively spliced transcripts *(14)*. Recently quantitative SNP analysis has been scaled up for genome-wide detection of loss of heterozygosity and copy number variation *(15)*.

The Tag-array minisequencing system utilizes generic capture oligonucleotides ("cTags") that are immobilized on a microarray. Multiplex cyclic minisequencing reactions with fluorescently labeled dideoxynucleotides (ddNTPs) are performed in solution using minisequencing primers designed to anneal immediately adjacent and upstream of the SNP site. The primers carry 5′-Tag-sequences complementary to one of the arrayed cTags, and the SNPs are separated by hybridizing the extended minisequencing primers to their corresponding cTags with known locations on the array. The incorporated fluorescently labeled ddNTPs allow deduction of the genotypes of each SNP based on measurement of the signal intensities by fluorescence scanning of the arrays *(11, 16)*. The use of generic Tag-sequences enables universal, non-SNP specific array designs. The "array of arrays" format described below is accomplished by a silicon rubber grid creating separate reaction chambers for multiple samples on a single microscope slide. Either 80 or 16 samples can be simultaneously analyzed for up to 200 or 600 SNPs, respectively *(10, 17, 18)* (**Fig. 14.1**).

The procedure described in detail in **Section 3** outlines (1) selection of appropriate SNPs, (2) design of oligonucleotides for PCR, immobilization on the microarray, and SNP genotyping, (3) preparation of microarrays, (4) manufacturing of the silicon

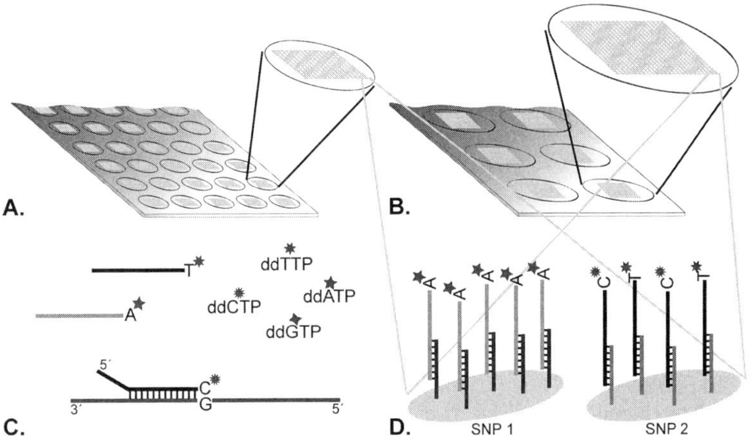

Fig. 14.1. Principle of Tag-array minisequencing. Schematic cut views of two arrayed slides with the subarrays in either 384-well (**A**) or 96-well (**B**) format. One of the subarrays for each slide containing up to 200 (A) or 600 (B) cTags is showed enlarged. The principle of the minisequencing reaction is illustrated with a minisequencing primer carrying a 5′-Tag-sequence annealed to its target, and extended with a labeled ddCTP at the position of the SNP (**C**). The Tag-sequences of the extended minisequencing primers are allowed to hybridize to their complementary cTags arrayed as spots in the subarrays (**D**). The genotypes are deduced by measuring the fluorescence of the incorporated nucleotides. Part of one subarray, with the result for two SNPs, is shown. This sample is homozygous (A/A) for SNP 1 and heterozygous (C/T) for SNP 2.

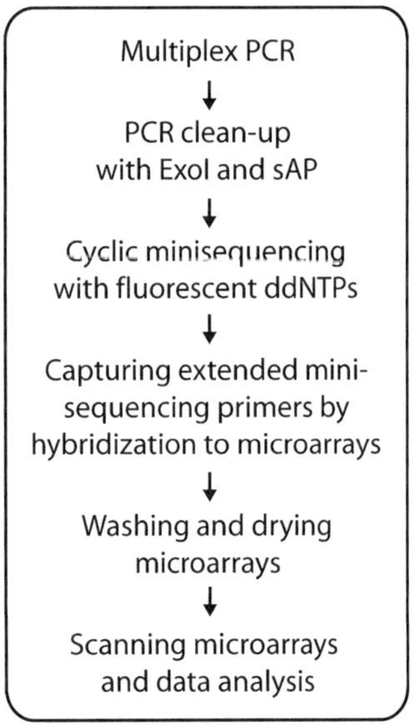

Fig. 14.2. Flow chart illustrating the main steps of the procedure for genotyping SNPs by minisequencing using Tag-arrays.

rubber grid, (5) the genotyping reaction, and (6) data analysis and genotype calling. The main steps of the assay are illustrated in **Fig. 14.2**. Several alterations and modifications of the method are possible and a number of suggestions are given in **Section 4**. The protocol is provided under the assumption that the instrumentation, reagents, and consumables specified in **Section 2** are to be used, but other equivalent procedures are also feasible.

2. Materials

2.1. Instrumentation and Software

1. Access to arraying instrument or purchased customized arrayed slides. We use a ProSys 5510A instrument (Cartesian Technologies Inc., Huntingdon, UK) with Stealth Micro Spotting Pins (TeleChem International Inc., Sunnyvale, CA).
2. PCR instrument.
3. Multichannel pipette and/or a pipetting robot (optional).
4. Centrifuge for microtiter plates (recommended).
5. Minisequencing reaction rack (**Fig. 14.3**).
6. Hybridization oven at 42°C.

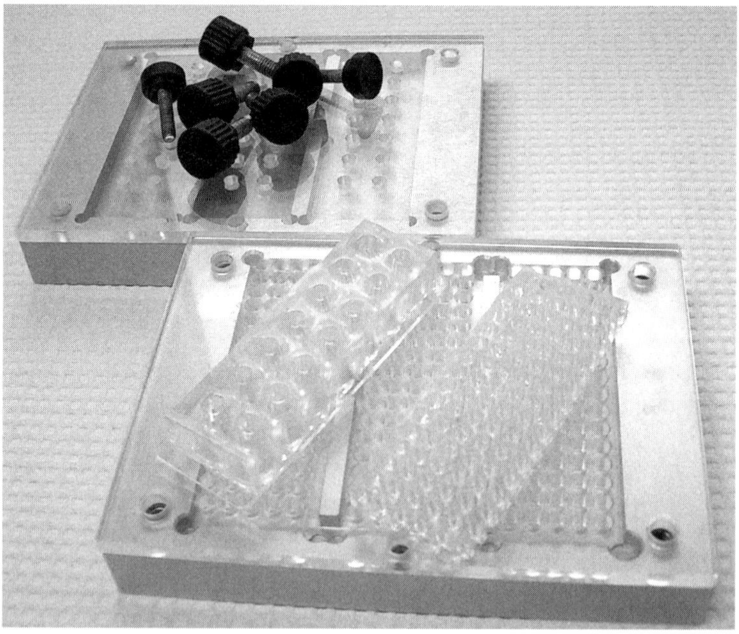

Fig. 14.3. The arrayed slide is covered with a silicon rubber grid to give separate reaction chambers and placed in a custom-made heat conducting aluminum rack. A 384-well (*right*) and a 96-well (*left*) format silicon rubber grid is shown on top of a reaction rack. A plexiglas cover with drilled holes through which the reaction chambers are accessible is tightly screwed on top of the assembly, thus securing correct positioning of the silicon grid during hybridization.

7. Array scanner and software for signal quantification. We use the ScanArray Express system (PerkinElmer Lifesciences, Boston, MA).
8. Software for cluster analysis and genotype calling. We use the SNPSnapper software (Juha Saharinen, http://www.bioinfo.helsinki.fi/SNPSnapper/).

2.2. Reagents and Consumables

All reagents and consumables should be of standard molecular biology grade. Use sterile distilled or deionized water.

2.2.1. Oligonucleotides Primers and Nucleotides

1. PCR primers.
2. Minisequencing primers with 5′-Tag-sequences.
3. cTags with a 3′ 15-T residue spacer and a 3′-amino group.
4. Reaction control – four oligonucleotides differing at one internal nucleotide position and a minisequencing primer (with a 5′-Tag-sequence) complementary to the oligonucleotide templates.
5. Spot control – fluorescently labeled cTag oligonucleotide.
6. Hybridization control – two fluorescently labeled Tag-oligonucleotides.
7. Print control – fluorescently labeled oligonucleotide designed to hybridize to any cTag.
8. Fluorescent dideoxynucleotides (Texas Red-ddATP, TAMRA-ddCTP, R110-ddGTP and Cy5-ddUTP; PerkinElmer Life Sciences). Keep light-protected working aliquots at +4°C and store stock solutions at –20°C.

2.2.2. Enzymes

1. DNA polymerase. We use Smart-Taq Hot DNA polymerase (Naxo, Tartu, Estonia).
2. Exonuclease I (Fermentas, Vilnius, Lithuania).
3. Shrimp Alkaline Phosphatase (GE Healthcare, Uppsala, Sweden).
4. DNA polymerase compatible with fluorescently labeled ddNTPs. We use KlenThermase DNA polymerase (GENECRAFT, Lüdinghausen, Germany).

2.2.3. Buffer Solutions

1. Standard PCR reagents or reagents optimized for multiplex PCR.
2. 2x Printing buffer: 300 mM phosphate buffer pH 8.5. Store at room temperature up to 1 month.
3. Blocking solution: 50 mM ethanolamine, 100 mM Tris–HCl, pH 9.0 and 0.1 % SDS. Prepare directly before use. Ethanolamine is highly corrosive and should be handled according to safety instructions.

4. Washing solutions; (I) 4xSSC, (II) 2xSSC and 0.1% SDS, and (III) 0.2xSSC. (20xSSC: 3 M NaCl, 300 mM sodium citrate, pH 7.0. 10% SDS: 10% w/v sodium dodecyl sulfate)
5. 1 M Tris–HCl, pH 9.5.
6. 50 mM $MgCl_2$.
7. 1 % v/v Triton X-100.
8. Hybridization solution: 6.25x SSC

2.2.4. Consumables

1. 384- or 96-well v-bottomed microtiter plates (ABgene, Epsom, UK).
2. CodeLink^{TM} Activated Slides (GE Healthcare, Uppsala, Sweden).
3. Elastosil RT 625 A and B (polydimethyl siloxan; Wacker-Chemie, München, Germany).

3. Methods

3.1. SNP Selection

SNPs can be identified either experimentally or in databases, for example dbSNP (http://www.ncbi.nlm.nih.gov/projects/SNP). Database searches for SNPs may be aimed at genes of interest, candidate chromosomal regions or randomly distributed SNPs with known allele frequencies, depending on the aim of the project, *see also* **Note 1**.

3.2. Oligonucleotide Design

3.2.1. PCR Primers

The length of the PCR fragments should be short, optimally around 100–150 bp. Design PCR primers flanking the SNPs of interest using available software. Primer3 is freely available on the internet (http://frodo.wi.mit.edu/primer3/input.htm). *See also* **Note 2**.

Each PCR primer pair should be tested in an *in silico* PCR to verify that a specific DNA fragment is being amplified (e.g. http://genome.brc.mcw.edu/cgi-bin/hgPcr). *See also* **Note 3**.

3.2.2. Minisequencing Primers

Minisequencing primers are designed to anneal immediately adjacent and upstream of the SNP position. The minisequencing primers should be approximately 20 bases long, and have a melting temperature of 55–60°C to ensure specificity in the cyclic primer extension reaction. The 5′ end of each minisequencing primer should contain Tag-sequences complementary to the cTags that are printed on the microarray. The Tags should be 20 bases long, have similar melting temperature and not be complementary to either each other, to the gene specific part of the minisequencing primers or to the human genome *(16)*. The

Affymetrix GeneChip®Tag Collection can be used as source for Tag-sequences (Affymetrix, Santa Clara, CA). Minisequencing primers for both forward and reverse DNA strands are often helpful as controls for the genotyping results (*see* **Note 4**).

3.2.3. Complementary Tag-Sequences

The complementary Tag-sequences (cTags) have 15 3′ T-residues as a spacer and a 3′-amino group to enable covalent attachment of the cTags to the microarray slides.

3.2.4. Control Oligonucleotides

We recommend the use of control oligonucleotides for each step of the genotyping procedure *(18)*.

To control for the spotting procedure, a fluorescently labeled cTag must be included on the array. A control oligonucleotide designed to hybridize to any cTag (5′-AAA AAA AAA ANN NNN NNN NN– Fluorophore -3′) is recommended for testing the printing of some subarrays or microarrays from each batch.

As minisequencing reaction control, a minisequencing primer that is complementary to four synthesized single-stranded oligonucleotide templates differing at one nucleotide position to mimic the four possible alleles of a SNP is useful. Add the control templates to the minisequencing reaction mixture at a final concentration of 1.5 nM. A corresponding cTag must be included on the array. To control the hybridization reaction, a fluorescently labeled oligonucleotide complementary to an arrayed cTag is used. Optimally, use two differently labeled hybridization control oligonucleotides, and add them in an alternating pattern over the microarray to allow assessment that no leaking between wells has occurred.

3.3. Preparation of Microarrays

3.3.1. Microarray Printing

Dissolve the cTags in Printing buffer to a final concentration of 25 μM, *see also* **Note 5**. If not used directly or if to be reused, store the cTags at –20°C, but limit freeze–thawing cycles to a maximum of ten. Prepare the arrays by contact printing of the cTag oligonucleotides onto CodeLink™ Activated slides using the ProSys 5510A instrument with SMP3 pins. These pins deliver 1 nl of the cTag-solution to the slides to form spots with a diameter of 125–150 μm and with a center-to-center distance of for example 200 μm (*see* **Note 6**). For the possibility of using the "array of arrays" format, print spots in a subarray pattern corresponding to the spacing of wells in a 384-well microtiter plate, (*see* **Note 7**; **Fig. 14.1**). After arraying, mark the position of some of the subarrays on the back side of the slides using a diamond-pen.

3.3.2. Post-Printing Processing of the Microarray Slides

Process the slides according to the instructions of the manufacturer. The protocol for CodeLink™ Activated Slides is given below.

1. Prepare an incubation chamber with 75% relative humidity. Add as much solid NaCl to water as needed to form a 1-cm deep slurry at the bottom of a plastic container with an air-tight lid.
2. After printing, keep the arrays in the incubation chamber for 4–72 h.
3. Prepare the blocking solution and preheat it to 50°C.
4. Deactivate the excess of amine-reactive groups by immersing the arrayed slides into the blocking solution for 30 min at 50°C.
5. Rinse twice with dH_2O. Immerse the slides in washing solution I for 30 min at 50°C (at least 10 ml per slide should be used). Rinse again with dH_2O.
6. Spin dry the slides for 5 min at 900 rpm. Store the slides desiccated at room temperature until use.

3.3.3. Quality Control of Printing Procedure

For each batch of printed slides it is useful to analyze a few subarrays by hybridization to control the quality of the spots. After blocking the slides, hybridize the 3′-fluorescently labeled print-control oligonucleotide to some subarrays at 300 nM concentration in 6.25x SSC for 5 min with subsequent washing and scanning as described below.

3.4. Preparation of Reusable Silicon Rubber Grid

A grid of silicon rubber reaction chambers is prepared using inverted v-bottomed microtiter plates as mould (**Fig. 14.3**).

1. Add the two Elastosil RT 625 components in a 50 ml Falcon tube in a mass ratio of 9:1 (i.e. 46.8 g of A and 5.2 g of B) and rotate and turn the tube manually until the components are fully mixed ($\sim$ 30 min; *see* **Note 8**).
2. Pour the mixture onto an inverted v-bottomed 384-well microtiter plate, leaving about 1–2 mm of the tip of the wells uncovered. Allow the silicon rubber to harden at least over night at room temperature (*see* **Note 9**).
3. Remove the silicon rubber grid from the plate, and use a scalpel to cut the silicon rubber into pieces of the same size as microscope slides, with the wells matching the printed subarrays.
4. The silicon rubber grid is reusable, wash it in 10% chlorine, rinse with water and allow to dry after each use.

3.5. Genotyping

3.5.1. Multiplex PCR and Clean-Up

1. Amplify DNA samples and non-template control samples by multiplex PCR according to an optimized protocol. The success of the amplification may be verified on a 2% agarose gel for a subset of the samples.
2. For each sample, pool the multiplex PCR products (*see* **Note 10**).
3. Prepare a master-mix of the exonuclease (ExoI) and alkaline phosphatase (sAP) reagents for clean-up of the PCR products (**Table 14.1**; *see* **Note 11**).

Table 14.1
PCR clean-up reagents

Reagent	Volume per reaction (μl)	Final concentration
PCR products	7.1	
50 mM $MgCl_2$	1.6	7.61 mM[a]
1 M Tris–HCl pH 9.5	0.5	0.05 M
20 U/μL Exonuclease I	0.3	0.57 U/μl
1 U/μL Shrimp Alkaline Phosphatase	1.0	0.10 U/μl
Total volume	10.5	

[a] The true final concentration of $MgCl_2$ is higher depending on the contribution from the PCR products.

4. Add 3.4 μl of the clean-up mixture in a final volume of 10.5 μl.
5. Incubate at 37°C for 45 min.
6. Inactivate the enzymes by heating to 85°C for 15 min.

3.5.2. Cyclic Minisequencing

1. Prepare a master-mix with minisequencing reagents (**Table 14.2**). The fluorophores in this mixture are light-sensitive (*see* **Note 12**).

Table 14.2
Minisequencing reagents

Reagent	Volume per reaction (μl)	Final concentration
PCR products after clean-up	10.50	
100 nM each pooled Minisequencing primers	1.50	10 nM
100 μM Fluorescently labeled ddNTPs[a]	4 × 0.015	0.10 μM
1 % Triton X-100	0.30	0.02 %
25 U/μl KlenThermase	0.04	0.067 U/μl
H_2O	2.60	
Total volume	15.00	

[a]Texas Red-ddATP, TAMRA-ddCTP, R110-ddGTP, Cy5-ddUTP.

2. After the clean-up step, add 4.5 μl of minisequencing reaction mixture in a final volume of 15 μl.
3. Perform the minisequencing reactions in a thermal cycler using an initial 3 min denaturation step at 96°C followed by for example 33 cycles of 20 s at 95°C and 20 s of 55°C in a thermocycler (*see* **Note 13**).

3.5.3. Capture by Hybridization

1. Position a silicon rubber grid over the arrayed slide according to the diamond-pen markings. Place the arrayed slides into the custom-made aluminum reaction rack and tighten the Polymethyl methacrylate (PMMA or "plexiglas") cover (**Fig. 14.3**). Preheat the assembly to 42°C on a heat block (*see* **Note 14**).
2. Add 7 μl of the hybridization solution to each minisequencing reaction to a final volume of 22 μl. It is recommended to include hybridization control oligonucleotides at 0.25 nM concentrations in the hybridization mixture.
3. Transfer 20 μl of each sample to a separate reaction chamber on the microscope slide. A multichannel pipette is convenient for this step.
4. Hybridize for 2 h at 42°C in a humid and dark environment accomplished for example by placing a wet tissue on the plexiglas lid and covering it with plastic film and aluminum foil.

3.5.4. Washing

1. Prepare the three washing solutions. Preheat solution II to 42°C.
2. After hybridization, take the slides from the reaction rack and immediately rinse with Solution I, at room temperature.
3. Wash the slides twice for 5 min with Solution II at 42°C, and twice for 1 min with Solution III, at room temperature, in 50 ml Falcon tubes.
4. Spin dry the slides for 5 min at 900 rpm, and store them protected from light.

3.5.5. Fluorescence Scanning

If allowed by the used scanner, balance the signal intensity from each laser channel so that no signals are saturated and the signals from the four fluorophores are as equal as possible. Balancing is feasible if a reaction control with signals from all four fluorophores has been included. **Figure 14.4** shows an example of a scanned array.

3.6. Data Analysis and Genotype Assignment

A quantification program such as the one supplied with the ScanArray Express instrument handles the scanning images and quantifies the signals from each spot. The raw-data is collected in an Excel sheet. Subtract the background, measured either around the spots or at negative control spots, i.e., spotted cTags

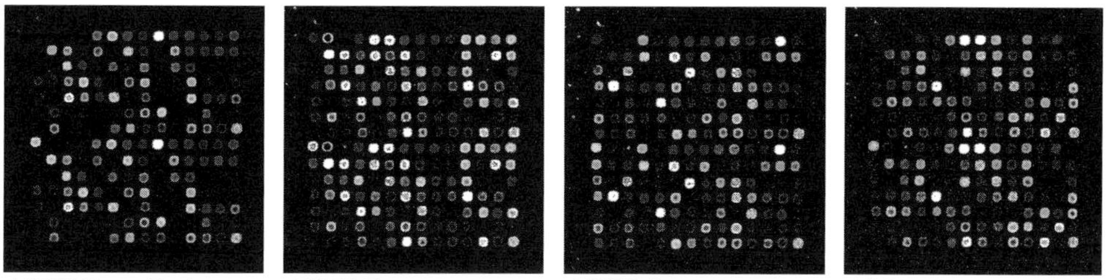

Fig. 14.4. Scanning results from genotyping one sample for 45 SNPs in one subarray, after cyclic minisequencing with ddATP, ddCTP, ddGTP, and ddUTP labeled with Texas Red, TAMRA, R110, and Cy5, respectively. Each cTag was spotted as horizontal duplicates and both polarities of the SNPs were analyzed, which together with the spots for the controls resulted in 196 spots *(18)*.

without corresponding tagged primers, from the signals measured in each channel.

Assign the genotypes of the SNPs in each sample by calculating the ratios between the signals from one of the alleles and the sum of the signals from both the alleles; Signal Allele1/ (Signal Allele 1 + Signal Allele2). A scatter plot with this ratio on the horizontal axis and the sum of the signals from both alleles on the vertical axis may be used for assigning the genotypes (**Fig. 14.5**). This scatter plot should give three distinct genotype clusters, with the homozygote samples clustering at each side and the heterozygotes in the middle. The ratios may vary between SNPs, depending on the sequence surrounding it, the type of nucleotide incorporated, and the signal intensity of the fluorophores, *see also* **Note 15**.

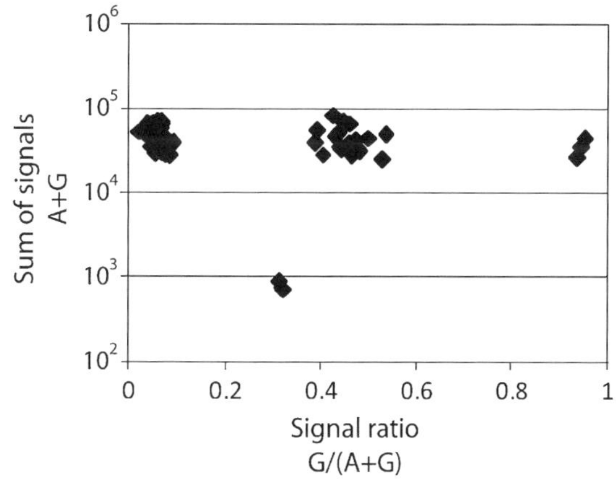

Fig. 14.5. Scatter plot for one SNP with a G/A variation analyzed in 80 samples, i.e., one slide with a 384-well format and 5 × 16 subarrays. The sum of the fluorescence signals from both alleles in each sample is plotted on the vertical axis. The ratio between the signal from one allele divided by the sum of the signals from both alleles is plotted on the horizontal axis. The three distinct clusters represent the three genotypes, where in this example three non-template control samples fall below the clusters.

When using the ScanArray Express or QuantArray program for signal analysis, or if the signal quantitation output files have been converted to fit their format, the genotyping results can be visualized using the SNPSnapper software customized for this method. The SNPSnapper software can be used for exporting the SNP genotypes together with allele fractions and ratios to a text file for further analysis.

4. Notes

1. Some of the SNPs in the databases may not be polymorphic in the population from which the study samples originate. The SNP allele frequencies in a particular population may be determined by analyzing pooled DNA samples using quantitative minisequencing in microtiter plates, or in the microarray format *(11, 19)*.
2. A touchdown PCR procedure may be used *(20)*. One strategy when designing primers for multiplex PCR is to aim at as similar primer melting temperature and G/C content as possible. Complementary 3′ sequences in the primers can be avoided by designing primers with the same 3′ terminal nucleotides *(21)*. Other options are to introduce common tails on the 5′ ends of all PCR primers, and subsequent amplification with one common primer for all the fragments at an elevated temperature *(22)*, or the use of universal 5′-sequences making the PCR primers eligible for the same reaction conditions *(23)*.
3. We recommend excluding SNPs located in repetitive elements identified by the RepeatMasker program (http://www.repeatmasker.org).
4. To avoid strong hairpin-loop structures, evaluate the complete minisequencing primer, including the Tag-sequence. Secondary structures that involve the 3′ end of a primer may lead to mis-incorporation of nucleotides. A primer design software that predicts secondary structures (mfold: http://mfold.bioinfo.rpi.edu/ or NetPrimer http://www.premierbiosoft.com/netprimer/netprimer.html) can be used.
5. The array may also be manufactured with immobilized minisequencing primers. In this assay variant the genotyping reaction is performed directly on the array surface *(24)*. This may be useful when SNPs located less than 20 bp apart are to be genotyped.
6. Microarrays may be purchased from a commercial supplier or manufactured in-house. There are several different slide types

and attachment chemistries. Some of them have been tested in our system *(25)*.

7. The number of spots in each subarray can be varied by changing the subarray pattern from a 384- to a 96-wells format, thus the number of SNPs to be analyzed per subarray is increased, but the maximum number of samples that can be analyzed simultaneously is decreased *(26)*.
8. Elastosil RT601 may be used instead of RT625 to give a slightly harder silicon rubber to decrease deformation of the wells when the rack lid is tightened. If large subarrays, utilizing all available surface, are printed, deformation of the wells may cause the cTags at the corners of the subarray to be covered by the silicon. The softer, RT625, silicon sticks better to the glass surface, and decreases the risk of leakage between wells.
9. Depending on the number of SNPs to be interrogated, an inverted 96-well microtiter plate may be used as silicon rubber mould to allow larger subarrays *(26)*.
10. Instead of multiplex PCR, single fragment PCR can be used with subsequent pooling of the amplified fragments, possibly after concentration using ethanol precipitation or spin dialysis. This is especially recommended when genotyping SNPs in cDNA (RNA) for detection of allele-specific gene expression *(12)*. Also if a large number of multiplex PCR products are pooled, it may be advantageous to concentrate the pool prior to the subsequent steps.
11. Alkaline phosphatase inactivates the remaining dNTPs and exonuclease I degrades the single-stranded PCR primers, which would disturb the subsequent minisequencing reactions.
12. Cy5-ddUTP can be used at a 1.5–2-fold higher concentration than the other ddNTPs to compensate for its lower incorporation efficiency. Instead of using four differently labeled nucleotides in the same reaction, depending on the available microarray scanner, a single label or two labels may be used in four or two separate reactions respectively *(27)*.
13. If the obtained fluorescent signals are weak, the number of cycles may be increased. We have used up to 99 cycles, but increasing the number of cycles often also increases the background signals.
14. Background problems can arise if the hybridization chamber is not kept humid which causes drying out of the samples on the slide.
15. The flanking sequence as well as the fluorophores attached to the dideoxynucleotides affect the efficiency and sequence

specificity of nucleotide incorporation by the DNA polymerase. The different properties of the fluorophores, such as molar extinction coefficients, emission spectra and quantum yield, as well as unspecific background may also affect the obtained signal intensities and signal ratios *(11)*.

Acknowledgments

This protolcol is the result of the combined work effort of both former and present members of the research group in Molecular Medicine at the Department of Medical Sciences at Uppsala University. The group has received financial support from the Swedish Research Council, the Wallenberg Foundation, the Swedish Cancer Foundation, and the European Commission.

References

1. The International Human Genome Sequencing Consortium (2004) Finishing the euchromatic sequence of the human genome. *Nature*, **431**, 931–945.
2. Lander, E.S., Linton, L.M., Birren, B., Nusbaum, C., Zody, M.C., Baldwin, J., Devon, K., Dewar, K., Doyle, M., FitzHugh, W. et al. (2001) Initial sequencing and analysis of the human genome. *Nature*, **409**, 860–921.
3. Sachidanandam, R., Weissman, D., Schmidt, S.C., Kakol, J.M., Stein, L.D., Marth, G., Sherry, S., Mullikin, J.C., Mortimore, B.J., Willey, D.L. et al. (2001) A map of human genome sequence variation containing 1.42 million single nucleotide polymorphisms. *Nature*, **409**, 928–933.
4. The International HapMap Consortium (2005) A haplotype map of the human genome. *Nature*, **437**, 1299–1320.
5. Frazer, K.A., Ballinger, D.G., Cox, D.R., Hinds, D.A., Stuve, L.L., Gibbs, R.A., Belmont, J.W., Boudreau, A., Hardenbol, P., Leal, S.M. et al. (2007) A second generation human haplotype map of over 3.1 million SNPs. *Nature*, **449**, 851–861.
6. Syvanen, A.C. (2005) Toward genome-wide SNP genotyping. *Nat Genet*, **37 Suppl**, S5–S10.
7. Syvanen, A.C., Aalto-Setala, K., Harju, L., Kontula, K. and Soderlund, H. (1990) A primer-guided nucleotide incorporation assay in the genotyping of apolipoprotein E. *Genomics*, **8**, 684–692.
8. Steemers, F.J., Chang, W., Lee, G., Barker, D.L., Shen, R. and Gunderson, K.L. (2006) Whole-genome genotyping with the single-base extension assay. *Nat Methods*, **3**, 31–33.
9. Andres, O., Ronn, A.-C., Bonhomme, M., Kellermann, T., Crouau-Roy, B., Doxiadis, G., Verschoor, E.J., Goossens, B., Domingo-Roura, X., Bruford, M.W. et al. (2008) A microarray system for Y chromosomal and mitochondrial single nucleotide polymorphism analysis in chimpanzee populations. *Mol Ecol* Resources, 8, 529–539.
10. Chen, D., Ahlford, A., Schnorrer, F., Kalchhauser, I., Fellner, M., Viràgh, E., Kiss, I., Syvanen, A.C. and Dickson, B.J. (2007) High-resolution high-throughput SNP mapping in Drosophila melanogaster. *Nature Methods*, 5, 323–329.
11. Lindroos, K., Sigurdsson, S., Johansson, K., Ronnblom, L. and Syvanen, A.C. (2002) Multiplex SNP genotyping in pooled DNA samples by a four-colour microarray system. *Nucleic Acids Res*, **30**, e70.
12. Milani, L., Gupta, M., Andersen, M., Dhar, S., Fryknas, M., Isaksson, A., Larsson, R. and Syvanen, A.C. (2007) Allelic imbalance in gene expression as a guide to cis-acting regulatory single nucleotide polymorphisms in cancer cells. *Nucleic Acids Res*, 35, e34.
13. Liljedahl, U., Fredriksson, M., Dahlgren, A. and Syvanen, A.C. (2004) Detecting imbalanced expression of SNP alleles by

minisequencing on microarrays. *BMC Biotechnol*, **4**, 24.
14. Milani, L., Fredriksson, M. and Syvanen, A.C. (2006) Detection of alternatively spliced transcripts in leukemia cell lines by minisequencing on microarrays. *Clin Chem*, **52**, 202–211.
15. Peiffer, D.A., Le, J.M., Steemers, F.J., Chang, W., Jenniges, T., Garcia, F., Haden, K., Li, J., Shaw, C.A., Belmont, J. et al. (2006) High-resolution genomic profiling of chromosomal aberrations using Infinium whole-genome genotyping. *Genome Res*, **16**, 1136–1148.
16. Hirschhorn, J.N., Sklar, P., Lindblad-Toh, K., Lim, Y.M., Ruiz-Gutierrez, M., Bolk, S., Langhorst, B., Schaffner, S., Winchester, E. and Lander, E.S. (2000) SBE-TAGS: an array-based method for efficient single-nucleotide polymorphism genotyping. *Proc Natl Acad Sci U S A*, **97**, 12164–12169.
17. Lindroos, K., Liljedahl, U. and Syvanen, A.C. (2003) Genotyping SNPs by minisequencing primer extension using oligonucleotide microarrays. *Methods Mol Biol*, **212**, 149–165.
18. Lovmar, L., Fredriksson, M., Liljedahl, U., Sigurdsson, S. and Syvanen, A.C. (2003) Quantitative evaluation by minisequencing and microarrays reveals accurate multiplexed SNP genotyping of whole genome amplified DNA. *Nucleic Acids Res*, **31**, e129.
19. Lagerstrom-Fermer, M., Olsson, C., Forsgren, L. and Syvanen, A.C. (2001) Heteroplasmy of the human mtDNA control region remains constant during life. *Am J Hum Genet*, **68**, 1299–1301.
20. Don, R.H., Cox, P.T., Wainwright, B.J., Baker, K. and Mattick, J.S. (1991) 'Touchdown' PCR to circumvent spurious priming during gene amplification. *Nucleic Acids Res*, **19**, 4008.
21. Zangenberg, A.P., Saiki, R.K. and Reynolds, R. (1999) In Innis, M. A., Gelfand, D. H. and Sninsky, J. J. (eds.), *PCR Applications*. Academic Press, London, UK, pp. 73–94.
22. Brownie, J., Shawcross, S., Theaker, J., Whitcombe, D., Ferrie, R., Newton, C. and Little, S. (1997) The elimination of primer-dimer accumulation in PCR. *Nucleic Acids Res*, **25**, 3235–3241.
23. Shuber, A.P., Grondin, V.J. and Klinger, K.W. (1995) A simplified procedure for developing multiplex PCRs. *Genome Res*, **5**, 488–493.
24. Liljedahl, U., Karlsson, J., Melhus, H., Kurland, L., Lindersson, M., Kahan, T., Nystrom, F., Lind, L. and Syvanen, A.C. (2003) A microarray minisequencing system for pharmacogenetic profiling of antihypertensive drug response. *Pharmacogenetics*, **13**, 7–17.
25. Lindroos, K., Liljedahl, U., Raitio, M. and Syvanen, A.C. (2001) Minisequencing on oligonucleotide microarrays: comparison of immobilisation chemistries. *Nucleic Acids Res*, **29**, E69–69.
26. Fredriksson, M., Barbany, G., Liljedahl, U., Hermanson, M., Kataja, M. and Syvanen, A.C. (2004) Assessing hematopoietic chimerism after allogeneic stem cell transplantation by multiplexed SNP genotyping using microarrays and quantitative analysis of SNP alleles. *Leukemia*, **18**, 255–266.
27. Li, J.G., Liljedahl, U. and Heng, C.K. (2006) Tag/anti-tag liquid-phase primer extension array: a flexible and versatile genotyping platform. *Genomics*, **87**, 151–157.

Chapter 15

Resequencing Arrays for Diagnostics of Respiratory Pathogens

Baochuan Lin and Anthony P. Malanoski

Abstract

Microarray technology has revolutionized the detection and analysis of microbial pathogens. The success of this technology is evident from the various microarrays that have been developed for this purpose, variation in the density of probes, and the time ranges required for assay completion. Among these, high-density re-sequencing microarrays have demonstrated great potential for detecting bacterial, viral pathogens, and virulence markers. Resequencing microarrays use closely overlapping probe sets to determine a target organism's nucleotide sequence. Hybridization to a series of perfect matched probes provides confirmatory presence/absence information, while hybridization to mismatched probes reveals strain-specific single nucleotide polymorphism (SNP) data. This approach provides sequence information of the diagnostic regions of detected organisms that is considerably more informative over that provided from other microarray techniques.

Key words: Resequencing microarray, pathogen detection, probe sets, respiratory pathogens, single nucleotide polymorphism, febrile respiratory illness.

1. Introduction

High-density resequencing microarrays were developed to detect single nucleotide polymorphisms (SNP) and so produce detailed genetic sequence reads. A resequencing microarray comprises "probe sets" – high-density arrangements of short highly specific oligonucleotide probes (25 and 29 used currently) where each base in a reference sequence is queried by four probes. One probe is an exact match of the reference sequence and the other three represent the same section of reference sequence with the central base position replaced by one of the possible SNP variants. In practical use, the number of probes is doubled so that for each

Martin Dufva (ed.), *DNA Microarrays for Biomedical Research: Methods and Protocols, vol. 529*

DOI 10.1007/978-1-59745-538-1_15 Springerprotocols.com

base both the forward (sense) and reverse (antisense) directions are contained in a probe set. It is possible to completely "resequence," resolve every base in the sequence of an unknown sample, the reference sequence itself or any other sequence that differs from the reference by one mismatch or fewer per 25 base pair (bp) *(1)*. At higher mismatch rates, a large number of bases can still be identified, but an increasing number will fail as the differences increase. This array-based format, combined with specific PCR, has proven ideal for SNP genotyping and phylogenetic analysis *(2–6)*. Initial work demonstrated the advantages of using a resequencing array with many short reference sequences to detect multiple bacterial and viral pathogens *(7–11)*. Taking full advantage of the sequential base resolution capability of resequencing microarrays, similarity searches of DNA databases have been incorporated into the analysis allowing for fine detailed discrimination of closely related pathogens and tracking mutations within the targeted pathogen even with only partial base call resolution in a reference sequence *(8,10–12)*.

The effective use of resequencing microarrays for respiratory pathogen detection or any large collection of organisms relies on the integration of several components. The overall design for the resequencing microarray and selection of primers for amplification must occur first. This consists of several tasks: First, selection of organisms and desired level of discrimination for each organism and whether specific nucleic acid markers must be tested for; second, determination from known sequence data of sequence regions to choose reference sequences from; third, selection of reference sequences and check for possible conflicts; fourth, primer selection. The order of several of these steps can be interchanged and refinements consist of repeating several of these steps after making changes. Once fabricated, an amplification method is required in order to achieve the sensitivity required for diagnosis/surveillance applications, so that any of the target pathogens can be detected directly from collected samples. Finally, because so many potential organism detection events are to be dealt with, a standardized algorithm is applied to determine if pathogens are detected and report the maximum level of detail possible using the resolved base sequence information from the multiple-pathogen resequencing microarrays.

2. Materials

2.1. Controls

1. IQ-Ex (0.2 pg/μL) control template, control forward 1.0 kb PCR primers (20 μM), control reverse primers (20 μM), and oligonucleotide control reagent (3.2 nM). These are part of

GeneChip® Resequencing Assay Kit (Affymetrix Inc., Santa Clara, CA). These reagents can be stored at –20°C, for up to 6 months. Set up 100 μl PCR for 1.0 kb IQ-EX containing 20 mM Tris–HCl (pH 8.4), 50 mM KCl, 2.5 mM $MgCl_2$, 200 μM dNTPs, 1 U of Platinum *Taq* DNA polymerase (Invitrogen Life Technologies, Carlsbad, CA), 3 μl each of control forward 1.0 kb PCR primers (20 μM) and control reverse primers (20 μM), and 5 μl IQ-Ex (0.2 pg/μl) control template. The amplification reaction is carried out with initial denaturation at 95°C for 10 min, followed by 30 cycles of: 94°C for 30 s, 68°C for 30 s, 72°C for 60 s, and a final extension at 72°C for 5 min (**Note 1**).

2. Internal controls for reverse transcription (RT) and PCR. Plasmid vector containing DNA fragment from non-related organism, such as *Arabidopsis thaliana* plant genes, will be suitable for this purpose. Two *Arabidopsis thaliana* plant genes, corresponding to NAC1 and triosphosphate isomerase (TIM), were chosen as internal controls for reverse transcription (RT) and PCR.
3. MEGAscript™ High Yield Transcription Kit (Ambion, Austin, TX). This kit is stored at –20°C, for up to 3 months (**Note 2**).

2.2. Sample Extraction

1. MasterPure™ DNA and RNA purification kit (Epicentre Technologies, Madison, WI). Proteinase K is stored at −20°C frost-free freezer. The 2X T&C lysis solution and MPC protein precipitation reagents are stored at room temperature (**Note 3**).

2.3. RT-PCR

1. PCR primers are dissolved in TE buffer to make 100 μM stock and stored at −80°C for up to 1 year. 1–10 μM working solutions are prepared by dilution in nuclease-free H_2O and stored at −20°C for up to 6 months.
2. RT master mix: mix 4 μL 5X first strand RT buffer and 2 μL 0.1 M DTT, 1 μL RNaseOUT™ (40 U/μL), and 1 μL SuperScript™ III reverse transcriptase (200 U/μL; Invitrogen Life Technologies, Carlsbad, CA). Add 8 μL RT master mix to each RT reaction. The final RT reaction should contain 50 mM Tris−HCl (pH 8.3), 75 mM KCl, 3 mM $MgCl_2$, 500 μM dNTPs, 40 U of RNaseOUT™, 10 mM DTT, 2 μM primer LN (5′-CGA TAC GAC GGG CGT ACT AGC GNN NNN NNN N-3′), 200 U of Superscript™ III reverse transcriptase.
3. Platinum Taq (Invitrogen) or GoTaq® Flexi DNA polymerase (Promega Corporation, Madison, WI) are stored at −20°C for up to 6 months (**Note 4**).
4. 10 mM dNTPs (Invitrogen) is stored in 100 μL aliquots at −20°C for up to 6 months.
5. 50X dNTPs (ACGU, Sigma-Aldrich, St. Louis, MO) is stored in 100 μL aliquots at −20°C for up to 6 months (**Note 5**).

6. Heat-labile uracil-DNA glycosylase (1 U/μL, USB Corporation, Cleveland, OH) is stored at −20°C for up to 6 months (**Note 5**).
7. PCR master mix: mix 5 μL 10X PCR buffer (200 mM Tris–HCl, pH 8.4, 500 mM KCl), with 2 μL $MgCl_2$ (50 mM), 2 μL 50x dNTPs (ACGU), 1 μL Uracil-DNA glycosylase, heat-labile (USB; 1 U/μL), 2 μL primer L (100 μM, 5′-CGA TAC GAC GGG CGT ACT AGC G-3′), 2 μL of primers mix A (1 μM, **Table 15.1**) or 2.5 μL of primer mix B (1 μM, **Table 15.2**), 2 μL Platinum *Taq* DNA (5 U/μL) polymerase. Add 40 μL PCR master mix to 10 μL of RT product.

2.4. Fragmentation and Labeling of Amplified Product

1. QIAquick PCR purification kit (Qiagen, Valencia, CA), store at room temperature (**Note 6**).
2. Fragmentation solution: For each reaction, mix 4.3 μL 10X GeneChip® fragmentation buffer (Affymetrix), 3.2 μL nuclease-free H_2O, 0.1 μL fragmentation reagent (3 U/μL; Affymetrix) on ice. The solution can be stored at 4°C for up to 24 h before using (**Note 7**). Individual components can be stored at −20°C for up to 6 months.
3. Labeling solution: For each reaction, mix 12 μL 5X terminal deoxynucleotidyl transferase buffer (Affymetrix), 2 μL GeneChip DNA labeling reagent (5 mM), and 3.4 μL terminal deoxynucleotidyl transferase (30 U/μL) on ice. The solution can be stored at 4°C for up to 24 h before using (**Note 7**). Individual components can be stored at −20°C for up to 6 months.

2.5. Hybridization, Washing, and Staining

1. Anti-streptavidin antibody (goat), biotinylated (Vector Laboratories, Burlingame, CA) is dissolved at 0.5 mg/mL in water and stored in 50 μL aliquots at 4°C for up to 6 months.
2. Pre-hybridization buffer: 10 mM Tris–HCl, pH 7.8 and 0.01% Tween 20. Store the solution at room temperature.
3. Hybridization buffer: Mix 138.6 μL TMAC (5 M), 2.3 μL each of 1 M Tris–HCl, pH 7.8, 1% Tween 20, 50 mg/ml BSA, 10 mg/mL Herring sperm DNA (Promega), and 1.8 μL of oligonucleotide control reagent (Affymetrix), 5.4 μL fragmented and labeled IQ-EX PCR product, and 13 μL water. The hybridization master mix can be stored at –20°C for up to 8 weeks.
4. 12X MES stock buffer (1.22 M MES, 0.89 M [Na^+], dissolve 70.4 g of MES hydrate and 193.3 g of MES sodium salt in 800 mL of molecular biology grade water. Mix and adjust the volume to 1,000 mL, the pH of the solution should be between 6.5 and 6.7. Filter the solution through a 0.2 μm

Table 15.1
List of PCR primers in primer mix A used for multiplex PCR

Primer name	Sequence (5′→3′)	Organism/gene	Amplicon size
FluAHA1-F2	CGA TAC GAC GGG CGT ACT AGC GGC CAA CAA CTC AAC CGA CAC	**Influenza A**	810 bp
FluAHA1-R2	CGA TAC GAC GGG CGT ACT AGC GAC ACT TCG CAT CAC ATT CAT CC	*hemaggultinin*	
FluAHA3-F6	CGA TAC GAC GGG CGT ACT AGC GAC TTC CCG GAA ATG ACA ACA	**Influenza A**	873 bp
FluAHA3-R7	CGA TAC GAC GGG CGT ACT AGC GGG TTT GTC ATT GGG AAT GCT	*hemaggultinin*	
FluAHA5-F2	CGA TAC GAC GGG CGT ACT AGC GGC CAT TCC ACA ACA TAC ACC	**Influenza A**	736 bp
FluAHA5-R2	CGA TAC GAC GGG CGT ACT AGC GAG CTA CCA TGA TTG CCA GTG	*hemaggultinin*	
FluANA1-F5	CGA TAC GAC GGG CGT ACT AGC GAC GTT GTT GCT GGA AAG GAC	**Influenza A**	1,000 bp
FluANA1-R6	CGA TAC GAC GGG CGT ACT AGC GAA ACT TCC GCT GTA CCC TGA	*neuraminidase*	
FluANA2-F6	CGA TAC GAC GGG CGT ACT AGC GGG AAA TAT GCC CCA AAC TAG C	**Influenza A**	1,029 bp
FluANA2-R6	CGA TAC GAC GGG CGT ACT AGC GAT GCA GCT TTT GCC TTC AAC	*neuraminidase*	
FluAMA-F4	CGA TAC GAC GGG CGT ACT AGC GTT CTA ACC GAG GTC GAA ACG	**Influenza A**	891 bp
FluAMA-R5	CGA TAC GAC GGG CGT ACT AGC GCT CTG GCA CTC CTT CCG TAG	*matrix*	
FluBHA-F5	CGA TAC GAC GGG CGT ACT AGC GGG GAG GTC AAT GTG ACT GGT	**Influenza B**	898 bp
FluBHA-R5	CGA TAC GAC GGG CGT ACT AGC GGG GCA ATT TCC TAT GGC TTT	*hemaggultinin*	
FluBNA-F4	CGA TAC GAC GGG CGT ACT AGC GGT GAA CCG TTC TGC AAC AAA	**Influenza B**	899 bp
FluBNA-R3	CGA TAC GAC GGG CGT ACT AGC GCC AAT CTT GGA TGC CAT TCT	*neuraminidase*	

(continued)

Table 15.1 (continued)

Primer name	Sequence (5′→3′)	Organism/gene	Amplicon size
FluBMA-F2	CGA TAC GAC GGG CGT ACT AGC GCA TTG ACA GAA GAT GGA GAA GG	**Influenza B**	411 bp
FluBMA-R2	CGA TAC GAC GGG CGT ACT AGC GAA GCA CAG AGC GTT CCT AG	*matrix*	
Ad5 hexon-F2	CGA TAC GAC GGG CGT ACT AGC GCT GTG GAC CGT GAG GAT ACT	**Adenovirus 5** *hexon*	1,768 bp
Ad5 hexon-R2	CGA TAC GAC GGG CGT ACT AGC GTT GGC GGG TAT AGG GTA GAG C		
Ad5 fiber-F2	CGA TAC GAC GGG CGT ACT AGC GTT ATT CAG CAG CAC CTC CTT G	**Adenovirus 5**	2,046 bp
Ad5fiber-R2	CGA TAC GAC GGG CGT ACT AGC GGG TGG CAG GTT GAA TAC TAG	*fiber*	
Ad5 E1A-F3	CGA TAC GAC GGG CGT ACT AGC GGG CTG ATA ATC TTC CAC CTC C	**Adenovirus 5**	808 bp
Ad5 E1A-R3	CGA TAC GAC GGG CGT ACT AGC GCT CTC ACG GCA ACT GGT TTA A	*E1A*	
Ad4 hexon-F3	CGA TAC GAC GGG CGT ACT AGC GGA CAG GAC GCT TCG GAG TAC	**Adenovirus 4** *hexon*	1,334 bp
Ad4 hexon-R3	CGA TAC GAC GGG CGT ACT AGC GGG CAA CAT TGG CAT AGA GGA AG		
Ad4 fiber-F2	CGA TAC GAC GGG CGT ACT AGC GGG TGG AGT GAT GGC TTC G	**Adenovirus 4**	1,245 bp
Ad4 fiber-R2	CGA TAC GAC GGG CGT ACT AGC GAG TGC CAT CTA TGC TAT CTC C	*fiber*	
Ad4F E1A-F1	CGA TAC GAC GGG CGT ACT AGC GGC CGT GGA GTA AAT GGC TAA	**Adenovirus 4**	1,506 bp
Ad4F E1A-R1	CGA TAC GAC GGG CGT ACT AGC GAG TCT TCC AAG ACC GTC CAA	*E1A*	
Ad7 hexon-F2	CGA TAC GAC GGG CGT ACT AGC GAT GTG ACC ACC GAC CGT AG	**Adenovirus 7** *hexon*	2,417 bp
Ad7 hexon-R2	CGA TAC GAC GGG CGT ACT AGC GGT TGC TGG AGA ACG GTA TG		

(continued)

Table 15.1 (continued)

Primer name	Sequence (5′→3′)	Organism/gene	Amplicon size
Ad7 fiber-F1	CGA TAC GAC GGG CGT ACT AGC GTC TAC CCC TAT GAA GAT GAA AGC	**Adenovirus 7**	688 bp
Ad7fiber-R1	CGA TAC GAC GGG CGT ACT AGC GGG ATA GGC AGT TGT GCT GGG CAT	*fiber*	
Ad7 E1A-F2	CGA TAC GAC GGG CGT ACT AGC GTG AGT GCC AGC GAG AAG AG	**Adenovirus 7**	786 bp
Ad7 E1A-R2	CGA TAC GAC GGG CGT ACT AGC GCA GGA GGT GAG GTA GTT GAA TC	*E1A*	
A tha TIM-F2	CGA TAC GAC GGG CGT ACT AGC GTC AAA TCC TCG TTG ACA GAC	***A. thaliana***	503 bp
A tha TIM-R2	CGA TAC GAC GGG CGT ACT AGC GTG CAC TGT TGC CTC CAT TGA	*TIM*	

Table 15.2
List of PCR primers in primer mix B used for multiplex PCR

Primer name	Sequence (5′ → 3′)	Organism/gene	Amplicon size
PIV I HN-F2	CGA TAC GAC GGG CGT ACT AGC GAC AGG AAT TGG CTC AGA TAT G	**Parainfluenza 1** *hemagglutinin-neuraminidase*	382 bp
PIV I HN-R2	CGA TAC GAC GGG CGT ACT AGC GAC ATG ATC TCC TGT TGT CGT		
PIV III HN-F2	CGA TAC GAC GGG CGT ACT AGC GTC GAG GTT GCC AGG ATA TAG G	**Parainfluenza 3** *hemagglutinin- neuraminidase*	477 bp
PIV III HN-R2	CGA TAC GAC GGG CGT ACT AGC GGG ACT ATG AGA TGC CTG ATT GC		
PIV III 5′ND-F2	CGA TAC GAC GGG CGT ACT AGC GCA ACT ATT AGC AGT CAC ACT CG	**Parainfluenza 1** *5' noncoding region*	180 bp
PIV III 5′ND-R2	CGA TAC GAC GGG CGT ACT AGC GAA GTT GGC ATT GTG TTC AGT G		
HRhino 5'ND-F2	CGA TAC GAC GGG CGT ACT AGC GTC ATC CAG ACT GTC AAA GG	**Rhinovirus 89** 5' noncoding region	423 bp
HRhino 5'ND-R2	CGA TAC GAC GGG CGT ACT AGC GAA ACA GGA AAC ACG GAC ACC		
RSV Lpol-F2	CGA TAC GAC GGG CGT ACT AGC GCT CTA TCA TCA CAG ATC TCA GC	**RSV*-A** *L-polymerase*	388 bp
RSV Lpol-R2	CGA TAC GAC GGG CGT ACT AGC GCA TGA GTC TGA CTG GTT TGC		

(continued)

Table 15.2 (continued)

Primer name	Sequence (5′→ 3′)	Organism/gene	Amplicon size
RSVA MNN-F2	CGA TAC GAC GGG CGT ACT AGC GAC AAA GAT GGC TCT TAG CAA AG	**RSV*-A** *major nucleocapsid*	196 bp
RSVA MNN-R2	CGA TAC GAC GGG CGT ACT AGC GAC CCA GTG AAT TTA TGA TTA GC		
RSVB MNN-F2	CGA TAC GAC GGG CGT ACT AGC GAA AAC CAA CCC AAC CAA ACC	**RSV*-B** *major nucleocapsid*	248 bp
RSVB MNN-R2	CGA TAC GAC GGG CGT ACT AGC GGC ACA TCA TAA TTG GGA GTG TC		
WNV C-F 2	CGA TAC GAC GGG CGT ACT AGC GGC TCT CTT GGC GTT CTT CAG	**West Nile virus** C and prM	407 bp
WNV C-R2	CGA TAC GAC GGG CGT ACT AGC GTC ATT ACC AGC CGA CAG CAC		
WNV E-F2	CGA TAC GAC GGG CGT ACT AGC GCC GTC AGC GAT CTC TCC AC	**West Nile Virus** E	107 bp
WNV E-R2	CGA TAC GAC GGG CGT ACT AGC GCC TGT CCA CCA CTC CTT GTC		
WNV NS1-F2	CGA TAC GAC GGG CGT ACT AGC GGC TTG AAA GGG CAG TTC TGG	**West Nile Virus** NS1	150 bp
WNV NS1-R2	CGA TAC GAC GGG CGT ACT AGC GCA GGT CTC CGA TTG TGA TTG C		

(continued)

Table 15.2 (continued)

Primer name	Sequence (5′→ 3′)	Organism/gene	Amplicon size
coron229E MG-F2	CGA TAC GAC GGG CGT ACT AGC GCT CTG GTG TGT GGT GCT TAT A	**Coronavirus 229E** *membrane glycoprotein*	718 bp
coron229E MG-R2	CGA TAC GAC GGG CGT ACT AGC GCT CGG CAC GGC AAC TGT C		
coronOC43 MG-F2	CGA TAC GAC GGG CGT ACT AGC GAT GTG GAT GAC GTT TAG GTA	**Coronavirus OC43** *Membrane glycoprotein*	676 bp
coronOC43 MG-R2	CGA TAC GAC GGG CGT ACT AGC GGG TTG ATG GCA GTC GGT AA		
S pne lytA-F2	CGA TAC GAC GGG CGT ACT AGC GAA GAA GAG TTC ATG ACG GAC	***S. pneumoniae*** *Autolysin*	148 bp
S pne lytA-R2	CGA TAC GAC GGG CGT ACT AGC GTG GTT GTT TGG TTG GTT ATT CG		
S pne ply-F2	CGA TAC GAC GGG CGT ACT AGC GCC GAT GAC TTA TAG TAT TGA	***S. pneumoniae*** *pneumolysin*	129 bp
S pne ply-R2	CGA TAC GAC GGG CGT ACT AGC GAT AAT CTT GAT GCC ACT TAG C		
M pne CytP1-F2	CGA TAC GAC GGG CGT ACT AGC GGT TCT TCA GGC TCA GGT CAA TC	***M. pneumoniae*** *Cytadhesin P1 protein*	390 bp
M pne CytP1-R2	CGA TAC GAC GGG CGT ACT AGC GAC AGC GGT ATG TAC TGG TCA T		

(continued)

Table 15.2 (continued)

Primer name	Sequence (5′→ 3′)	Organism/gene	Amplicon size
N men ctrA-F2	CGA TAC GAC GGG CGT ACT AGC GTG GGA ATA GTG TGC GTA TGC	***N. meningitidis***	195 bp
N men ctrA-R2	CGA TAC GAC GGG CGT ACT AGC GAC ATC ACC GCG ACG CAG CAA	*capsular transport protein*	
N men crgA-F2	CGA TAC GAC GGG CGT ACT AGC GGA TTC CGC GAT GCC GAT G	***N. meningitidis***	318 bp
N men crgA-R2	CGA TAC GAC GGG CGT ACT AGC GCG CCC ATG TAT TTA GAG AAC CG	*regularoty protein, crgA*	
B per PTXP-F2	CGA TAC GAC GGG CGT ACT AGC GCC GGC GTC GTG CGC GAA A	***B. pertussis***	361 bp
B per PTXP-R2	CGA TAC GAC GGG CGT ACT AGC GCA GCC ACG TCA GCC AGC C	*pertussis toxin promoter region*	
B per ptxS1-F3	CGA TAC GAC GGG CGT ACT AGC GGA GCG AAT ATC TGG CAC ACC	***B. pertussis***	337 bp
B per ptxS1-R3	CGA TAC GAC GGG CGT ACT AGC GGG GCC AGG TCT AGA ACG AAT	*pertussis toxin S1 subunit*	
C pne VD4-F2	CGA TAC GAC GGG CGT ACT AGC GTG GAG TAC AAT GGT CTC GAG C	***C. pneumoniae***	161 bp
C pne VD4-R2	CGA TAC GAC GGG CGT ACT AGC GTT TGC ATG AAG TCT GAG AAC GA	*major outer membrane protein VD4*	

(continued)

Table 15.2 (continued)

Primer name	Sequence (5′→ 3′)	Organism/gene	Amplicon size
C pne rpoB-F2	CGA TAC GAC GGG CGT ACT AGC GAC GGC ATT ACA ACG GCT AG	***C. pneumoniae***	406 bp
C pne rpoB-R2	CGA TAC GAC GGG CGT ACT AGC GCA TCT TCT GGT AAT CCC TGT TC	*DNA directed RNA polymerase*	
C pne VD2-F2	CGA TAC GAC GGG CGT ACT AGC GAC AGC GTT CAA TCT CGT TGG	***C. pneumoniae***	249 bp
C pne VD2-R2	CGA TAC GAC GGG CGT ACT AGC GAG AGA ATT GCG ATA CGT TAC AG	*major outer membrane protein VD2*	
S pyo speB-F2	CGA TAC GAC GGG CGT ACT AGC GCC TTA CAA CCT ATT GAC ACC TG	***S. pyogenes***	371 bp
S pyo speB-R2	CGA TAC GAC GGG CGT ACT AGC GAC ACG AGA GCT ACC TGC AGA	*pyrogenic exotoxin B*	
S pyo mef-F2	CGA TAC GAC GGG CGT ACT AGC GTT TAT ACA ATA TGG GCA GGG	***S. pyogenes***	381 bp
S pyo mef-R2	CGA TAC GAC GGG CGT ACT AGC GTC GTA AGC TGT TCT TCT GGT AC	*macrolide-efflux determinant (mefA, mefE)*	
S pyo ermB-F2	CGA TAC GAC GGG CGT ACT AGC GTC ATT GCT TGA TGA AAC TGA T	***S. pyogenes***	244 bp
S pyo ermB-R2	CGA TAC GAC GGG CGT ACT AGC GTT GGA TAT TCA CCG AAC ACT AG	*erythromycin resistance methylase (ermB)*	

(continued)

Table 15.2 (continued)

Primer name	Sequence (5′→ 3′)	Organism/gene	Amplicon size
S pyo ermTR-F2	CGA TAC GAC GGG CGT ACT AGC GCT TGT GGA AAT GAG TCA ACG G	***S. pyogenes***	233 bp
S pyo ermTR-R2	CGA TAC GAC GGG CGT ACT AGC GAG GTA GCT ATA TTT CGC TTG AC	*erm(TR)*	
B ant rpoB-F2	CGA TAC GAC GGG CGT ACT AGC GGA GCG TCT ACG TCC TGG TGA	***B. anthracis***	291 bp
B ant rpoB-R2	CGA TAC GAC GGG CGT ACT AGC GCA TTG GTT TCG CTG TTT TGA	*RNA polymerase beta-subunit*	
B ant pag-F2	CGA TAC GAC GGG CGT ACT AGC GTG GAA GAG TGA GGG TGG ATA C	***B. anthracis***	486 bp
B ant pag-R2	CGA TAC GAC GGG CGT ACT AGC GAA TAA TCC CTC TGT TGA CGA A	*protective antigen*	
B ant capB-F2	CGA TAC GAC GGG CGT ACT AGC GAG GAG CAA TGA GAA TTA CAC G	***B. anthracis***	311 bp
B ant capB-R2	CGA TAC GAC GGG CGT ACT AGC GCT AAG TTC CAA TAC TCT TGC	*poly(D-glutamic acid) capsule*	
VMVHA-F3	CGA TAC GAC GGG CGT ACT AGC GGC CGG TAC TTA TGT ATG TGC ATT	**Variola Major Virus**	439 bp
VMVHA-R3	CGA TAC GAC GGG CGT ACT AGC GCA TCA TTG GCG GTT GAT TTA	*hemagglutinin*	

(continued)

Table 15.2 (continued)

Primer name	Sequence (5′ → 3′)	Organism/gene	Amplicon size
VMVcrmB-F3	CGA TAC GAC GGG CGT ACT AGC GGG GAA CAT ACG CTT CCA GAT	**Variola Major Virus**	257 bp
VMVcrmB-R3	CGA TAC GAC GGG CGT ACT AGC GTT CCA CAT TTT GTT TGG GAA A	*cytokine response modifier B*	
EVLGV-F3	CGA TAC GAC GGG CGT ACT AGC GCC TTA TCC GAC TCG CAA TGT	**Ebola Virus**	366 bp
EVLGV-R3	CGA TAC GAC GGG CGT ACT AGC GCA GTG TGA GGT TAT GTG GTG GA	*L gene*	
LVGPC-F4	CGA TAC GAC GGG CGT ACT AGC GTT GGT TGC GCA ATT CAA GT	**Lassa Virus**	302 bp
LVGPC-R4	CGA TAC GAC GGG CGT ACT AGC GTG TTG TTC TTT GTG CAG GAG A	*GPC gene*	
F tul 13Kd-F2	CGA TAC GAC GGG CGT ACT AGC GTC GTA ATG TTA GCT GTA TCA TC	***F. tularensis***	513 bp
F tul 13Kd-R2	CGA TAC GAC GGG CGT ACT AGC GTA CAT TAG CTG TCC ACT TAC CG	*13-kDa lipoprotein*	
F tul FopA-F2	CGA TAC GAC GGG CGT ACT AGC GGT GGG TGG TGG TCT TAA GTT T	***F. tularensis***	192 bp
F tul FopA-R2	CGA TAC GAC GGG CGT ACT AGC GCT GGA TAT TAC CAG TGT CAT T	*FopA*	

(continued)

Table 15.2 (continued)

Primer name	Sequence (5′→ 3′)	Organism/gene	Amplicon size
Y pes cve-F2	CGA TAC GAC GGG CGT ACT AGC GAC TGA TAA AGG GGA GTG GAT A	*Y. pestis* *cve2155 sequence*	388 bp
Y pes cve-R2	CGA TAC GAC GGG CGT ACT AGC GCT CGC CTT GCT CTT TGA GC		
Y pes caf1-F2	CGA TAC GAC GGG CGT ACT AGC GGG ACA CAA GCC CTC TCT ACG	*Y. pestis* *Caf1*	564 bp
Y pes caf1-R2	CGA TAC GAC GGG CGT ACT AGC GTA GAT ACG GTT ACG GTT ACA G		
AthaNAC1-F5	CGA TAC GAC GGG CGT ACT AGC GCA TGG GAA GCT GTT TTG ATG	***A. thaliana*** *NAC1*	497 bp
AthaNAC1-R5	CGA TAC GAC GGG CGT ACT AGC GCC CGA AGA ATT GTT CCA ATC		

*RSV: Respiratory syncytial virus

filter, and store at 2–8°C shielded from light. The solution should be clear, discard solution if it turns yellow.

5. Wash A (non-stringent wash buffer): 6X SSPE, 0.01% Tween-20. Filter the solution through a 0.2 μm filter and store at room temperature for up to 6 months.
6. Wash B (stringent wash buffer): 0.6X SSPE, 0.01% Tween-20. Filter the solution through a 0.2 μm filter, and store at room temperature for up to 6 months.
7. 1X array holding buffer (100 mM MES, 1 M [Na^+], 0.01% Tween-20). Mix 8.3 mL of 12X MES stock buffer with 18.5 mL of 5 M NaCl, 0.1 mL of 10% Tween 20, and 73.1 mL of water. Store the buffer at 2–8°C and shield from light.
8. SAPE stain solution: 6X SSPE, 0.01% Tween-20, 1X Denhardts solution (Sigma-Aldrich), 10 μg/mL of Streptavidin Phycoerythrin (SAPE, Molecular Probes). The solution can be stored at 4°C shielded from light for up to 1 week.
9. Antibody stain solution: 6X SSPE, 0.01% Tween-20, 1X Denhardts solution (Sigma-Aldrich), 5 μg/mL of biotinylated anti-streptavidin antibody (goat, Vector Labs). The solution can be stored at 4°C for up to 1 week.

3. Methods

The success of using resequencing microarray for multi-organism detection, as in broad spectrum detection of various respiratory pathogens, relies on resolving two issues before the assay can be applied to samples. First the chip must be designed by selecting appropriate reference sequences to answer the questions that will be asked. The second consideration is multiplex primer selection since the assay outlined, rapid analysis of samples with large amounts of background nucleic acid material, requires the use of specific or semi-specific primers.

3.1. Chip Design

Selection of partial genomic sequences from pathogens (reference or target sequences) for placement on a resequencing microarray to provide direct sequence-based identification of multiple pathogens depends on what specific knowledge is required for the various pathogens. For example, the respiratory pathogen microarray v.1 (RPM v.1) chip design includes 57 target genes, partial sequences from the genes containing diagnostic regions of each pathogen (i.e., *E1A*, *hexon*, and *fiber* for human adenoviruses (HAdVs); *hemagglutinin*, *neuraminidase* and the *matrix* genes for influenza A viruses). The targets for both HAdVs and influenza are both long enough that RPM v.1 not only allows identification but also produce strain-specific sequence data at the same time. The remaining respiratory pathogens only required detection, so fewer and shorter partial sequences were selected allowing resequencing of 29.7 kb of sequences to provide at least species level identification of 26 distinct organisms *(10)* (**Note 8**). For the RPM v.1 chip, selection of partial sequences to generate probe sets on the microarray was based on the same rules used in selecting probes for long-oligonucleotide spotted microarrays even though such rules do not account for the strengths and weaknesses of a resequencing microarray. Overall, the detection and discrimination performance of such sequences was good. In fact, probes that for a spotted array would only discern to a particular level such as serotype will at least give the same level of discrimination on a resequencing array and often provide more detailed discrimination such as strain differentiation. Using these selection rules can however lead to wasted space on the resequencing microarray because in some cases where two probes were required on a spotted microarray, the information from only one is sufficient to provide equivalent or greater detection and discrimination on a resequencing microarray. Design methods have since been refined to reduce redundancy and better incorporate the advantages of resequencing arrays into probe selection. Once selected, the sequence file was sent to Affymetrix for fabrication (**Note 9**).

3.2. Multiplex Primer Design

The gene-specific primer pairs for all targets on the RPM v.1 chip *(8)* were designed according to the following criteria to meet minimum amplification efficiency requirement.

1. From our work we have established a gross predictor that hybridization will occur for an organism on the array and that at least 70% of the bases match between the sequence used on the microarray and the organism when aligned (BLAST) *(12)*. A list of sequences that may potentially hybridize to the reference sequence is constructed using a BLAST query. Primers are selected from consensus sequences of well-conserved regions flanking the reference sequence from the list. All potential primers that are 18–24 bases in length with ~50% GC content, with no repetitive sequences and have annealing temperature range from 55 to 60°C without potential for self annealing and hairpin formation are considered. This list is further filtered to ensure uniqueness with respect to the other pathogens and human genome by using a full search of the GenBank database with the BLAST program. This insured that the potential primers for an organism have a number of mismatches with these two groups of sequences and would not mis-prime on a sequence region not of interest in the assay (**Note 10**).

2. Once selected, all primers in the same primer cocktails are checked for potential hybridization to other primers to reduce the potential of primer-dimer formation. The primers that form conceivable primer dimers, 8 or more contiguous base matches between the primers, are replaced with new ones until all potential primer dimers are removed. Also, we adapt a method developed by Shuber et al. and Brownie et al. *(13,14)* to further suppress primer-dimer formation by adding a linker sequence of 22 bp (primer L) to the 5′-end of primers used (**Note 11**).
3. To minimize the possibility of intra-primer interactions, the number of primers in a mix is kept to no more than 100. For RPM v.1, the primers were divided into two independent reactions to satisfy this requirement. Fine-tuning adjustments to both mixtures (swapping primers that amplified poorly for new ones) were carried out to ensure all target genes from the 26 targeted pathogens (West Nile Virus is included on the array but not in this amplification scheme) would amplify sufficiently to generate detectable hybridization. Primer sequences are listed in **Tables 15.1** and **15.2** *(8)*.

3.3. Sample Preparation

1. Mix 150 µL of the fluid samples (nasal washes or throat swabs in storage media) with 150 µL of 2X T&C lysis solution premixed with 1 µl of 50 µg/µL proteinase K thoroughly by vortexing. The sample mixture is incubated at 65°C for 15 min with vortex mixing every 5 min. After incubation, place the sample on ice for 3–5 min.
2. Add 150 µL of MPC protein precipitation reagent to the sample mixture and vortex vigorously for 10 s. At this point, sample mixture should appear cloudy. Pellet the debris by centrifugation for 10 min at 13,000 rpm at room temperature using microcentrifuge. If the pellet is clear, small or loose, add an additional 25µL of MPC protein precipitation reagent, mix, and spin again.
3. Transfer the supernatant to a clean 1.5 mL tubes and discard the pellet (**Note 12**). Add 500 µL of isopropanol, then invert tube several times to mix thoroughly. Pellet the DNA by centrifugation at 4°C for 10 min at 13,000 rpm using microcentrifuge. Pour off the isopropanol, be careful not to lose the DNA pellet. Rinse twice with 75% ethanol, centrifuge briefly if the pellet is dislodged. Remove all the residual ethanol with a pipette and air dry the pellet for 5–10 min. Resuspend the total nucleic acids in 25 µl of nuclease-free water and store at –20°C until further use (**Note 13**).

3.4. Internal Controls

Two *Arabidopsis thaliana* plant genes, corresponding to NAC1 and TIM, were chosen as internal controls for reverse transcription

(RT) and PCR as they are unlikely to occur naturally in clinical samples (**Note 14**). Two plasmids, pSP64poly(A)-NAC1 and pSP64poly(A)-TIM, containing ~500 bp of the two genes were kindly provided by Dr. Norman H. Lee at The Institute for Genome Research (Rockville, MD) *(15)*.

1. NAC1 fragment is amplified by PCR with SP6 (5′-GAT TTA GGT GAC ACT ATA-3′ and M13R (5′-CAG GAA ACA GCT ATG AC-3′) primers, and the PCR products are purified using QIAquick PCR Purification Kit (Qiagen).
2. To generate RNA from pSP64poly(A)-TIM, the plasmids were linearized with *EcoRI* and in vitro transcribed from the SP6 promoter using the MEGAscript™ High Yield Transcription Kit (Ambion). 60 fg each of NAC1 and TIM are used as internal controls for checking the amplification efficiency and the presence of inhibitors in the specimens.

3.5. Multiplex RT-PCR Amplification

1. Reverse transcription (RT) reactions are performed in 20 μL volumes. Mix 5–8 μL of extracted DNA with 1 μL 10 mM dNTPs (**Note 15**), 2 μL 40 μM primer LN, 60 fg/μL each of NAC1 and TIM, and bring the volume up to 12 μL. The reaction is incubated at 65°C for 5 min, then chilled on ice for at least 1 min. Add 8 μL RT master mix and incubate at 25°C for 10 min, 50°C for 50 min, then denature the enzyme at 85°C for 5 min.
2. The RT reaction products are split up into two 10 μl volumes and used in two different multiplex PCR. Primer mix A contains 19 primer pairs and amplifies 18 gene targets from three different influenza A viruses, 1 influenza B virus, 3 serotypes of HAdVs, and one internal control (TIM; **Table 15.1**). Primer mix B contains 38 primer pairs and amplifies the remaining 37 gene targets and the other internal control (NAC1; **Table 15.2**) (8). The amplification reaction is carried out with an initial incubation at 25°C for 10 min., preliminary denaturation at 94°C for 3 min., followed by 5 cycles of: 94°C for 30 s, 50°C for 90 s, 72°C for 120 s, then 35 cycles of: 94°C for 30 s, 64°C for 120 s, and a final extension at 72°C for 5 min (**Note 16**). The amplified products from both PCR are combined into a single volume and subjected to purification and processing prior to hybridizing to the RPM v.1 chips.

3.6. Microarray Hybridization and Processing

1. Combine PCR product into one tube, then add 5 volumes of PB buffer to one volume of the PCR samples (500 μL PB for 100 μL PCR product). Mix the reaction mixture by pipetting up and down. The color of the mixture should not change at this point. If the color turns orange or purple, add 10 μL of 3 M sodium acetate, pH 5.0 and mix again.

2. Pipette 700 μL of the mixture into QIAquick spin column in a 2 mL collection tube and centrifuge for 30–60 s at 13,000 rpm using microcentrifuge at room temperature. Discard flow-through and repeat the process if the mixture volume is larger than 700 μL.
3. Wash the spin column with 750 μL PE buffer with the indicated amount of ethanol added, and centrifuge for 30–60 s at 13,000 rpm using microcentrifuge at room temperature. Discard the flow-through, then place column back into the collection tubes. Centrifuge for 60 s at 13,000 rpm using microcentrifuge at room temperature to remove residual ethanol.
4. Elute the DNA by placing the spin column in a new 1.5 mL tube, and add 50 μL EB buffer (10 mM Tris–HCl, pH 8.5) to the center of the spin column and centrifuge for 60 s at 13,000 rpm using microcentrifuge at room temperature (**Note 17**).
5. Purify the IQ-EX PCR products as described in step 1. Determine the concentration of the IQ-EX using UV spectrophotometry.
6. Add 7.6 μL of the fragmentation solution to 35 μL of eluted DNA and 3 μg of IQ-EX PCR product in a final volume of 35 μL. Incubate the reaction mixture at 37°C for 5 min (**Note 18**), then denature the enzyme activity at 95°C for 15 min. Store at 4°C after incubation. At this point, you can store the sample at 4°C for up to 1 week before labeling.
7. Add 17.4 μL of the labeling solution to 35 μL of fragmented DNA and IQ-EX PCR product from step 3. Incubate the reaction mixture at 37°C for 30 min (**Note 18**), and then denature the enzyme activity at 95°C for 15 min. Store at 4°C. Use fragmented and labeled IQ-EX PCR product to prepare the hybridization buffer.
8. Add 160 μL of hybridization buffer to 60 μL of fragmented and labeled PCR products. At this point, you can store the sample at –20°C for up to 1 month before hybridization.
9. Add 200 μL pre-hybridization buffer to each chip, and incubate the chip in the hybridization oven at 49°C at 60 rpm for 15 min. In the meantime, denature the samples from step 5 at 95°C for 5 min, and then equilibrate the tubes at 49°C for 5 min (**Note 19**).
10. Remove the arrays from the hybridization oven, and remove and discard the pre-hybridization buffer. Add 200 μL of denatured samples. At this point, you should see a small bubble inside the chip which serves as a mixing mechanism for the microarray, so ensure such a bubble is present. Incubate the chip at the hybridization oven at 49°C at 60 rpm for

4–16 h (**Note 20**). Remove the hybridization mixture from the array, and fill with 250 μL of array holding buffer. At this point, you can store the array at 4°C for up to 3 h before washing and staining (**Note 21**).

11. Prime the GeneChip® fluidic stations (Affymetrix) with Wash A and B. Register a new experiment in GeneChip Operating Software Service (GCOS). Load the array, SAPE and antibody stain solution, into the designed fluidic module. Start the washing and staining protocol "DNAArray_WS5_450." Remove the array when the protocol is complete; make sure at this time that there is not a bubble. If there are visible bubbles, manually fill the array with array holding buffer using a pipette. Apply two tough spots to each of the two septa on the back of the array. The array can be stored at 4°C for up to 24 h before scanning (**Note 22**). Flush the fluidic stations with DI water and shut down.
12. Turn on the GeneChip Scanner 3000 at least 10 min before use. If the array was stored at 4°C, allow to warm to room temperature before scanning. Insert the array into scanner. Use GCOS to start scanning the array by selecting the corresponding experiment. GCOS will process the image file of the scanned microarray and create cell intensity data. An example of the results is shown in **Fig. 15.1**.

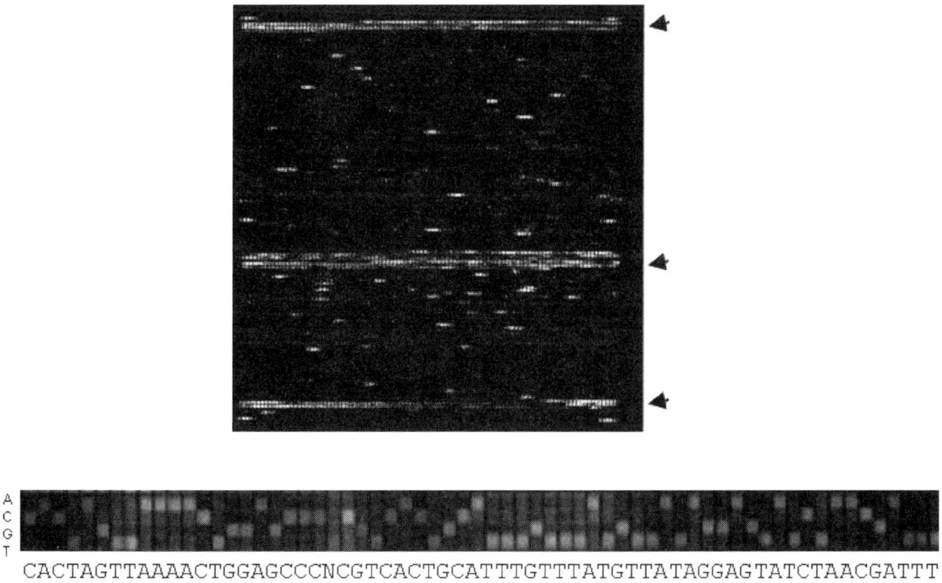

Fig. 15.1. *Top panel*: The hybridization image of the HAdV-7 strain NRRC 1315. *Lower panel*: magnification of a portion of Ad7EIA probe sets showing an example of the primary sequence data generated by the hybridization of amplified targets to RPM V.1. The primary sequence generated can be read from left to right. The *top arrow* indicates Affymetrix hybridization control-IQ-EX; the *middle arrow* indicates the regions that HAdV-7 hybridize; the *lower arrow* indicates MAC1 and TIM, internal amplification control.

13. Use GeneChip® Sequence Analysis Software (GSEQ) to analysis the cell intensity data and generate the base call (**Note 23**). Export the sequence information to FASTA file.

3.7. Pathogen Identification Algorithm

1. Sending entire FASTA files to be searched by BLAST is wasteful of time and potentially misleading (**Note 24**). The set of resolved bases resulting from hybridization is instead subjected to a filtering process. Each references sequence is examined by itself and split into possible subsequences (SubSeqs) suitable for BLAST search. SubSeqs are found by finding seed locations within the sequence that have at least 18 of 20 bases resolved. One of these locations is increased in size while the total called base percentage stays above 40% unless a contiguous stretch of at least 18 or 19 bases of N calls is encountered. The section is marked as its own SubSeq if its length is at least 30 bases and the remaining seed locations are examined in a similar manner.
2. BLAST is used to perform a similarity search of the NCBI nr database using the SubSeqs as the queries. The BLAST program used is the NCBI Blastall –p blastn with a defined set of parameters. Masking of low complex regions is performed for the seeding phase; however, such regions are included in the actual scoring. The default gap penalty and nucleotide match score are used. The nucleotide mismatch penalty, –q, parameter is set to –1 rather than the default. The results of any BLAST query with an expected value <0.0001 are returned in tabular format from the blastall program. If any SubSeq has a value of 10^{-6} or less then it is considered positive for identification of whatever organism is reported in the return.
3. A SubSeq might return many records from the database with the same score. The identified organism is whatever taxonomic classification encompasses all tied best scoring returns when there are more than one (**Note 25**). If only a single return has the best score then that is considered the closest specific strain to the organism in the sample.
4. Different SubSeq for the same reference sequence are required to result in a single pathogen identification for that reference sequence. If one SubSeq has a significantly better score and more detailed identification than others that is taken as the identification of the reference sequence. If all SubSeqs have similar scores, then the taxonomic classification that is consistent for all of them is considered the best identification that can be made.
5. A final examination can be made for the results from reference sequences that targeted the same organism to insure that they are reporting consistent results. It is not required that all reference sequences identify an organism nor is strictly

required that they make the same exact identification. This process can be very laborious and time consuming considering the large number of reference sequences and automation of this process is possible (**Note 26**).

4. Notes

1. For amplification of IQ-EX, other Taq DNA polymerase can also be used. GeneChip® Resequencing Assay Kit also contains control forward 7.5 kb PCR primers (20 μM) which can replace control forward 1.0 kb PCR primers. However, the 7.5 kb IQ-EX amplification requires using Taq DNA polymerase designed for long-range PCR amplification.
2. Other transcription kits can also be used to generate RNA controls.
3. The MasterPure™ DNA and RNA purification kit gave us the highest yield of nucleic acids, but other nucleic acid extraction kits can also be used for this purpose.
4. For multiplex PCR, it is highly recommended that you test different Taq DNA polymerase. We tested several different Taq DNA polymerases from various companies; our experience suggests that Platinum Taq or GoTaq® Flexi DNA polymerase give the highest amplification yield.
5. Good laboratory practice, e.g., always use filtered pipette tips, clear gloves, and UV irradiate PCR hood after each PCR set up, is necessary to prevent contamination issue in the PCR. Additionally, 50x dNTPs (ACGU) and uracil-DNA glycosylase are used to prevent trace amount of carry-over contamination from the previous PCR.
6. Other PCR purification kit can also be used for cleaning up PCR products.
7. The fragmentation reagent is an equivalent of DNase I. Terminal deoxynucleotidyl transferase and biotinylated ddNTP from other sources can also be used for labeling reaction.
8. The resequencing microarray platform is generally more flexible at target selection. However, the design process can still be very complicated. Fortunately, the RPM v.1 chips are available through Tessarae Inc. (Potomac Falls, VA) with the permission from US Naval Research Laboratory. Newer version of RPM (RPM v.3.x) can be also obtained through Tessarae, Inc. (Potomac Falls, VA) with complete protocol. So initial use of the assay does not require design of a microarray.

9. The details of how to make Affymetrix CustomSeq resequencing microarray is described in GeneChip® CustomSeq Custom Resequencing Array Design Guide (http://www.affymetrix.com/support/technical/other/customseq_design_-manual.pdf). Briefly, the first step is to identify and generate the base sequences of interest in FASTA format. Once the content of the array has been selected, the total number of bases is calculated and an array format (different formats have different sequence capacities) is chosen. The sequence submission documents can then be prepared from this information (design request form, sequence and instruction files and purchase order). During the fabrication process at Affymetrix, their chip design group will perform a design clarification process which will check and suggest removing ambiguous, repetitive and homologous sequences. Upon completion of this step, the masks required to produce chips are fabricated and checked for quality. Finally, the arrays will be manufactured. The clarification process can take as little as 2 weeks if little feedback is required but can run longer and the production of chips depends upon scheduling constraints to meet delivery to all their customers.
10. Primers selected should have at least three base mismatches with human genome sequences to avoid non-specific amplification.
11. Short linker primers can also be used, but the linker primers must be unique with higher melting temperature and unrelated to the target pathogens and background genome sequences that the samples may contain.
12. Try to avoid lipid and small white powdery protein substances in the supernatant.
13. If you air dry the pellet too long, you will need to add nuclease-free water to the pellet and store at 4°C overnight to ensure the complete suspension of the nucleic acids.
14. Internal control is not absolutely necessary for the reaction, but it is good to ensure there is no false negative result due to the RT or PCR step. Other genes besides NAC1 and TIM can also be used for internal controls provided the custom design chip has reference sequences to hybridize with the control genes.
15. Do not use 50x dNTPs (ACGU, Sigma) at this step.
16. Two-step PCR are used here to shorten the cycling time, alternatively, a three-step PCR can be used as long as the annealing temperature is raised enough to switch to linker only priming.
17. Smaller amount of elution buffer can be used; EB buffer can be used to bring up the volume for the next step.

18. Longer incubation time is recommended for both fragmentation (30 min) and labeling (2 h) by Affymetrix. However, our experience suggested that a shorter incubation time is sufficient.
19. It is recommended that the chip should be warm up to room temperature before performing pre-hybridization. Our experiences suggested that this is not a critical step.
20. It is recommended that the hybridization should be carried out for 16 h in order to reach equilibrium. Our experience indicates that 4 h hybridization can generate sufficient base call for pathogen identification.
21. Wash A can also be used to fill the chip at this stage. Chips can be stored for up to 24 h before washing.
22. Chips can still be scanned after storing for more than 24 h and less than 1 week, although the fluorescence signal will be weaker.
23. The base calling algorithm for resequencing microarray is changed to allow for the most "permissive" base calling to be executed compared to "conservative" settings that are defaults for GDAS. The parameters are listed below:

    "Permissive" Base Calling Algorithm Settings –

    a. Filter Conditions
        i. No Signal threshold = 0.500 (default = 1.000000)
        ii. Weak Signal Fold threshold = 20000.000 (default = 20.000000)
        iii. Large SNR threshold = 20.000000 (default = 20.000000)
    b. Algorithm Parameters
        i. Strand Quality Threshold = 0.000 (default = 0.000000)
        ii. Total Quality Threshold = 25.0000 (default = 75.000000)
        iii. Maximum Fraction of Heterozygote Calls = 0.99000 (default = 0.900000)
        iv. Model Type (0 = Heterozygote, 1 = Homozygote) = 1
        v. Perfect Call Quality Threshold = 0.500 (default = 2.000000)
    c. Final Reliability Rules
        i. Min Fraction of Calls in Neighboring Probes = 1.0000 (disables filter)
        ii. Min Fraction of Calls of Samples = 1.0000 (disables filter)
24. Resequencing arrays provide positional information which allows for the use of similarity searches; however, because of how the information is obtained they can potentially bias for or against variants with insertions or deletions depending on

the reference sequence selected. Splitting regions that are separated by large sections of N calls reduces this bias.

25. The organism identified may not necessarily be the primary organism the reference sequence is intended to identify but a near-neighbor species.

26. The algorithm as described was built into a new software program, Computer-Implemented Biological Sequence-based Identifier system, version 2 (CIBSI 2.0) to automate the pathogen identification process for the RPM v.1 array *(12)*. CIBSI 2.0 besides determining what each reference sequence detects furthermore, whether the identifications from separate targets support a common organism identification and determine whether detected organisms belong to the target set that the assay was designed to detect or are related to close genetic near neighbors. Target pathogens are the organisms the assay was specifically designed to detect.

References

1. Hacia, J.G. (1999) Resequencing and mutational analysis using oligonucleotide microarrays. *Nat Genet*, **21**, 42–47.
2. Gingeras, T.R., Ghandour, G., Wang, E., Berno, A., Small, P.M., Drobniewski, F., Alland, D., Desmond, E., Holodniy, M. and Drenkow, J. (1998) Simultaneous genotyping and species identification using hybridization pattern recognition analysis of generic Mycobacterium DNA arrays. *Genome Res*, **8**, 435–448.
3. Kozal, M.J., Shah, N., Shen, N., Yang, R., Fucini, R., Merigan, T.C., Richman, D.D., Morris, D., Hubbell, E., Chee, M. et al. (1996) Extensive polymorphisms observed in HIV-1 clade B protease gene using high-density oligonucleotide arrays. *Nat Med*, **2**, 753–759.
4. Wong, C.W., Albert, T.J., Vega, V.B., Norton, J.E., Cutler, D.J., Richmond, T.A., Stanton, L.W., Liu, E.T. and Miller, L.D. (2004) Tracking the evolution of the SARS coronavirus using high-throughput, high-density resequencing arrays. *Genome Res*, **14**, 398–405.
5. Wilson, K.H., Wilson, W.J., Radosevich, J.L., DeSantis, T.Z., Viswanathan, V.S., Kuczmarski, T.A. and Andersen, G.L. (2002) High-density microarray of small-subunit ribosomal DNA probes. *Appl Environ Microbiol*, **68**, 2535–2541.
6. Wilson, W.J., Strout, C.L., DeSantis, T.Z., Stilwell, J.L., Carrano, A.V. and Andersen, G.L. (2002) Sequence-specific identification of 18 pathogenic microorganisms using microarray technology. *Mol Cell Probes*, **16**, 119–127.
7. Davignon, L., Walter, E.A., Mueller, K.M., Barrozo, C.P., Stenger, D.A. and Lin, B. (2005) Use of resequencing oligonucleotide microarrays for identification of Streptococcus pyogenes and associated antibiotic resistance determinants. *J Clin Microbiol*, **43**, 5690–5695.
8. Lin, B., Blaney, K.M., Malanoski, A.P., Ligler, A.G., Schnur, J.M., Metzgar, D., Russell, K.L. and Stenger, D.A. (2007) Using a resequencing microarray as a multiple respiratory pathogen detection assay. *J Clin Microbiol*, **45**, 443–452.
9. Lin, B., Malanoski, A.P., Wang, Z., Blaney, K.M., Ligler, A.G., Rowley, R.K., Hanson, E.H., von Rosenvinge, E., Ligler, F.S., Kusterbeck, A.W. et al. (2007) Application of broad-spectrum, sequence-based pathogen identification in an urban population. *PLoS ONE*, **2**, e419.
10. Lin, B., Wang, Z., Vora, G.J., Thornton, J.A., Schnur, J.M., Thach, D.C., Blaney, K.M., Ligler, A.G., Malanoski, A.P., Santiago, J. et al. (2006) Broad-spectrum

respiratory tract pathogen identification using resequencing DNA microarrays. *Genome Res*, **16**, 527–535.

11. Wang, Z., Daum, L.T., Vora, G.J., Metzgar, D., Walter, E.A., Canas, L.C., Malanoski, A.P., Lin, B. and Stenger, D.A. (2006) Identifying influenza viruses with resequencing microarrays. *Emerg Infect Dis*, **12**, 638–646.
12. Malanoski, A.P., Lin, B., Wang, Z., Schnur, J.M. and Stenger, D.A. (2006) Automated identification of multiple micro-organisms from resequencing DNA microarrays. *Nucleic Acids Res*, **34**, 5300–5311.
13. Brownie, J., Shawcross, S., Theaker, J., Whitcombe, D., Ferrie, R., Newton, C. and Little, S. (1997) The elimination of primer-dimer accumulation in PCR. *Nucleic Acids Res*, **25**, 3235–3241.
14. Shuber, A.P., Grondin, V.J. and Klinger, K.W. (1995) A simplified procedure for developing multiplex PCRs. *Genome Res*, **5**, 488–493.
15. Wang, H.Y., Malek, R.L., Kwitek, A.E., Greene, A.S., Luu, T.V., Behbahani, B., Frank, B., Quackenbush, J. and Lee, N.H. (2003) Assessing unmodified 70-mer oligonucleotide probe performance on glass-slide microarrays. *Genome Biol*, **4**, R5.

Chapter 16

Comparative Genomic Hybridization: DNA Preparation for Microarray Fabrication

Richard Redon, Diane Rigler and Nigel P. Carter

Abstract

The spatial resolution of microarray-based comparative genomic hybridization (array-CGH) is dependent on the length and density of target DNA sequences covering the chromosomal region of interest. Here we describe the methods developed at the Wellcome Trust Sanger Institute (Cambridge, UK) to construct microarrays comprising large-insert clones available through genome sequencing projects. These methods are applicable to Bacterial and Phage Artificial Chromosomes (BAC and PAC) as well as fosmid and cosmid clones. The protocols are scalable for the construction of microarrays composed of several hundreds up to several ten thousands clones.

Key words: Microarray fabrication, large-insert clones, BAC, PAC, fosmid, cosmid, DOP-PCR, Comparative Genomic Hybridization, array-CGH.

1. Introduction

Constructing large-insert clone arrays originally involved extraction of DNA inserts from large volumes of bacterial cultures, resulting in only small quantities of DNA available for spotting *(1–3)*. Although this strategy is practical for the construction of small microarrays composed of tens to hundreds of clones, it is not scalable to arrays composed of thousands of clones, such as arrays covering the whole genome with one clone every megabase or with tiling path coverage.

In order to amplify with high fidelity small amounts of large-insert clone DNA obtained from high throughput extraction of small bacterial cultures, several strategies have been developed, such as strand-displacement rolling circle amplification *(4)* or ligation-mediated

Martin Dufva (ed.), *DNA Microarrays for Biomedical Research: Methods and Protocols, vol. 529*

DOI 10.1007/978-1-59745-538-1_16 Springerprotocols.com

PCR amplification *(5)*. Here we describe the protocols for the use of degenerate oligonucleotide-primed PCR (DOP-PCR) *(6)*.

DOP-PCR was originally designed for species-independent general DNA amplification *(7)*. Used with a cycling protocol incorporating a small number of initial low temperature annealing cycles, DOP-PCR allows a general amplification of any target DNA i.e., a chromosomal segment, whole chromosome, or whole genome. The target DNA is amplified by PCR with a mix of primers comprising unique 5′ and 3′ end sequences flanking a random variable hexanucleotide sequence *(7)*.

Although DOP-PCR was designed for general amplification of any DNA fragment, the representation of the target DNA by the DOP-PCR products relies on the six fixed nucleotides located at the 3′ end of the primer sequence. In our protocols, each clone is amplified with three different sets of degenerate primers to improve the representation of large-insert clone sequence after DOP-PCR amplification. Each degenerate primer is designed with a different fixed 3′ hexanucleotide sequence, which has been selected for its presence at high frequency in the human genome *(6)*.

In addition, because DNA preparations of human large-insert clones from bacterial cultures are contaminated with *Escherichia coli* genomic DNA *(8)*, the three primer sets have been selected to be inefficient in the amplification of *Escherichia coli* DNA. In consequence, the PCR products using these three primer sets contain mostly human insert sequences, which will greatly increase the quantity of printed DNA able to hybridize with the labeled test and reference genomic probes *(6)*.

The three degenerate primers contain the same 5′ decanucleotide sequence. This property enables a secondary PCR amplification with one universal 5′-modified primer whose 3′ end matches the 5′ end of all DOP primers. The secondary PCR presents two advantages: (i) it increases exponentially the quantity of large-insert clone DNA available for array printing; (ii) it facilitates the incorporation of reactive groups for covalent attachment of the amplified DNA to specially coated glass slides. In our protocol, we use 5′ amino-linked secondary primers such that amplified large-insert clone DNA will bind covalently to CodeLink® activated slides *(6)*.

2. Materials

2.1. Extraction of Large-Insert Clone DNA

1. 1 ml deep-well microtiter plates (Beckman).
2. ‘U’ bottom microtiter plate (Greiner).
3. Multiscreen 0.2 μM filter plate (Millipore).
4. 2 × TY medium: Tryptone 16 g/l, Yeast Extract 10 g/l, NaCl 5 g/l, pH 7.4.

5. Solution I: 50 mM glucose, 5 mM Tris–HCl pH 8.0, 10 mM EDTA, store at 4°C.
6. Solution II: 0.2 M NaOH, 1% SDS.
7. Solution III: 3 M KOAc, store at 4°C.
8. Isopropanol.
9. 70% Ethanol.
10. T0.1E buffer: 10 mM Tris–HCl, 0.1 mM EDTA, pH 8.
11. RNAse A (ICN Biochemicals).

2.2. PCR Amplification of Large-Insert Clone DNA

1. Sequences of DOP Primers:
   (a) DOP1: CCGACTCGAGNNNNNNCTAGAA;
   (b) DOP2: CCGACTCGAGNNNNNNTAGGAG;
   (c) DOP3: CCGACTCGAGNNNNNNTTCTAG;
   (d) Aminolink primer: GGAAACAGCCCGACTCGAG.
2. TAPS solution: 250 mM TAPS (Sigma), 166 mM $(NH_4)_2SO_4$, 25 mM $MgCl_2$, pH 9.3 (adjusted with 5 M KOH). Store at −20°C.
3. TAPS2-buffer: Add 33 µl Bovine Serum Albumin (5%) and 7 µl β-mercaptoethanol to 960 µl TAPS solution. Prepare and sterilize under a UV hood freshly before use.
4. 1% W1: polyethylene glycol ether W-1 (or Brij® 58, Sigma), 1% (w/v) in water.
5. AmpliTaq polymerase (Roche).
6. dNTP mix (each at 2.5 mM in water; available from Amersham Biosciences in 100 mM solutions).
7. Aminolink buffer: 500 mM KCl, 25 mM $MgCl_2$, 50 mM Tris–HCl, pH 8.5. Store at room temperature for 1 week only.

2.3. Array Printing

1. CodeLink™ activated slides (GE Healthcare UK Limited, UK).
2. Spotting buffer (4 ×): 1 M sodium phosphate, 0.001% (w/v) sarkosyl, 0.4% (w/v) sodium azide, pH 8.5. Store at room temperature.
3. Ammonium hydroxide: 1% (w/v) in water.
4. 0.1% SDS: Sodium dodecyl sulfate, 0.1% (w/v) in water.

3. Methods

3.1. Extraction of Large-Insert Clone DNA

1. For each clone, set up 500 µl of bacterial cultures in 2 × TY medium with the appropriate antibiotic in a separate well of a 1 ml deep-well microtiter plate. Grow for 18 h with agitation at 37°C.

2. Transfer 250 μl of the culture into a 'U' bottom microtiter plate and centrifuge at 1,000 *g* for 4 min when extracting BAC/PAC clones, or for 2 min when extracting cosmids and fosmids.
3. Discard the supernatant, then add 25 μl of solution I to each well and mix gently by tapping the side of the plate.
4. Add 25 μl of solution II, mix as before and leave at room temperature for 5 min until the solution clears.
5. Add 25 μl of cold solution III, mix as before and leave at room temperature for 5 min.
6. After incubation, transfer the content of each well to a 0.2 μM filter plate. Filter the sample into a 'U' bottom 96-well plate containing 100 μl of isopropanol per well by centrifugation at 1,000 *g* for 2 min at 20°C.
7. Remove the filter and incubate the samples for 30 min at room temperature before spinning the samples at 1,600 *g* for 20 min at 20°C.
8. Remove the supernatant, wash the DNA pellet in 70% ethanol and centrifuge at 1,600 *g* for 10 min at 20°C.
9. Remove the supernatant, dry the DNA pellet then leave it to dissolve at 4°C overnight in 5 μl of T0.1E (pH8) with RNAseA (10 μl of 1 mg/ml RNAseA per 1 ml T0.1E). Store at −20°C until required.

3.2. PCR Amplification of Large-Insert Clone DNA

1. For each clone, three separate DOP-PCR are required, with three different DOP primers. For each DOP primer, all reactions are prepared in 96-well plates, each well corresponding to the amplification of one clone in a final reaction volume of 50 μl (**Note 1**). The reaction is prepared by mixing: 5 μl of TAPS2-buffer, 5 μl DOP primer (20 μM); 4 μl dNTP mix; 2.5 μl 1% W1; 0.5 μl of AmpliTaq polymerase; 5 μl clone template at 1 ng/μl (**Note 2**) and 28 μl water.
2. Place the PCR in a thermocycler and use the following program: (i) initial denaturing: 3 min at 95°C; (ii) 10 cycles: 1.5 min at 95°C, 2.5 min at 30°C, ramp up to 72°C by 0.1°C/s, 3 min at 72°C; (iii) 30 cycles: 1 min at 95°C, 1.5 min at 62°C, 2 min at 72°C; (iv) last extension: 8 min at 72°C
3. Run 5 μl of each PCR product on a 2.5% agarose gel (200 V, 30 min). The average size of the product ranges from 0.2 to 2 kb (**Fig. 16.1A**). Ensure you have no signals in the negative controls (**Note 3**). PCR products can be stored at −20°C until required.
4. Combine together the three different DOP-PCR products from each clone into a single 96-well plate position.

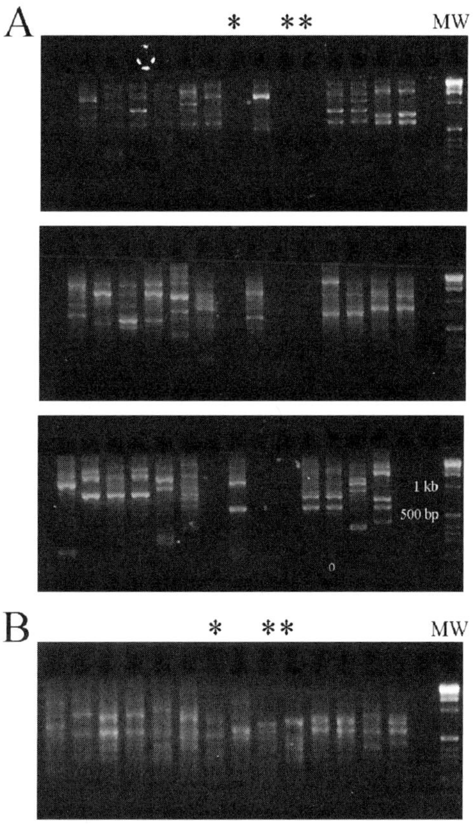

Fig. 16.1. Banding patterns of degenerate oligonucleotide-primed PCR (DOP-PCR) and aminolink-PCR products (**A**) Banding patterns of DNA products after PCR amplification of fosmids using DOP1 (*top*), DOP2 (*middle*), and DOP3 (*bottom*) primers. Note the specific banding pattern obtained with each primer. The asterisks (*) mark the absence of any band on negative controls (DOP-PCR without template DNA). (**B**) Banding pattern of DNA products after aminolink-PCR on combined DOP-PCR products. Specific patterns are still visible at this stage. However, patterns tend to be more homogenous with the presence of more bands for each product. Note that all positions corresponding to aminolink-PCR negative controls (*) show the presence of smaller bands: these bands are due to the use of combined products from DOP-PCR negative controls (**Note 2**). MW: molecular weight marker.

5. Prepare aminolink-PCR in 96-well plates, each well corresponding to the amplification of one clone in a final reaction volume of 90 μl, using the combined DOP-PCR products as a template. The reaction is prepared by mixing: 9 μl aminolink buffer, 9 μl dNTP mix, 4.5 μl aminolink primer (200 ng/μl), 0.9 μl AmpliTaq polymerase, 3 μl combined DOP-PCR products and 63.6 μl water.
6. Place the PCR in a thermocycler and amplify using program: (i) initial denaturing: 10 min at 95°C; (ii) 35 cycles: 1 min at 95°C, 1.5 min at 60°C, 7 min at 72°C; (iv) last extension: 10 min at 72°C.

7. Run 2 μl of the PCR products on a 2.5% agarose gel (200 V, 30 min). The average size of the product ranges from 0.2 to 2 kb (**Note 3** and **Fig. 16.1B**).
8. Transfer 88 μl of amino-linked DOP-PCR product mix into a Millipore Multiscreen 0.2 μm filter plate and add 29 μl 4 × spotting buffer (**Note 4**).
9. Filter the samples by centrifugation at 600 *g* for 10 min and transfer into Genetix 384-well plates. PCR products can be stored at −20°C until required.

3.3. Array Printing

The protocol in this section was developed at the Microarray Facility of the Sanger Institute (http://www.sanger.ac.uk/Projects/Microarrays/arraylab/methods.shtml) where it is routinely used to array DOP-PCR amplified DNA onto CodeLink^{TM} activated slides (GE Healthcare), using Microgrid II stations (Biorobotics). Alternative strategies of microarray printing are described in more details in **Chapters 5–7**.

1. Array DNA elements onto CodeLink^{TM} activated slides at 20–25°C, 40–50% relative humidity.
2. Transfer slides containing arrayed elements into a microscope slide rack, place in a humid chamber containing a saturated NaCl solution and incubate for 24–72 h at room temperature (**Note 5**).
3. Remove the slides from humid chamber, block reactive groups by immersing slides in 1% solution of ammonium hydroxide and incubate for 5 min with gentle shaking.
4. Transfer slides into a solution of 0.1% SDS and incubate for 5 min with gentle shaking.
5. Briefly rinse slides in water at room temperature and then place in 95°C water for 2 min to denature the bound DNA elements.
6. Transfer slides to ice-cold water for 1 min and then briefly rinse twice in water at room temperature.
7. Dry the slides by spinning in a centrifuge for 10 min at 150 *g*. Store slides at room temperature in a dark and dry place until use (**Note 6**).

4. Notes

1. Since DOP-PCR is a very efficient method for the amplification of template DNA from any organism, it is very sensitive to any trace of contaminating DNA. Always ensure that you are working in a sterile and DNA-free environment, i.e., in a PCR hood. Always work with materials and reagents (water, TAPS2 buffer and 1% W1) that have been UV-treated prior to use; do not UV treat the AmpliTaq polymerase, dNTPs, and primers.

2. After DNA extraction, each well should contain ~200 ng of DNA dissolved in 5 μl of T0.1E. Add 195 μl of water in each well before setting up PCR.
3. Negative controls should be included in every PCR set up. For each DOP primer, template DNA should be replaced by equivalent volume of water at several positions in 96-well plates to ensure that there is no unexpected DNA contamination. For aminolink-PCR, if combined products from primary DOP-PCR negative controls are used as templates in negative controls, some DNA bands will be detected at corresponding positions after gel electrophoresis (**Fig. 16.1B**). However, these DNA fragments are smaller in size than those detected after using DOP-PCR products as templates: their presence is due to the formation of primer oligomers during the first PCR in negative controls, which are then amplified and extended during the secondary PCR. Also include negative controls with only water as template in every secondary PCR set-up and check that there is no DNA product for these reactions.
4. The 4 × spotting buffer is specifically designed to print arrays on CodeLink® activated slides with Microgrid II stations. To print array on slides with another surface chemistry and/or using another printing station, refer to the documentation from the manufacturers.
5. Long incubations with high humidity levels are used to facilitate covalent binding between the activated surface of the slide and the 5′ amino groups of printed DNA elements.
6. Slides can be stored in the dark under dehumidified atmosphere for at least 3 months without loss of performance in hybridizations. We recommend the use of a sealed desiccation cabinet.

Acknowledgments

The authors would like to thank Heike Fiegler as well as Cordelia Langford and David Vetrie for their strong contributions in establishing the methods described in this chapter. This work was supported by the Wellcome Trust.

References

1. Pinkel, D., Segraves, R., Sudar, D., Clark, S., Poole, I., Kowbel, D., Collins, C., Kuo, W.L., Chen, C., Zhai, Y. *et al.* (1998) High resolution analysis of DNA copy number variation using comparative genomic hybridization to microarrays. *Nat Genet*, **20**, 207–211.
2. Redon, R., Hussenet, T., Bour, G., Caulee, K., Jost, B., Muller, D., Abecassis, J. and du Manoir, S. (2002) Amplicon mapping and transcriptional analysis pinpoint cyclin L as a candidate oncogene in head and neck cancer. *Cancer Res*, **62**, 6211–6217.

3. Solinas-Toldo, S., Lampel, S., Stilgenbauer, S., Nickolenko, J., Benner, A., Dohner, H., Cremer, T. and Lichter, P. (1997) Matrix-based comparative genomic hybridization: biochips to screen for genomic imbalances. *Genes Chromosomes Cancer*, **20**, 399–407.
4. Smirnov, D.A., Burdick, J.T., Morley, M. and Cheung, V.G. (2004) Method for manufacturing whole-genome microarrays by rolling circle amplification. *Genes Chromosomes Cancer*, **40**, 72–77.
5. Ishkanian, A.S., Malloff, C.A., Watson, S.K., DeLeeuw, R.J., Chi, B., Coe, B.P., Snijders, A., Albertson, D.G., Pinkel, D., Marra, M.A. *et al.* (2004) A tiling resolution DNA microarray with complete coverage of the human genome. *Nat Genet*, **36**, 299–303.
6. Fiegler, H., Carr, P., Douglas, E.J., Burford, D.C., Hunt, S., Scott, C.E., Smith, J., Vetrie, D., Gorman, P., Tomlinson, I.P. *et al.* (2003) DNA microarrays for comparative genomic hybridization based on DOP-PCR amplification of BAC and PAC clones. *Genes Chromosomes Cancer*, **36**, 361–374.
7. Telenius, H., Carter, N.P., Bebb, C.E., Nordenskjold, M., Ponder, B.A. and Tunnacliffe, A. (1992) Degenerate oligonucleotide-primed PCR: general amplification of target DNA by a single degenerate primer. *Genomics*, **13**, 718–725.
8. Foreman, P.K. and Davis, R.W. (2000) Real-time PCR-based method for assaying the purity of bacterial artificial chromosome preparations. *Biotechniques*, **29**, 410–412.

Chapter 17

Comparative Genomic Hybridization: DNA Labeling, Hybridization and Detection

Richard Redon, Tomas Fitzgerald and Nigel P. Carter

Abstract

Array-CGH involves the comparison of a test to a reference genome using a microarray composed of target sequences with known chromosomal coordinates. The test and reference DNA samples are used as templates to generate two probe DNAs labeled with distinct fluorescent dyes. The two probe DNAs are co-hybridized on a microarray in the presence of Cot-1 DNA to suppress unspecific hybridization of repeat sequences. After slide washes and drying, microarray images are acquired on a laser scanner and fluorescent intensities from every target sequence spot on the array are extracted using dedicated computer programs. Intensity ratios are calculated and normalized to enable data interpretation. Although the protocols explained in this chapter correspond primarily to the use of large-insert clone microarrays in either manual or automated fashion, necessary adaptations for hybridization on microarrays comprising shorter target DNA sequences are also briefly described.

Key words: Probe labeling, random priming, hybridization, detection, comparative genomic hybridization, array-CGH.

1. Introduction

Array-CGH was developed in the late 1990s *(1, 2)* to detect DNA copy number changes at high resolution along the genome or locus of interest (**Chapter 3** *for a general introduction on the method*). This chapter details the methods employed to label, hybridize, and detect genomic DNA probes for array-CGH.

As a first step, two genomic DNA samples – one test and one reference – are labeled separately using two different fluorescent dyes: generally Cyanine 3 (Cy3; excitation/emission wavelength maxima: 550/568 nm) and Cyanine 5 (Cy5; 650/668 nm). The probe DNA is commonly generated by enzymatic incorporation

Martin Dufva (ed.), *DNA Microarrays for Biomedical Research: Methods and Protocols, vol. 529*

DOI 10.1007/978-1-59745-538-1_17 Springerprotocols.com

of Cy3 or Cy5 labeled nucleotides (usually dCTP or dUTP). Several strategies are possible for genomic DNA labeling, such as nick translation *(1, 3)*, direct labeling PCR *(4)* or random priming *(5)*. Nick translation requires relatively high quantities of input DNA (usually 2 μg), which can be problematic when working with small and/or precious DNA samples. PCR labeling overcomes this problem, as DNA is amplified and labeled in the same reaction. However, the exponential amplification of genomic DNA can bias the representation of the target by the amplified probe, thus leading to artifactual ratio variations after hybridization *(4)*. Random primed labeling represents a good compromise between the two first techniques: (1) it requires moderate quantities of input DNA (usually 50–500 ng); (2) the representation of target sequences is not biased in the resulting probe, as the reaction consists of a moderate linear amplification of the template DNA by the Klenow fragment using a mix of random hexanucleotides as primers. The probe DNA can also be labeled by chemical techniques such as *cis*-platinum labeling *(6)*, which is based on the capability of monoreactive cisplatin derivatives to react at the N7 position of guanine moieties in DNA *(7)*. However, *cis*-platinum labeling requires more than 1 μg of template DNA as the chemical reaction does not create new DNA molecules.

After DNA labeling, both probes are co-precipitated and dissolved in hybridization buffer. The composition of this buffer, combined with the temperature of the hybridization, is critical to enable efficient and specific hybridization of the probes to the target DNA printed on the microarray. The hybridization buffer used in these protocols contains 50% formamide, 0.1% Tween20, and 5–10% of dextran sulfate. Formamide acts as a denaturing agent, which increases the hybridization stringency and prevents unspecific hybridization at lower temperatures such as 37°C. Tween20 is a nonionic detergent which minimizes non-specific fluorescence background on the surface of the slide. Dextran sulfate is a neutral component which consists of polymers of anhydroglucose in aqueous solutions: in homogeneous solution, it excludes DNA from the volume occupied by the polymer. In consequence, DNA concentration is 'artificially' increased, which improves hybridization kinetics *(8)*.

Hybridization to the array is performed at 37°C in the presence of Cot-1 DNA. Cot-1 DNA is the fraction of genomic DNA consisting largely of highly repetitive sequences. It is obtained from total genomic DNA by selecting for the most rapidly reassociating DNA fragments after denaturing. A large excess of Cot-1 DNA suppresses the hybridization of high-copy repeat sequences that are present in labeled DNA probes and thus prevents their hybridization to the corresponding repeat sequences that are also present in the target DNA *(9, 10)*.

After hybridization, slides are washed several times in PBS/0.05% Tween20 to remove excess hybridization buffer and reduce nonspecific background signal at the surface of the slide. The most critical wash is performed at 42°C in 50% formamide/ 2 × SCC (or at 54°C in 0.1 × SSC in the automated protocol): this stringent wash is essential for the selective and efficient elimination of probe fragments that are not hybridized specifically to the target arrayed DNA.

After washes and slide drying, microarray images are acquired for data analysis. Array-CGH was originally developed from CGH on chromosomes, a method in molecular cytogenetics using fluorescence microscopy *(11, 12)*. Consequently, the first acquisition systems used a CCD camera coupled to 0.5 × or 1 × magnification optical system *(1)* or a confocal laser scanning microscope *(2)*. Today, with the huge expansion of microarray technologies, many scanners specific for DNA microarrays have become available that enable quick and easy image acquisition. The resolution of these scanners is typically 5–10 μm per pixel. Images are usually saved under Tagged Image File Format (TIFF), which is compatible with many distributed image analysis programs.

The method below describes in detail how to label probes by random priming and how to hybridize and detect them on CGH microarrays (in particular on large-insert clone microarrays constructed using methods in **Chapter 16**) by either manual or automated procedures. Array-CGH profiles can then be produced from array images and interpreted using strategies described in **Chapter 3.**

2. Materials

2.1. Probe Preparation

1. BioPrime Labeling Kit (Invitrogen): contains 2.5 × random primers solution, water, Klenow fragment and stop buffer.
2. dNTP mix (10 ×): 1 mM dCTP, 2 mM dATP, 2 mM dGTP and 2 mM dTTP in TE buffer.
3. 1 mM Cy3-dCTP (NEN Life Science).
4. 1 mM Cy5-dCTP (NEN Life Science).
5. Microcon YM30 columns (Millipore/Amicon).

2.2. Hybridization and Detection; Manual Protocol

1. Hybridization buffer M: 50% formamide (deionized, available from Sigma), 10% dextran sulfate, 0.1% Tween 20, 10 mM Tris–HCl, 2 × SSC, pH 7.4. Store at −20°C.
2. 3 M Sodium acetate, pH 5.2.

3. Human Cot-1 DNA (Invitrogen).
4. Formamide.
5. 2 × SSC, pH 7.4.
6. PBS/0.05% Tween 20, pH 7.4.
7. 0.1 × SSC, pH 7.4.
8. Cover slip.
9. Hybridization chamber.

2.3. Hybridization and Detection, Automated Protocol

1. Hs.400PRO/4800PRO automated slide processing station with 51 × 20 mm hybridization chambers (Tecan, Inc).
2. Cysteamine (available from Sigma, **Note 1**).
3. Hybridization buffer A: 50% formamide (deionized), 7.5% dextran sulfate, 0.1% Tween 20, 10 mM Tris–HCl, 2 × SSC, pH 7.4. Store at −20°C.
4. 3 M Sodium acetate, pH 5.2.
5. Human Cot-1 DNA (Invitrogen).
6. PBS/0.05% Tween 20, pH 7.4.
7. 0.1 × SSC, pH 7.4.

3. Methods

Quantities and volumes described here correspond to the use of 2.4 × 3.6 cover slips or Tecan Hs. PRO 51 × 20 mm hybridization chambers: they should be rescaled when using cover slips or chambers with different dimensions. The protocols were developed for the use of large-insert clone microarrays, constructed following the protocols described in **Chapter 16**. However, this protocol has been applied successfully to microarrays comprising smaller target DNA sequences, such as small-insert clones (size ranging from 1 to 4 kb) or PCR products (150 bp–1 kb), by following the automated procedure with slight modifications as described in the footnote of **Table 17.1**.

3.1. Probe Preparation

1. Prepare two separate 1.5 ml tubes for the test and the reference DNA samples. In each tube, add 150 ng of test or reference genomic DNA, 60 μl of 2.5 × random primers solution, and water to a final volume of 130.5 μl (**Note 2**).
2. Denature the samples for 10 min at 100°C, and immediately cool on ice. Then add into each tube, still on ice:

Table 17. 1
Array-CGH program for Hs. PRO stations (Tecan, Inc)

			PROTOCOL
Step	**Type**	**Temp.**	**Options**
1	Wash	37°C	First: Yes, Channel: 1, Runs: 1, Wash time: 0:00:30, Soak time: 0:00:00
2	Injection	37°C	Agitation after injection: Yes
3	Hybridization	37°C	Agitation frequency: high, high viscosity mode: Yes, time: 0:45:00
4	Injection	37°C	Agitation after injection: Yes
5	Hybridization	37°C	Agitation frequency: high, high viscosity mode: Yes, time: 21:00:00 *
6	Wash	37°C	First: No, Channel: 1, Runs: 14, Wash time: 0:00:30, Soak time: 0:00:30
7	Wash	54°C*	First: No, Channel: 2, Runs: 5, Wash time: 0:01:00, Soak time: 0:02:00
8	Wash	25°C	First: No, Channel: 1, Runs: 7, Wash time: 0:00:30, Soak time: 0:00:30
9	Wash	25°C	First: No, Channel: 3, Runs: 1, Wash time: 0:00:30, Soak time: 0:00:00
10	Slide drying	25°C	Time: 0:02:30, Final manifold cleaning: Yes, Channel: 5
			LIQUID USAGE
Liquid channel			**Wash solution**
1			PBS / Tween 20, 0.05%
2			0.1 × SSC
3			water

*For the use of microarrays comprising short target DNA sequences: (1) the time of hybridization should be extended to 48–72 h; (2) the temperature for washes in step 7 should be 52°C instead of 54°C.

15 µl 10 × dNTP mix, 1.5 µl Cy3 or Cy5 labeled dCTP, and 3 µl Klenow fragment (**Note 1**).

3. Incubate the reactions at 37°C overnight.
4. Stop the reactions by adding 15 µl stop buffer.

5. The next step is the removal of unincorporated labeled nucleotides using Microcon YM30 columns. Put the columns onto the tubes provided by the suppliers. Then add the entire labeled test and reference samples to 140 μl of water on top of the columns (one column per sample).
6. Centrifuge for 5 min at 12,000 *g*. Discard the flow-through, put the columns (which should contain the labeled probes) back onto the tubes and add 300 μl water to the columns.
7. Spin again for 5 min at 12,000 *g*. Discard flow-through, add 50 μl HPLC water to the columns and place filter upside-down onto a fresh tube.
8. Centrifuge for 2 min at 400 *g*. The solution collected at this step should be colored in blue for Cy5 labeling, in pink for Cy3 labeling (**Note 3**).

3.2. Hybridization and Detection; Manual Protocol

1. Add the following solutions into a 1.5 ml tube: Cy3 and Cy5 labeled and cleaned test and reference DNA samples (the volume should be ~100 μl per sample); 135 μl human Cot1 DNA (1 μg/μl); 35 μl Sodium acetate 3 M; 1 ml ethanol (**Note 4**).
2. Mix and precipitate at −20°C overnight or at −70°C for 30 min. Centrifuge precipitated DNA for 30 min at 4°C, at maximum speed (~16,000 *g*).
3. Wash pellets in 1 ml 80%ethanol and centrifuge again for 5 min at maximum speed. Remove supernatant and dry the pellets. Re-dissolve the pellets in 35 μl hybridization buffer M.
4. Denature the labeled probes for 10 min at 70°C then incubate for 60 min at 37°C.
5. Apply 30 μl of the hybridization solution onto the array and cover with a 2.4 × 3.6 cm cover slip. Place the slide into a hybridization chamber previously humidified with 2 × SSC/ 20% formamide. Seal the chamber and incubate at 37°C for 24–48 h.
6. Remove the cover slip by placing the slide into a tall glass trough containing PBS/0.05% Tween 20 to wash off the hybridization solution
7. Transfer the slide into a new glass trough and wash in PBS/ 0.05% Tween 20 for 10 min at room temperature under quick agitation.
8. Place the slide into a pre-heated 50% formamide/2 × SSC solution and incubate for 30 min at 42°C under slow agitation.

9. Transfer the slide into fresh PBS/0.05% Tween 20 and wash again for 10 min at room temperature under quick agitation.
10. Dry the slide by centrifugation at 150 *g* for 1 min, then store it in a light-proof box and scan as soon as possible (**Note 6**).

3.3. Hybridization and Detection, Automated Protocol

The manual protocol described in **Section 2.2** enables the processing of up to 8 slides per person per day. For high-throughput array-CGH analysis, we have adapted the protocol for the use of automated hybridization stations (Hs400PRO/Hs4800PRO; Tecan, Inc). The Hs4800PRO station enables one person to process 12 slides per unit per day, and can be configured with up to 4 independent 12-position units.

There are two main changes in the automated procedure: (1) The hybridization buffer A contains only 7.5% of dextran sulfate: this reduces the viscosity of the solution and makes it compatible with injection and mixing in the hybridization chambers; (2) the second wash – with 50% formamide/2 × SSC at 42°C for 30 min – has been replaced by washes with 0.1 × SSC at 54°C, to preserve the components of the hybridization station. The Hs. PRO station should be programmed for array-CGH as described in **Table 17.1**, before starting Step 4 of the protocol below (**Note 5**).

1. Add the following solutions into a 2 ml tube (named Tube A): Cy3 and Cy5 labeled and cleaned test and reference DNA samples (the volume should be ~100 μl per sample), 135 μl human Cot1 DNA (1 μg/μl), and 35 μl Sodium acetate 3 M, and 1 ml ethanol.
2. Into another 2 ml tube (named Tube B), add 100 μl Herring Sperm DNA (10 μg/μl), 10 μl Sodium acetate 3 M, and 300 μl ethanol.
3. Mix both tubes and precipitate DNA at −70°C for 30 min or at −20°C overnight.
4. Centrifuge for 30 min at 4°C, at maximum speed (~16,000 *g*).
5. Remove supernatants and wash pellets in 1 ml 80%ethanol. Centrifuge again for 5 min at 16,000 *g*. Remove supernatants, dry pellets and re-dissolve them in 120 μl hybridization buffer A.
6. Denature DNA in Tubes A and B for 10 min at 70°C.
7. Keep Tube A at 37°C. Start the array-CGH program on the Tecan station and inject 100 μl of the hybridization solution from Tube B into the corresponding slide position following the instructions displayed on the machine.

8. After ~45 min, inject 100 μl of the hybridization solution from Tube A into the corresponding slide position following the instructions displayed on the machine.
9. When the array-CGH program is completed, place the slide in a light-proof box and scan as soon as possible (**Note 6**).

4. Notes

1. Cy3 and Cy5 are the most commonly used fluorochromes for DNA labeling. However, both molecules are very sensitive to environmental conditions such as light, high ozone, and humidity levels *(13)*. To avoid any problem of dye degradation or fading, all experiments from DNA labeling to slide scanning should be performed in a laboratory with controlled temperature (20–25°C), relative humidity (25–35%), and ozone level (<0.02 ppm). Alternatively, to limit premature degradation of Cy3 and Cy5, an anti-oxidant, such as cysteamine can be added to hybridization buffers A and M (cysteamine 10 mM) as well as PBS/0.01% Tween 20 (cysteamine 2 mM).
2. The quality and purity of DNA samples used as templates for DNA labeling should be carefully monitored. Genomic DNA should show no degradation (by electrophoresis on agarose gel; **Fig. 17.1A**) and no protein contamination (on spectrophotometer, 280/260 ratio should be greater than 1.8; **Fig. 17.1B and C**). The use of DNA samples which do not fulfill these quality criteria may result in failure of the labeling reaction or low quality of the array-CGH results (i.e. higher technical variability of the array-CGH profile impairing the detection of copy number changes).
3. DNA probe quantity and quality should be controlled after removal of unincorporated labeled nucleotides. First, 3 μl of each sample collected in step 8 should be run on a 2.5% agarose gel to check for the presence of a DNA smear with most fragments below 500 bp (**Fig. 17.1A**). In addition, probe DNA concentration as well as Cy3 or Cy5 incorporation can be measured by using only 1 μl of each collected sample with a NanoDrop Spectrophotometer (**Fig. 17.1B and C**).
4. One critical factor for the success of array-CGH is the efficient suppression of repeated sequences by the Cot-1 DNA during hybridization (**Section 1**). We have noticed that commercially available Cot-1 DNA tends to show batch to batch variations in terms of suppression efficiency. Suppressive hybridization with high quality Cot-1 DNA results in log2ratio values close to 0.6 for single copy gains and –1 for single copy losses. The

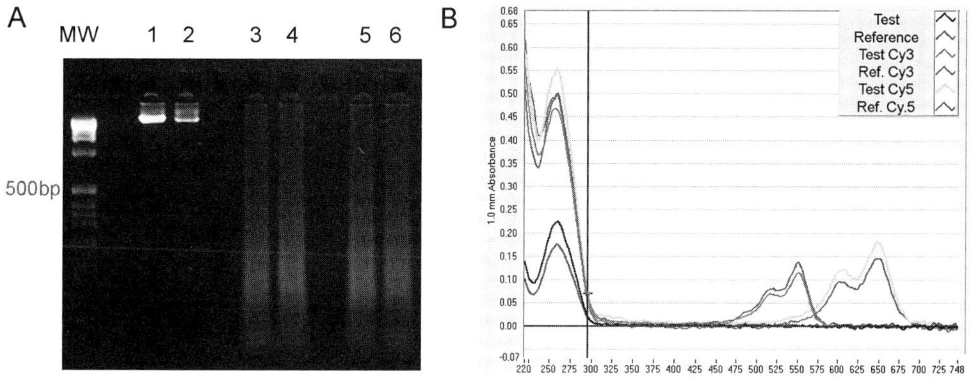

C

No.	Sample ID	ng/ul	260/280	Dye name	Dye pmol/ul	PSA pmol/ug	Yield ug
1	Test	73	1.88	None	NaN	NaN	NaN
2	Reference	57	1.86	None	NaN	NaN	NaN
3	Test Cy3	149	1.83	Cy3	7.47	50.1	15
4	Reference Cy3	162	1.82	Cy3	8.97	55.5	16
5	Test Cy5	176	1.9	Cy5	7.05	40.1	18
6	Reference Cy5	162	1.88	Cy5	5.85	36.1	16

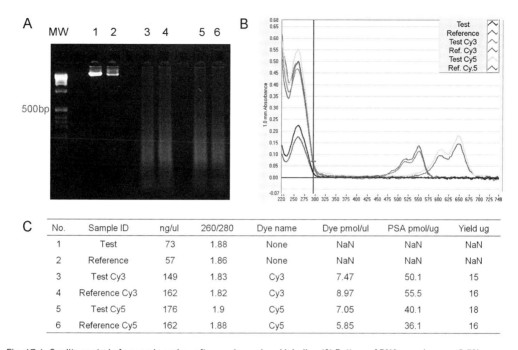

Fig. 17.1. Quality control of genomic probes after random primed labeling (**A**) Pattern of DNA samples on a 2.5% agarose gel, before and after labeling by random primed labeling. The test and Reference DNA (Positions 1–2) samples show no sign of degradation before random priming. After the labeling reaction, each probe labeled with Cy3 (3–4) or Cy5 (5–6) consists of a DNA smear with most fragments below 500 bp. MW: molecular weight marker. (**B** and **C**) Measurement of DNA concentration and dye incorporation on a NanoDrop spectrophotometer. All samples give a 260/280 ratio above 1.8, indicating that there is no residual protein contamination before and after labeling and cleanup. The graph (B) displays the 1 mm absorbance of each sample according to the light wavelength in nm. Test and reference genomic DNA show one single pick at 260 nm, while each probe labeled with Cy3 or Cy5 shows a second pick with maximum absorbance at 550 and 650 nm, respectively. The table (C) reports Cy3 or Cy5 incorporation rates for each labeled probe. PSA is the probe specific activity: PSA = (pmol of dye per μl)/(μg of DNA per μl). It reports the dye incorporation rate and should be greater than 25 pmol/μg of DNA with a DNA yield (which is the quantity of synthesized probe) greater than 8 μg.

use of lower quality Cot-1 DNA may result in incomplete repeat suppression, compressed abnormal ratios, and increased background ratio variability.

5. The hybridization stations Hs.400PRO/4800PRO replace two other stations, which are still available: Hs400 and Hs4800. These two non-PRO stations can also be used for array-CGH. In this case, as the mixing system is different in the two non-PRO stations, the hybridization buffer A should contain only 5% of dextran sulfate instead of 7.5% in order to reduce the viscosity of the solution.

6. Array images are generally acquired using a commercial DNA microarray scanner. We routinely use the scanner from Agilent technologies: this fully automated system (with a 48-slide loading carousel) uses dynamic auto focus to keep features in focus while scanning. Scanners can also be purchased from companies such as Perkin Elmer, Tecan or Axon Instruments.

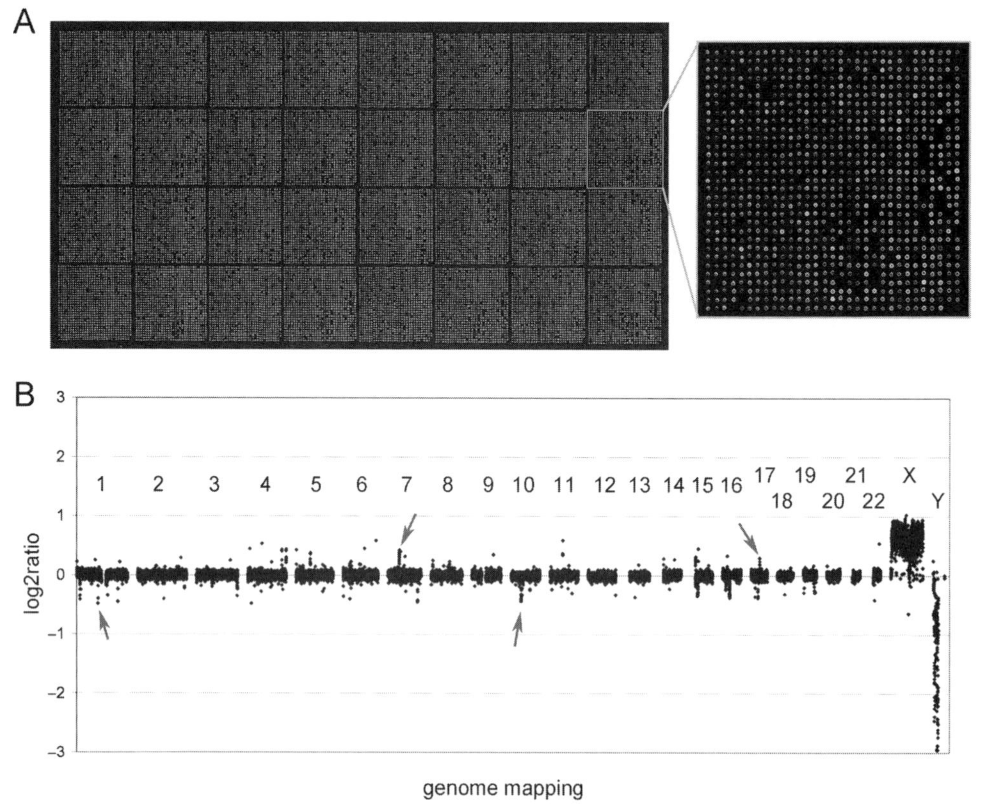

Fig. 17.2. Example of microarray image and genome profile after array-CGH (**A**) Merged images displaying signal intensities after comparative genomic hybridization of one test female DNA (labeled with Cy3, in *green*) versus one reference male DNA (labeled with Cy5, in *red*) on a microarray covering the entire human genome with ~30,000 large-insert clones *(16)*. The microarray (*left panel*) is divided in 8 × 4 blocks, each of them comprising 31 × 31 spots (*right panel*). As both DNA samples are from healthy anonymous individuals, most spots (clones) show no copy number changes and appear in *yellow* (green = red = 2 DNA copies). However, on each block (*right panel*), a few clones appear in green or in red: most of them are clones mapped on chromosome X (green = 2, red = 1) or on chromosome Y (green = 0, red = 1). (**B**) Whole genome array-CGH profile derived from the images merged in (A). For each spot, the log2ratio value is calculated, normalized by the median of all log2ratio values and plotted against the genome position of the corresponding clone. The profile shows log2ratio values that are close to zero on autosomes, increased (up to +1) on chromosome X and decreased (down to less than –3) on chromosome Y. Some local log2ratio variations indicate the presence of small copy number changes between the two individuals (examples indicated by *red arrows*).

There are several software solutions available for image quantification (*see example of microarray image in* **Fig. 17.2A**). Although spot intensity extraction programs are usually supplied with commercially available scanners, they can be purchased separately. The usual commercial programs include GenePix (Axon Instruments), ScanArray (Perkin Elmer), Agilent Feature Extraction software (Agilent Technologies), and BlueFuse for microarrays (BlueGnome Ltd). Other programs are freely available, such as TIGR Spotfinder *(14)* (http://www.tm4.org/spotfinder.html) or UCSF Spot *(15)* (http://www.jainlab.org/downloads.html). For further data analysis (*described in* **Chapter 3**), the ratio values between the fluorescent intensities in both channels on every spot on the array are calculated (usually after the subtraction of local background fluorescence). Intensity ratios are then normalized, for example by dividing each individual ratio by the median ratio value of all clones. The normalized ratio for each clone is then plotted against its position along the genome (*see example of normalized array-CGH profile in* **Fig. 17.2B**).

Acknowledgments

The authors would like to thank Heike Fiegler who developed some of the methods described in this chapter. This work was supported by the Wellcome Trust.

References

1. Pinkel, D., Segraves, R., Sudar, D., Clark, S., Poole, I., Kowbel, D., Collins, C., Kuo, W.L., Chen, C., Zhai, Y. et al. (1998) High resolution analysis of DNA copy number variation using comparative genomic hybridization to microarrays. *Nat Genet*, **20**, 207–211.
2. Solinas-Toldo, S., Lampel, S., Stilgenbauer, S., Nickolenko, J., Benner, A., Dohner, H., Cremer, T. and Lichter, P. (1997) Matrix-based comparative genomic hybridization: biochips to screen for genomic imbalances. *Genes Chromosomes Cancer*, **20**, 399–407.
3. Redon, R., Hussenet, T., Bour, G., Caulee, K., Jost, B., Muller, D., Abecassis, J. and du Manoir, S. (2002) Amplicon mapping and transcriptional analysis pinpoint cyclin L as a candidate oncogene in head and neck cancer. *Cancer Res*, **62**, 6211–6217.
4. Tsubosa, Y., Sugihara, H., Mukaisho, K., Kamitani, S., Peng, D.F., Ling, Z.Q., Tani, T. and Hattori, T. (2005) Effects of degenerate oligonucleotide-primed polymerase chain reaction amplification and labeling methods on the sensitivity and specificity of metaphase- and array-based comparative genomic hybridization. *Cancer Genet Cytogenet*, **158**, 156–166.
5. Fiegler, H., Carr, P., Douglas, E.J., Burford, D.C., Hunt, S., Scott, C.E., Smith, J., Vetrie, D., Gorman, P., Tomlinson, I.P. et al. (2003) DNA microarrays for comparative genomic hybridization based on DOP-PCR amplification of BAC and PAC clones. *Genes Chromosomes Cancer*, **36**, 361–374.
6. Raap, A.K., van der Burg, M.J., Knijnenburg, J., Meershoek, E., Rosenberg, C., Gray, J.W., Wiegant, J., Hodgson, J.G. and Tanke, H.J. (2004) Array comparative genomic hybridization with cyanin cis-platinum-labeled DNAs. *Biotechniques*, **37**, 130–134.
7. Wiegant, J.C., van Gijlswijk, R.P., Heetebrij, R.J., Bezrookove, V., Raap, A.K. and Tanke, H.J. (1999) ULS: a versatile method of labeling nucleic acids for FISH based on a monofunctional reaction of cisplatin derivatives with guanine moieties. *Cytogenet Cell Genet*, **87**, 47–52.
8. Wahl, G.M., Stern, M. and Stark, G.R. (1979) Efficient transfer of large DNA fragments from agarose gels to diazobenzyloxymethyl-paper and rapid hybridization by using dextran sulfate. *Proc Natl Acad Sci U S A*, **76**, 3683–3687.
9. Sealey, P.G., Whittaker, P.A. and Southern, E.M. (1985) Removal of repeated sequences from hybridisation probes. *Nucleic Acids Res*, **13**, 1905–1922.
10. Landegent, J.E., Jansen in de Wal, N., Dirks, R.W., Baao, F. and van der Ploeg, M. (1987) Use of whole cosmid cloned genomic sequences for chromosomal localization by non-radioactive in situ hybridization. *Hum Genet*, **77**, 366–370.

11. Lichter, P., Joos, S., Bentz, M. and Lampel, S. (2000) Comparative genomic hybridization: uses and limitations. *Semin Hematol*, **37**, 348–357.
12. du Manoir, S., Speicher, M.R., Joos, S., Schrock, E., Popp, S., Dohner, H., Kovacs, G., Robert-Nicoud, M., Lichter, P. and Cremer, T. (1993) Detection of complete and partial chromosome gains and losses by comparative genomic in situ hybridization. *Hum Genet*, **90**, 590–610.
13. Fare, T.L., Coffey, E.M., Dai, H., He, Y.D., Kessler, D.A., Kilian, K.A., Koch, J.E., LeProust, E., Marton, M.J., Meyer, M.R. et al. (2003) Effects of atmospheric ozone on microarray data quality. *Anal Chem*, **75**, 4672–4675.
14. Saeed, A.I., Sharov, V., White, J., Li, J., Liang, W., Bhagabati, N., Braisted, J., Klapa, M., Currier, T., Thiagarajan, M. et al. (2003) TM4: a free, open-source system for microarray data management and analysis. *Biotechniques*, **34**, 374–378.
15. Jain, A.N., Tokuyasu, T.A., Snijders, A.M., Segraves, R., Albertson, D.G. and Pinkel, D. (2002) Fully automatic quantification of microarray image data. *Genome Res*, **12**, 325–332.
16. Fiegler, H., Redon, R., Andrews, D., Scott, C., Andrews, R., Carder, C., Clark, R., Dovey, O., Ellis, P., Feuk, L. et al. (2006) Accurate and reliable high-throughput detection of copy number variation in the human genome. *Genome Res*, **16**, 1566–1574.

Chapter 18

Chromatin Immunoprecipitation Using Microarrays

Mickaël Durand-Dubief and Karl Ekwall

Abstract

Chromatin immunoprecipitation (ChIP) is a powerful procedure to investigate the interactions between proteins and DNA. ChIP-chip combines chromatin immunoprecipitation and DNA microarray analysis to identify protein–DNA interactions that occur *in vivo*. This genome-wide analysis of protein–DNA association is carried out in several steps including chemical cross-linking, cell lysis, DNA fragmentation and immunoaffinity purification that allow the identification of DNA interactions and provide a powerful tool for genome-wide investigations. Immunoprecipitated DNA fragments associated with the desired protein are amplified, labelled and hybridized to DNA microarrays to detect enriched signals compared to a labelled reference sample.

Key words: Chromatin immunoprecipitation, arrays, affymetrix, eurogentec, ChIP, ChIP-chip.

1. Introduction

Chromatin immunoprecipitation has become a popular method for investigating how proteins interact within the genome. When combined with DNA microarray analysis, the ChIP technique can provide a whole-genome view of protein–DNA interactions. The basic protocol presented here has been established in the eukaryotic model organism, fission yeast (*Schizosaccharomyces pombe*), for the purpose of detecting histone modifications, histone modifying enzymes, transcription factors or remodelling factors *(1–6)*. The protocol has been adapted for *S. pombe* from several other protocols originally developed for budding yeast (*Saccharomyces cerevisiae*) *(7–9)*. The main differences from the budding yeast protocols are different cell lysis methods due to the more rigid cell walls of *S. pombe* cells, and of course the establishment of *S. pombe* specific genomic microarray platforms *i.e.* Eurogenetec and

Martin Dufva (ed.), *DNA Microarrays for Biomedical Research: Methods and Protocols, vol. 529*

DOI 10.1007/978-1-59745-538-1_18 Springerprotocols.com

Affymetrix arrays. Although this method is described for fission yeast, adaptation to other organisms with small genomes is possible, provided that microarray platforms are available. In the case of larger genomes, e.g., mammals and plants, the newly developed ChIP methods combined with high throughput DNA sequencing such as '454' and 'Solexa' are cheaper alternatives to the ChIP-chip approach *(10, 11)*.

Briefly, in a ChIP experiment protein–DNA interactions are first captured *in vivo* by treating cells with paraformaldehyde, a reversible crosslinker. Cells are then lysed, DNA fragmented *via* sonication and the protein–DNA complexes enriched by immunoaffinity capture of the desired protein. Next template and reference are reverse cross-linked and DNA extracted, amplified, labelled and hybridized onto the microarrays. This coupling of chromatin immunoprecipitation and whole genome DNA microarrays (ChIP-chip) allows researchers to answer specific questions by generating high or low resolution genome-wide maps of in vivo interactions between DNA-associated proteins and DNA.

The parameters used for ChIP depend to a large extent on the abundance of the antigen and the specificity of the antibody. The small amount of DNA recovered after the immunoprecipitation is not sufficient for labelling and hybridization onto microarrays. A new DNA amplification step is therefore required. In this chapter, we describe how to amplify and label ChIP samples both for spotted microarrays on glass slides and for high resolution tiling microarrays in fission yeast. The different microarray platforms are *S. pombe* Intergenic Region (IGR)+ Open Reading Frame (ORF) microarrays (Eurogentec) and GeneChip® *S. pombe* Tiling Array (Affymetrix).

2. Materials

2.1. Cell Culture and Formaldehyde Fixation of Fission Yeast Cells

1. YES media filter sterilized: 5 g/l Yeast Extract (Difco, BD); 0.1 g/l uracil, adenine, l-leucine, l-glutamine, l-arginine, l-histidine and l-lysine, 30 g/l glucose.
2. Phosphate buffered saline (PBS): Prepare a 10 × stock solution with 1.37 *M* NaCl, 27 mM KCl, 100 mM Na_2HPO_4, 18 mM KH_2PO_4 (adjust to pH 7.4 with HCl if necessary) and autoclave before storage at room temperature. Prepare a working solution by dilution of one part with nine parts water.
3. Formaldehyde solution 37% (Sigma; Cat. no: 252549; hazardous).
4. 2.5 *M* glycine filter sterilized.

2.2. Preparation of a Sonicated Chromatin Extract

1. Acid-washed glass beads (425–600 μm; Sigma, Cat no. G-9268)
2. 2 ml Screw Cap 72693 Sarstedt tubes.
3. Lysis buffer: 150 mM NaCl, 50 mM Hepes–KOH, pH 7.5, 0.1% (w/v) sodium dodecyl sulphate (SDS), 1% (w/v) Triton X-100, 0.1% (w/v) sodium deoxycholate, 1 mM ethylenediamine tetraacetic acid (EDTA).
4. Complete, EDTA-free; Protease Inhibitor Cocktail Tablets, 11873580001, Roche. (Hazardous). One tablet of Protease Inhibitor Cocktail (PI) in 2 ml lysis buffer gives a 25 × stock solution (store at –20°C). For 1 ml of lysis buffer with PI, add 40 μl of 25 × stock solution of Protease Inhibitor Cocktail Tablets (PI) and 960 ml of lysis buffer.
5. FastPrep® machine, Thermo Electron.
6. Sonicator Vibracell VCX 130 with CV18 tip Model (Sonic and Material, Inc, Newtown. CT).

2.3. Chromatin Immunoprecipitation and Chromatin Recovery

1. Protein A beads (Sigma; Cat no: P3391). To prepare a slurry of beads, resuspend them in 1 × cold PBS in 50 ml tube, rotate 30 min on spinning wheel in cold room and centrifuge at 1,500 *g*, 4°C. Again wash beads twice in cold PBS. Then resuspend in 15 ml cold lysis buffer. Spin again and remove all the supernatant. For 1 volume of protein A bead add 1 volume of cold lysis buffer.
2. 3 *M* NaOAc, pH 5.2.
3. 96% Ethanol.
4. Proteinase K (Invitrogen; Cat. no: 25530049).
5. Glycogen (Roche; Cat. no: 10 901 393 001).
6. DNAase-free RNase (Roche; Cat. no: 1119915).
7. Lysis buffer (150 mM NaCl): 150 mM NaCl, 50 mM Hepes–KOH, pH 7.5, 1 mM EDTA, 0.1% (w/v) SDS, 1% (w/v) Triton X-100, 0.1% (w/v) sodium deoxycholate.
8. Deoxycholate buffer: 10 mM Tris–HCl, pH 8.0, 1 mM EDTA, 0.5% (w/v) sodium deoxycholate, 0.5% (w/v) NP-40, 0.25 M LiCl.
9. TES buffer (elution buffer): 50 mM Tris–HCl, pH 8.0, 1.5% (w/v) SDS, 10 mM EDTA.
10. TE buffer: 10 mM Tris–HCl, pH 8.0, 1 mM EDTA.
11. 0.45 μm filter unit (Millipore Ultrafree-MC; UFC30HV00).
12. Phenol: chloroform (e.g. Sigma). Caution phenol is hazardous.
13. Chloroform (e.g. Sigma).
14. Spinning wheel.

2.4. Verification of Enrichment of Recovered ChIP DNA, Amplification of Purified DNA (Round A and Round B)

1. 100 mM of dATP, dTTP, dGTP, dCTP (Invitrogen; cat no.10297-117). Make 20 mM dNTP stock (e.g. Roche).
2. 25 mM $MgCl_2$ solution.
3. 100 mM of dATP, dTTP, dGTP, dCTP (Invitrogen; cat no.10297-117).
4. AmpliTaq Taq polymerase 5 U/μl (Roche; cat no: N-808-0153).
5. 0.5 M EDTA solution.
6. Agarose for gel (e.g. Sigma).
7. Ethidium bromide solution.
8. Solution TBE 10 × : 890 mM Tris–HCl, 890 mM borate, 20 mM EDTA.
9. Gel image analysis software (e.g. GE HealthCare, ImageQuant).
10. Specific pairs of primers for the targets and controls (*see* **Note 2**).
11. 40 μM T-PCRA primer: 5′ – GTTTCCCAGTCAC-GATCNNNNNNNNN-3′
12. 364 mM T-PCRB primer: 5′ – GTTTCCCAGTCACGATC-3′
13. Bovine serum albumin (BSA; Sigma; Cat. no: A7906).
14. T7 Sequenase V.2.0, 13 U/μl (USB; Cat no: 70775 Z) including both Sequenase buffer 5 × and Sequenase dilution buffer 1 × .
15. Tris 10 mM, EDTA 1 mM; pH 8.0.
16. QIAquick PCR Purification Kit (Quiagen, Cat. no: 28104).
17. Buffer PB (500 ml; Quiagen, Cat. no: 19066).

2.5. Sample Preparation for Spotted Microarrays

1. BioPrime Labelling System (containing 2.5 × random primer solution and concentrated klenow fragment (40 U/μl; Invitrogen; Cat. no: 18094011).
2. Cy3-dCTP and Cy5-dCTP (Amersham; Cat no: PA55321).
3. UltraPure™ Salmon Sperm DNA Solution (Invitrogen; Cat no: 15632011).
4. Lifter Slip, 25 × 60 mm (Erie Scientific company, Portsmouth, NH, USA).
5. Hybri-Slips, 22 × 60 mm (Sigma; Cat no: Z370274).
6. Corning® hybridization chambers (Sigma; Cat. no: CLS40001).
7. Working solution of DIG Easy Hyb Granules (Roche; Cat no: 1796895) following the manufacturer's recommendations. Add 64 ml of ddH_20 to the plastic bottle dissolve by stirring immediately for 5 min at 37°C. Store the solution at 15–25°C.
8. Microcon-30 (Amicon Inc., Beverly, MA, U.S.A.).

9. 10 × dNTP mix for Cy3- or Cy5-dCTP klenow labelling: 1.2 mM each (dATP, dTTP, dGTP), 0.6 mM dCTP, 10 mM Tris–HCl pH 7.8, 1 mM EDTA.

2.6. Pre-Hybridization, Hybridization and Data Analysis on Eurogentec Spotted Arrays

1. Bench top centrifuge (e.g. Eppendorf centrifuge type 5804)
2. Glass/Slide Microarray adapter (e.g. CombiSlide™ glass slide/microarray adapter).
3. 20 × Salt Sodium Citrate (SSC), pH 7.0: 3.0 M NaCl and 0.3 M sodium citrate.
4. Sodium dodecyl sulphate (SDS) 10 % w/v.
5. Oven or incubator at 42°C.
6. Wet chamber as plastic box containing soaked paper.
7. Hybridization chamber (Corning Inc.; Cat no: 2551).
8. DNA microarray scanner (Agilent or Biorad).
9. Biorad Image Quant software version 4.2 (Biorad).
10. Genespring software (Agilent).

2.7. DNAase I Fragmentation of Amplified Targets, Hybridization, Washing and Scanning for GeneChip® Pombe Tiling Arrays

1. 5 μg Round B DNA.
2. 10 × One-Phor-All Buffer (GE Healthcare, Cat no. 27-0901-02).
3. Deoxyribonuclease I, Amplification Grade (Invitrogen; cat no: 18068-015).
4. Terminal Deoxy Transferase kit (TdT) containing 25 mM $CoCl_2$ (Roche, Cat. no: 3 333 566)
5. TdT buffer 5 × (Roche; Cat no: 1 243 276).
6. GeneChip® Fluidics Station (Affymetrix).
7. GeneChip® Scanner 3000 7G (Affymetrix).
8. GeneChip® DNA Labelling Reagent containing Biotin-Ń-ddATP, 7.5 mM (Affymetrix; Cat no: 900542).
9. GeneChip® Hybridization Oven 640, Affymetrix, Cat. no: 800139).
10. Herring Sperm DNA, 10 mg/ml (Promega Corporation, Cat. no: D1811).
11. GeneChip® Eukaryotic Hybridization Control Kit (contains Control Oligo B2; Affymetrix, Cat. no: 900299).
12. 20X SSPE: 3 M NaCl, 0.2 M NaH_2PO_4, 0.02 M EDTA.
13. Triton® X-100 (Sigma, Cat no : T8787).
14. Filter sterile solution 12X MES stock (1.22 M MES, 0.89 M [Na+]). For 1000 ml: prepare 70.4 g MES free acid monohydrate, 193.3 g MES Sodium Salt, 800 ml ddH2O. Mix and adjust the volume to1,000 ml. The pH should be between 6.5 and 6.7. Filter through a 0.22 μm filter.

15. 2 × Hybridization Buffer (Final 1 × concentration): 100 mM MES, 1 M [Na+], 20 mM EDTA, 0.01% Tween-20. For 50 ml: 8.3 ml of 12X MES Stock Buffer, 17.7 ml of 5 M NaCl, 4.0 ml of 0.5 M EDTA, 0.1 ml of 10% Tween-20 and 19.9 ml of water. Store at 2–8°C and shield from light. Discard it if turned yellow.
16. Wash Buffer A: Non-Stringent Wash Buffer (included in P/N 900720; 6 × SSPE, 0.01% Tween-20). For 1 l: 300 ml of 20 × SSPE 1.0 ml of 10% Tween-20, 699 ml of water.
17. Wash Buffer B: Stringent Wash Buffer (included in P/N 900720; 100 mM MES, 0.1 M [Na+], 0.01% Tween-20) For 1,000 ml: 83.3 ml of 12X MES Stock Buffer, 5.2 ml of 5 M NaCl , 1.0 ml of 10% Tween-20 , 910.5 ml of water , Filter through a 0.2 μm filter. Store at 2–8°C and shield from light.
18. 2 × Stain Buffer (Final 1 × concentration: 100 mM MES, 1 M [Na+], 0.05% Tween-20) For 250 ml: 41.7 ml of 12 × MES Stock Buffer, 92.5 ml of 5 M NaCl, 2.5 ml of 10% Tween-20, 113.3 ml of water. Filter through a 0.2 μm filter. Store at 2–8°C and shield from light.
19. 10 mg/ml Goat IgG Stock Resuspend 50 mg in 5 ml of 150 mM NaCl. Store at 4°C.
20. 1 × Array Holding Buffer: 100 mM MES, 1 M [Na+], 0.01% Tween-20. For 100 ml: 8.3 ml of 12 × MES Stock Buffer, 18.5 ml of 5 *M* NaCl, 0.1 ml of 10% Tween-20, 73.1 ml of water. Store at 4°C and shield from light.
21. Acetylated bovine serum albumin (BSA) solution (50 mg/ml; Invitrogen; Cat. no: 15561-020).
22. Anti-streptavidin antibody (goat), biotinylated (Vector Laboratories; Cat. no: BA-0500).
23. Streptavidin Phycoerythrin (SAPE; Molecular Probes; Cat. no: S-866). Store in the dark at 4°C.
24. Dimethyl sulfoxide (DMSO; e.g. Sigma)

3. Methods

The quality of ChIP-Chip results is highly dependent on the antibody used for the immunoprecipitation. Since different antibodies behave significantly differently in ChIP, we recommend testing several different commercially available antibodies before starting your ChIP-chip experiment. If antibodies are not available for a specific protein, it is also possible to use epitope-tagged proteins, which are detected by commercially available monoclonal antibodies. In our laboratory we have successfully used MYC and HA tags and other laboratories have also used other tags *(1, 2, 4, 5)*.

The choice of reference sample is important since the ChIP signal is calculated as a relative enrichment based on a reference sample and not as absolute enrichment. Usually for ChIP of sonicated chromatin extracts, we use the same chromatin extract as a reference. However, other types of references may be required, such as in ChIP to detect histone modifications where nucleosome occupancy *i.e.* histone H3 has been used as a reference *(1, 3–5)*. Before starting the experiments, we also recommend that suitable negative and positive controls be chosen for the ChIP assay. Negative controls are usually mock samples (absence of antibody), strains carrying no epitope detected by the antibodies or the use of a non-specific antibody. Positive controls such as a high-quality antibody that always works in ChIP assay will also help to indicate if conditions need to be changed. After ChIP, the amount of DNA recovered is not sufficient for labelling and hybridization onto microarrays. Therefore a first PCR amplification step (round A) is required that involves a partially random oligonucleotide whose universal sequence is used for a second PCR amplification (round B).

3.1. Three Different S. pombe Microarray Platforms

1. IGR+ORF DNA microarray for *S. pombe* (Eurogentec SA custom microarray services; Belgium). These arrays are spotted on glass slides with 500 bp of PCR fragments for each region representing RNA pol II promoter regions (IGR) and the 3' end of each open reading frame (ORF). This resulted in about 10 thousand IGR and ORF fragments that also include several tRNA genes, fragments for different rDNA promoters and fragments for noncoding centromere repeats *(5)*.
2. The Gene Chips S. pombe Tiling Array (Affymetrix) is a high-density oligonucleotide microarray (resolution 250 bp) that covers chromosome II and half of chromosome III of the *S. pombe* genome *(12)*.
3. The GeneChip® *S. pombe* Tiling 1.0FR Array (Affymetrix) comprised of over 1.2 million perfect match/mismatch probe pairs tiled through the complete *S. pombe* genome. Probes are tiled for both strands of the genome at an average of 20 bp resolution, as measured from the central position of adjacent 25-mer oligonucleotides, creating an overlap of approximately 5 bp on adjacent probes. Website: http://www.affymetrix.com.

ChIP sample preparation starts with the same steps but differs at the end of the procedure before loading the samples on different platforms. For Eurogentec spotted arrays, 500 ng of amplified DNA is required before labelling, while for Affymetrix tiling arrays, 5 μg of amplified DNA is needed.

3.2. Culture and Formaldehyde Fixation of Fission Yeast Cells

1. Grow log-phase fission yeast (5×10^6 to 2×10^7 cells/ml) to get 2×10^8 cells.
2. Add paraformaldehyde directly to the cell culture to a final concentration of 1% and agitate slowly 30 min at room temperature (*see* **Note 1**).
3. Quench paraformaldehyde fixation by adding 2.5 ml 2.5 *M* glycine and agitating 5 min at room temperature.
4. Transfer the cells to a 50 ml centrifuge tube, and spin down the cells at 1,500 *g* for 5 min at 4°C.
5. Pour off the supernatant in the fume-hood and from now, keep everything on ice.
6. Wash cells twice in 25 ml ice-cold 1 × PBS and spin down again the cells at 1,500 *g* (*see* **Note 2**).

3.3. Preparation of Sonicated Chromatin Extract

1. On ice, resuspend the fixed cell pellet corresponding to 2×10^8 cells in 400 μl ice cold lysis buffer containing protease inhibitor (caution, harmful) and transfer it to pre-cooled Sarstedt tube containing one volume acid-washed glass beads.
2. In a cold-room at 4°C, lyse cells in the FastPrep machine 6 times for 30 s each at the maximum power 6.5, waiting at least 2 min on ice between the pulses to avoid warming the samples. Verify under a microscope that more than 95% of the cells are lysed.
3. Separate the chromatin extract from the glass beads by puncturing the bottom of the Sarstedt tube with a flame-heated needle. Place the punctured tube in 2 ml eppendorf tube with removed caps and insert those in 15 ml Falcon tubes or 10 ml culture tubes and centrifuge at 1,500 *g* for 5 min at 4°C.
4. Gently resuspend the chromatin using a big tip and transfer it, on ice, to a clean 1.5 ml eppendorf tube.
5. Set the Vibracell sonicator: pulse 10″ on/8″ off, amplitude 60%, time 2 min. Sonicate by inserting the tip in the lysate on water-ice mix to shear the chromatin to ~600 bp (*see* **Note 3**).
6. Spin down the cell debris for 5 min at 14,000 *g* at 4°C and transfer the supernatant to a fresh tube on ice.
7. Resuspend the cell debris in 400 μl of lysis buffer containing protease inhibitor cocktail and sonicate again.
8. Spin the cell debris down again for 5 min at 14,000 *g* at 4°C.
9. Pool the supernatants from step 5 and 7 and sonicate a third time.
10. Remove residual cell debris by spinning the total 800 μl of chromatin extract for 10 s at 14,000 *g* at 4°C.
11. Transfer the supernatant (lysate) to a fresh 1.5 ml tube on ice (*see* **Note 4**). Use right away for immunoprecipitation of the chromatin or store in aliquots at –80°C.

3.4. Immuno-precipitation of the Chromatin

1. We suggest making triplicates per antibody you will use. You also need two negative controls: one tube with negative antibody and one tube for the crude input (*see* **Note 5**). Remember you will also need crude input later on to get reference enrichment.
2. Add the right amount of antibody to 100 μl of the crude lysate (*see* **Note 6**). No antibody should be added to the last sample, this is an essential negative control. Incubate for 2 h on a spinning wheel at 4°C.
3. Cut the tip off of a 200 μl pipette tip and use it to pipette up the slurry beads. Add 25 μl of well-mixed protein A slurry beads equilibrated in lysis buffer to the samples including the negative control (*see* **Note** 7).
4. Continue the incubation on the spinning wheel 1 h at 4°C (*see* **Note 8**).

3.5. Washing of the Beads

1. In a 4°C cold room, spin tubes 30 s at 1,500 *g* (*see* **Note 9**).
2. Resuspend the beads with 250 μl cold lysis buffer. Carefully transfer all of the beads to 0.45 μm filter unit (*see* **Note 10**). Rinse the tubes with another 100 μl lysis buffer and add the remaining beads to the filter unit.
3. Spin columns 30 s at 1,500 *g* at 4°C and discard flow through.
4. Add to the beads 400 μl of lysis buffer and wash twice for 15 min at 4°C on a rotating wheel.
5. Repeat the washing step one more time with lysis buffer at room temperature (RT).
6. Wash twice with Deoxycholate Buffer for 15 min at RT.
7. Rinse the beads for 15 min with 400 μl TE, pH 8.0.
8. Elute ChIP material by adding 100 μl TES to the beads and incubating at 65°C for 15 min. Centrifuge at 1,500 *g* for 30 s at room temperature.
9. Repeat elution, combine the fractions in new 1.5 ml tubes and add 200 μl TES buffer to the control crude non-immunoprecipitated chromatin (Input).
10. Incubate the controls and immunoprecipitated chromatin samples in well-closed tubes at 65°C overnight to reverse the cross-linking.

3.6. Protein Removal and Chromatin Recovery

1. The samples are cooled to room temperature, spin briefly. Add to each sample 20 μg of proteinase K (2 μl) as protease and 20 μg glycogen (2 μl) as a DNA carrier in 50 μl TE pH 8.0 and incubate at 56°C for 2 h.
2. The samples are spun down briefly and transferred to fresh 1.5 ml Eppendorf tubes (*see* **Note 11**).

3. To extract the DNA, under the fume hood add 500 μl phenol:-chloroform solution to each tube. Vortex thoroughly and centrifuge for 5 min at 14,000 *g*.
4. Carefully pipette the upper phase and transfer in 1.5 ml tubes containing 500 μl of chloroform. Vortex thoroughly and centrifuge again 5 min at 14,000 *g*.
5. Transfer upper phase containing DNA to fresh 1.5 ml eppendorf tube and precipitate the DNA sample by addition of 1/11 volume of 3 M NaOAc pH 5.2 followed by 2.5 volumes of 96% ethanol. Mix the tubes well and precipitate at –20°C overnight.
6. Spin the samples for 30 s at 14,000 *g* at 4 °C and remove the supernatant.
7. Dry the DNA pellet at 40°C for about 15 min.
8. Resuspend the DNA in 40 ml TE containing 0.1 μg RNase A DNAase-free and incubate for 30 min at 37°C. Use right away for the next step or store at –20°C.

3.7. Verification of Enrichment of the Recovered DNA

1. Dilute the input DNA sample to 1:100, 1:250 and 1:500 in TE, pH 8.0.
2. Prepare on ice a PCR mix with 3 μl aliquot of each sample including control and the three dilutions of the crude extract in small PCR tubes (*see* **Note 13**).

DNA ChIP template		3 μl
ddH_2O		22.8 μl
10 × Taq buffer without $MgCl_2$		3.5 μl
$MgCl_2$	(25 mM)	5.6 μl
dNTP	(20 mM)	0.6 μl
Primer F target	(100 μ*M*)	0.3 μl
Primer R target	(100 μ*M*)	0.3 μl
Primer F control	(100 μ*M*)	0.3 μl
Primer R control	(100 μ*M*)	0.3 μl
Taq polymerase	(5 U/μl)	0.3 μl
Total		35 μl

3. Put PCR tubes on pre-heated PCR block at 95°C and start the PCR programme, 30 × (95°C 30″/45 to 55°C 30″/72°C 1′) and 72°C 10′.
4. Prepare 2 or 3 % agarose TBE 1 × gels containing Ethidium Bromide. Add loading buffer to the PCR tube and run 20 μl of each PCR mix. Quantify the relative enrichment of PCR-products compared to internal and external

controls using the EtBr quantification apparatus and gel image analysis software.

3.8. Round A: Tagging the Immunoprecipitated DNA

1. Prepare on ice two to four PCR tubes (*see* **Note 14**) of a mix of 7 µl ChIP DNA, 2 µl 5 × Sequenase buffer, 1 µl 40 µM oligo T-PCRA. Denature 2 min at 94°C and incubate 2 min at 8°C.
2. Prepare on ice Round A reaction mix for 5 rounds and add 5 µl to each PCR tube:

5 × Sequenase	5µl
25 mM dNTP	0.9µl
100 mM DTT	3.75µl
500 µg/ml BSA	7.5µl
T7 Sequenase v 2.0	1.5µl
ddH_20	6.35µl
Total	25µl

3. Use a gradient PCR programme to increase the temperature very slowly from 8 to 37°C over 8 min.
4. Incubate another 8 min at 37°C. Denature 2 min at 94°C and incubate again 2 min at 8°C.
5. Add 1 µl of 4 units of Sequenase v2.0 by using the specific Sequenase dilution buffer 1 × and repeat steps 3 and 4.
6. Add to each tube 35 µl of 10 mM, Tris and 1 mM EDTA.

3.9. Round B: Amplification of Round A DNA

1. With round A samples prepare a Round B mix on ice and split in 50 µl aliquots in ten PCR tubes (*see* **Note 15**).

Round A samples (*see* **Section 3.8**)	75 µl
10 × PCR buffer	50 µl
25 mM dNTP	5 µl
364 mM oligo T-PCRB	17 µl
25 mM $MgCl_2$	40 µl
AmpliTaq Taq polymerase	5 µl
ddH2O	308 µl
Total	500 µl

2. Run the following PCR programme: 24 × (92°C 30″/ 40°C 30″/50°C 30″/72°C 1′) and 72°C 10′ (*see* **Note 16**).
3. Purify all the amplified fragments using Quiagen PCR purification kit and the binding buffer PB according the recommendation of the manufacturer and elute with 40 µl of the kit EB buffer 10 mM Tris pH 8.5 (*see* **Note 17**).
4. Measure concentration with NanoDrop® ND-1000 Spectrophotometer or another spectrophotometer. If there is enough material load a few microlitres on 1.5% agarose gel to check the quality of amplified materials.

3.10. Sample Preparations and Hybridization on Eurogenetec Spotted Arrays

3.10.1. Fluorophore Incorporation in DNA Samples

1. Prepare sample and reference in Eppendorf tube on ice:

Round B material	x µl for 500 ng
2.5 × oligo mix	20 µl
ddH_2O	adjust to 41 µl

2. Heat the mix 5 min at 95°C and put immediately in ice-cold water for 2 min.
3. Add on ice 5 µl of dNTP Klenow mix and 3 µl Cy3-dCTP (or 3 µl Cy5-dCTP) to the ChIP sample. Add the same mix with the other dye for the reference sample. Add 1 µl concentrated Klenow to each sample and incubate at 37°C for 2 h (*see* **Note 18**).
4. After incubation, add 450 µl TE pH 8.0 and 1 µl DNA (salmon sperm DNA) and transfer to Microcon-30 column. Centrifuge columns at full speed for about 7 min and discard the supernatant. Again add 450 µl TE pH 8.0 and concentrate the samples to a maximum of 4.6 µl. The colour of combined probes should be purple and clearly visible.

3.10.2. Hybridization, Washing and Scanning for Eurogentec Spotted Arrays

1. Perform pre-hybridization by carefully mounting the Lifter Slip coverslip on the Eurogenetec spotted arrays. Carefully add between the array slide and the coverslip, 50 µl of pre-warmed mix (2.5 µl sonicated DNA salmon sperm + 47.5 µl DIG Easy Hyb solution) at 42°C (*see* **Note 19**).
2. Carefully insert microarrays in hybridization chambers, then put chambers in the pre-warmed wet chamber at 42°C for 1 h.
3. Carefully take slides mounted with the coverslip in falcon tube containing 50 ml of 0.1 × SSC. The coverslip should detach by itself. Remove it carefully with tweezers.
4. Dry slides using CombiSlide™ glass slide/microarray adapter in bench top centrifuge (e.g. Eppendorf centrifuge type 5804) at 60 *g* for 5 min at room temperature (*see* **Note 20**).
5. Prepare hybridization mix. Combine fluorophores Cy3/Cy5 probes mix (containing either template or reference), 50 µl DIG Easy Hyb Solution and 1 µl sonicated DNA salmon sperm (10 µg/µl). Denature probe for 5 min at 95°C, put immediately in ice-cold water for 1–2 min and pre-warm the hybridization mix at 42°C.
6. Carefully Mount the Lifter Slip on the microarrays slide, apply the hybridization mix and incubate in dark overnight in wet chamber at 42°C.
7. Wash the slide with 1 × SSC. The coverslip should fall off by itself – if not then remove it carefully.
8. Wash using a rotating wheel for 5 min in 50 ml 0.1 × SSC/0.03% SDS in falcon tube wrapped in aluminium foil. Rinse briefly in 50 ml 0.1 × SSC to remove SDS.

9. On rotating wheel repeat washing thrice for 5 min in 50 ml 0.1 × SSC in falcon tube wrapped in aluminium foil.
10. Spin the slides dry using glass slide/microarray adapter in bench top centrifuge as in step 4.
11. Store slides in dark and scan them quickly using a microarray scanner at two wavelengths, 543 nm for Cy3 and 594 nm for Cy5, to generate .tiff files.
12. Quantify .tiff images using image quantification software (such as Image Quant version 4.2), import data into a microarray analysis software (such as GeneSpring, Agilent Technologies) and apply your normalizations.

3.11. Hybridization, Washing and Scanning for Affymetrix Tiling Arrays

For hybridization, washing and scanning, we followed exactly the protocol and recommendations from Affymetrix that can be downloaded on : www.affymetrix.com. In the Affymetrix Chromatin Immunoprecipitation Assay Protocol, samples are not fragmented *via* DNAase I treatment (*i.e.* described bellow), instead another fragmentation method of ChIP DNA is used. Briefly dUTP is incorporated during the Round B PCR amplification step and the Round B DNA sample is then fragmented by the UDG and APE nuclease. In our hands the provided Affymetrix conditions for the incorporation of dUTP during round B did not work and amplified too little DNA. If you wish to follow this fragmentation protocol, we just recommend using our round B condition and replacing the dNTP mix with the following dNTP/dUTP mix: 25 mM each dCTP, dGTP, dATP; 20 mM of dTTP, 5 mM of dUTP.

3.11.1. DNAase I Fragmentation of Amplified Targets for GeneChip® Pombe Tiling Arrays (Affymetrix)

1. Prepare a dilution mix of DNAase I (*DNAase I diluted mix*) on ice:

ddH_2O	16 µl
10 × One-Phor-All buffer	2 µl
DNAase I (1 U/µl)	2 µl

2. Prepare the DNAase I fragmentation mix on ice:

5 µg of Round B DNA	x µl
10 × One-Phor-All buffer	5 µl
DNAase I diluted mix	1.5 µl (0.15 units)
ddH_20	add to 50 µl final

3. Incubate at 37°C for 30 s. Then immediately inactivate the DNAase I at 95°C, 15 min (*see* **Note 21**) and keep at room temperature to slowly cool down.

3.11.2. Biotinylation of DNAase I Fragmented Samples

1. After DNAase treatment of the round B DNA, mix on ice:

5 × TdT buffer	10 µl
Round B DNAaseI fragmented (5 µg)	50 µl
25 mM $CoCl_2$	2.9 µl
TdT	2 µl
DNA labelling reagent	2 µl
Total	67 µl

2. Incubate for 1 h at 37°C and stop the reaction with 2 μl of 0.5 M EDTA.

4. Notes

1. For fixation of non-histone target proteins this step may need to be prolonged 1—6 h however we have observed that longer incubation may significantly increase the non-specific background. Observe that all manipulations with paraformaldehyde should be performed in the fume-hood.
2. Cells can be stored on ice even overnight if necessary by adding 25 ml of ice cold PBS to cover the pellet. Before you continue the next day – do not forget to centrifuge and discard PBS 1 × .
3. Caution, ultrasounds can cause *permanent ear damage*! Wear ear protection during sonication.
4. At this step you can check the size of the cross-linked protein-DNA. Run a 0.8 % agarose gel containing EtBr for 40 min at 130 V. If the sonication worked you should observe a 'smear' migrating around 600–1,500 bp and at least under 3000 bp.
5. Negative control such as unspecific antibody or protein A beads is only required for estimating the enrichment compared to the background enrichment.
6. For ChIP, the amount of antibodies can vary a lot. If you use commercial ChIP antibody, required amount is generally indicated. In our conditions we usually add 1 to 3 μg of antibodies for 100 μl lysate (about 25 μg of chromatin).
7. Note that it is important to have same amount of beads in each sample. Others and We observed that more beads can give more background. You need to carefully resuspend each time before pipetting the right volume.
8. To avoid background we do not encourage extending the incubation time.
9. Do not centrifuge for too long or at higher speed since it could damage the beads and affect the outcome of your experiment.
10. Resuspending the beads with wash buffer and transferring to 0.45 μM filter units can take some time if you have a large amount of different samples and this might affect your results. Therefore for a beginner, we suggest not handling more than 20–25 samples at the same time.
11. The tubes are weakened by long incubation at 65°C and might break during the phenol: Chloroform extraction.

12. Check the size of the sonicated DNA using input sample. Run 5 µl of DNA input aliquot on 1.5% agarose gel containing EtBr for 40 min at 130 V. If the sonication worked fine you should observe a 'smear' migrating around 500–300 bp.
13. This procedure is critical to validate the relative enrichment of your ChIP sample. We may also use a quantitative method like real-time PCR. Ideally several targets should be checked for DNA enrichment. However, if your protein targets are not yet known we strongly suggest using in parallel a positive control that has been validated in previous work (e.g. transcription specific factor or specific histone modification). To successfully validate the ChIP enrichment, we also encourage choosing a standard set of primers having same melting temperature. We recommend a small PCR product size *i.e.* 200 and 300 bases length for amplification if you use EtBr quantification *via* gel image analysis software (e.g. ImageQuant, GE Healthcare). In order to save your precious ChIP material we also encourage checking several control targets.
14. This step allows the incorporation of the partially degenerated T-PCRA oligonucleotide. If you plan to set up many ChIP microarrays we recommend buying a large amount of this oligo in order to get a higher reproducibility. Secondly if you choose to use Eurogentec arrays, only 500 ng of amplified material is required thus you may only need to make two reaction tubes. However Affymetrix arrays need much more amplified material (such as 5 µg) thus we recommend using at least four reaction tubes for round A.
15. Generally 500 µl of round B reaction mix would be enough for labelling step. However, depending on the immunoprecipitated protein you choose (e.g. transcription factor) low amounts of DNA might be recovered from the ChIP, thus we suggest using 1 ml of mix for Round B reaction.
16. It is important to use a relative small number of PCR cycles to keep amplification linear.
17. Note that the new PBI binding buffer included in the kit is not recommended by the manufacturer to use with microarrays. The QUIAquick PCR purification membrane column can take up to 10 µg of DNA. For the purification and concentration use only one column and concentrate all the volume on it. To save time, we suggest using the vacuum instead of centrifugation method.
18. If you perform microarrays in duplicate, it is recommended to switch the dye between the two probes to ensure reproducibility of the results, especially because the respective incorporation of dyes may not be equal.

19. The Lifter Slip coverslip has two sides (first one is smooth, the other has two thick layers in each parts). Apply the pre-hybridization/hybridization mix slowly and carefully without air bubbles, which are undesirable in the mounting medium. If desired, as these coverslips are costly, they can be washed successively by water and ethanol and re-used.
20. Depending on the bench centrifuge, centrifugation vibrations may draw the slides outside the adaptor thus we recommend testing before with a normal glass slide.
21. If you start the DNAase I fragmentation for the first time, we recommend testing the activity of your DNAase I batch e.g. on your amplified input DNA, using different incubation times. Check on agarose 2–3% gel to obtain fragments around 100 bp size.

Acknowledgments

Our laboratory is supported by grants from the Swedish Cancer Society, Swedish Research Council (VR) M Bergvalls stiftelse, Östersjöstiftelsen and EU 'The Epigenome' NoE network. We thank Jenna M. Bernstein for critical reading of the manuscript.

References

1. Durand-Dubief, M., Sinha, I., Fagerstrom-Billai, F., Bonilla, C., Wright, A., Grunstein, M., and Ekwall, K. (2007) Specific functions for the fission yeast Sirtuins Hst2 and Hst4 in gene regulation and retrotransposon silencing. *Embo J* **26**, 2477–88.
2. Fagerstrom-Billai, F., Durand-Dubief, M., Ekwall, K., and Wright, A. P. (2007) Individual subunits of the Ssn6-Tup11/12 corepressor are selectively required for repression of different target genes. *Mol Cell Biol* **27**, 1069–82.
3. Sinha, I., Wiren, M., and Ekwall, K. (2006) Genome-wide patterns of histone modifications in fission yeast. *Chromosome Res* **14**, 95–105.
4. Walfridsson, J., Khorosjutina, O., Matikainen, P., Gustafsson, C. M., and Ekwall, K. (2007) A genome-wide role for CHD remodelling factors and Nap1 in nucleosome disassembly. *Embo J* **26**, 2868–79.
5. Wiren, M., Silverstein, R. A., Sinha, I., Walfridsson, J., Lee, H. M., Laurenson, P., Pillus, L., Robyr, D., Grunstein, M., and Ekwall, K. (2005) Genomewide analysis of nucleosome density histone acetylation and HDAC function in fission yeast. *Embo J* **24**, 2906–18.
6. Zhu, X., Wiren, M., Sinha, I., Rasmussen, N. N., Linder, T., Holmberg, S., Ekwall, K., and Gustafsson, C. M. (2006) Genome-wide occupancy profile of mediator and the Srb8-11 module reveals interactions with coding regions. *Mol Cell* **22**, 169–78.
7. Kurdistani, S. K., and Grunstein, M. (2003) In vivo protein-protein and protein-DNA crosslinking for genomewide binding microarray. *Methods* **31**, 90–5.
8. Robyr, D., and Grunstein, M. (2003) Genomewide histone acetylation microarrays. *Methods* **31**, 83–9.
9. Kurdistani, S. K., Robyr, D., Tavazoie, S., and Grunstein, M. (2002) Genome-wide binding map of the histone deacetylase Rpd3 in yeast. *Nat Genet* **31**, 248–54.
10. Barski, A., Cuddapah, S., Cui, K., Roh, T., Schones, D., Wang, Z., Wei, G., Chepelev, I., and Zhao, K. (2007) High-resolution profiling of histone methylations in the human genome. *Cell* **129**, 823–37.

11. Euskirchen, G. M., Rozowsky, J. S., Wei, C. L., Lee, W. H., Zhang, Z. D., Hartman, S., Emanuelsson, O., Stolc, V., Weissman, S., Gerstein, M. B., Ruan, Y., and Snyder, M. (2007) Mapping of transcription factor binding regions in mammalian cells by ChIP: comparison of array- and sequencing-based technologies *Genome Res* **17**, 898–909.

12. Hayashi, M., Katou, Y., Itoh, T., Tazumi, A., Yamada, Y., Takahashi, T., Nakagawa, T., Shirahige, K., and Masukata, H. (2007) Genome-wide localization of pre-RC sites and identification of replication origins in fission yeast. *Embo J* **26**, 1327–39.

INDEX

A

B

C

D

Q

R

S

T

U